Elementary Algebra

Without Trumpets or Drums

Martin M. Zuckerman

City College
of
The City University of New York

Elementary Algebra

Without Trumpets or Drums

Allyn and Bacon, Inc.
Boston
London
Sydney
Toronto

TO MY FATHER

Library of Congress Cataloging in Publication Data

Zuckerman, Martin M.
Elementary algebra.

Includes index.
1. Algebra. I. Title.
QA152.2.Z8 512.9'042 75-29061

ISBN 0-205-04895-1

Contents

Preface *ix*

CHAPTER 1 The Real Line 1

1.1 Geometric Background *1* 1.2 Rational Numbers *3*
1.3 Inequalities *8* 1.4 Inverse and Absolute Value *12*

What Have You Learned in Chapter 1? 15
Let's Review Chapter 1. 15
Try These Exam Questions for Practice. 16

CHAPTER 2 Arithmetic 18

2.1 Addition and Subtraction *18* 2.2 Commutative and Associative Laws *23* 2.3 Multiplication *27* 2.4 Division *31*
2.5 Exponents *37* 2.6 Order of Operations *40*

What Have You Learned in Chapter 2? 46
Let's Review Chapter 2. 46
Try These Exam Questions for Practice. 47

CHAPTER 3 Algebraic Expressions 49

3.1 Terms *49* 3.2 Addition of Terms *53* 3.3 Polynomials *56* 3.4 Evaluating Polynomials *60*

What Have You Learned in Chapter 3? 66
Let's Review Chapter 3. 66
Try These Exam Questions for Practice. 67

CHAPTER 4 Products and Factors 68

4.1 Monomial Products *68* 4.2 Distributive Laws for Polynomials *73* 4.3 Prime Factors *77* 4.4 Common Factors *82* 4.5 Binomial Products *86* 4.6 Difference of Squares *89* 4.7 Factoring Trinomials $x^2 + Mx + N$ *93* 4.8 Factoring Trinomials $Lx^2 + Mx + N$ *98*

What Have You Learned in Chapter 4? 103
Let's Review Chapter 4. 103
Try These Exam Questions for Practice. 105

CHAPTER 5 Division of Polynomials 106

5.1 Fractions *106* 5.2 Evaluating Rational Expressions *111* 5.3 Division of Monomials *115* 5.4 Monomial Divisors *120* 5.5 Factoring and Simplifying *123* 5.6 Polynomial Division *125* 5.7 Division With a Remainder *131*

What Have You Learned in Chapter 5? 135
Let's Review Chapter 5. 135
Try These Exam Questions for Practice. 137

CHAPTER 6 Rational Expressions and Their Arithmetic 138

6.1 Multiplication of Rational Expressions *138* 6.2 Division of Rational Expressions *143* 6.3 Addition of Rational Expressions with the Same Denominator *146* 6.4 Least Common Multiples *150* 6.5 Addition and Subtraction *154* 6.6 Complex Expressions *160* 6.7 Decimals *165*

What Have You Learned in Chapter 6? 173
Let's Review Chapter 6. 173
Try These Exam Questions for Practice. 175

CHAPTER 7 Equations 176

7.1 Roots *176* 7.2 Solving Simple Equations *178* 7.3 Equations With Parentheses *183* 7.4 Equations With Rational Expressions *185* 7.5 Literal Equations *191*

What Have You Learned in Chapter 7? 198
Let's Review Chapter 7. 198
Try These Exam Questions for Practice. 199

CHAPTER 8 Word Problems 200

8.1 Integer Problems *200* 8.2 Age Problems *206* 8.3 Distance Problems *209* 8.4 Interest Problems *216* 8.5 Commissions, Mark-Ups, and Discounts *221*

What Have You Learned in Chapter 8? **227**
Let's Review Chapter 8. **227**
Try These Exam Questions for Practice. **228**

CHAPTER 9 Functions 229

9.1 Rectangular Coordinates *229* 9.2 What Is a Function? *237* 9.3 Graphs of Functions *248*
What Have You Learned in Chapter 9? **258**
Let's Review Chapter 9. **258**
Try These Exam Questions for Practice. **260**

CHAPTER 10 Lines and their Equations 261

10.1 Slope *261* 10.2 Equation of a Line *274* 10.3 Intersection of Lines *284* 10.4 Systems of Linear Equations *291* 10.5 Variation *301*
What Have You Learned in Chapter 10? **310**
Let's Review Chapter 10. **310**
Try These Exam Questions for Practice. **312**

CHAPTER 11 Roots 313

11.1 Square Roots *313* 11.2 Irrational Square Roots *320* 11.3 Roots of Products and Quotients *328* 11.4 Addition and Subtraction of Roots *334* 11.5 Multiplication of Roots *337* 11.6 Division of Roots *342* 11.7 *n*th Roots *346*
What Have You Learned in Chapter 11? **352**
Let's Review Chapter 11. **352**
Try These Exam Questions for Practice. **354**

CHAPTER 12 Quadratic Equations 355

12.1 Solutions By Factoring *355* 12.2 Equations of the Form $x^2 = a$ *360* 12.3 The Quadratic Formula *364* 12.4 Word Problems *369*
What Have You Learned in Chapter 12? **375**
Let's Review Chapter 12. **375**
Try These Exam Questions for Practice. **376**

ANSWERS TO ODD-NUMBERED EXERCISES

INDEX

Preface

This book is written for the multitudes of college students who have not had the good fortune to understand the basic algebraic and arithmetic concepts upon which all scientific studies depend. In clear, simple, yet precise, language, ELEMENTARY ALGEBRA WITHOUT TRUMPETS OR DRUMS presents basically intuitive material. Common learning difficulties are dealt with. Set-theoretic notation is avoided. Ann Klein, a learning disabilities specialist, examined the entire manuscript for readability.

Most sections in the book are broken up into subsections so that the student can master one topic at a time. In this way he or she can obtain a sense of accomplishment and at the same time can pinpoint the trouble spots. The textual material is interspersed with an abundance of illustrative examples. There are, in fact, over 450 worked-out examples and sample exercises, many of these with multiple parts. There are extensive drill exercises at the end of each section, including a number of more challenging problems. At the end of each chapter there are review exercises, a few of which review material from earlier chapters. There is also a sample test. Altogether there are over 3300 student exercises. A WORKBOOK is available for the student who may need additional practice.

Chapter 1 presents real numbers from a geometric point of view. Chapter 2 consists of a thorough review of arithmetic. If not used in class, the instructor can assign this chapter for self-study together with the corresponding material in the WORKBOOK.

Throughout the book there is an emphasis on "word problems." Chapter 8, in particular, is devoted to practical applications of linear equations. The student is first shown how to translate a problem into algebraic symbolism, and then how to solve the problem by solving the resulting equation. Section 12.4 presents applications of quadratic equations.

Graphical techniques are stressed. There are almost 200 figures accompanying the textual material, exercises, illustrative examples and their solutions.

For a detailed analysis of the contents of each chapter and for suggestions on presenting the material, see the "Chapter-by-Chapter Comments" in the Instructor's Manual.

I would like to thank each of the following reviewers for their helpful suggestions and encouraging comments:

Ben P. Bockstege of Broward Community College
Thomas R. Butts of Michigan State University
Paul R. Fallone of the University of Connecticut at Hartford
Victor Klee of the University of Washington
Ann Klein of Arlington, Massachusetts
Robert D. Klein of Northeastern University
Ken Seydel of Skyline College
Fred Toxepeus of Kalamazoo Valley Community College
Frank Wright of Cerritos College

I am grateful to Garen R. Wickham, my editor at Allyn and Bacon, and to Paul Tasner, who was in charge of production, for their cooperation throughout this endeavor.

Martin M. Zuckerman

1

The Real Line

1.1 GEOMETRIC BACKGROUND

The numbers you study in algebra are known as **real numbers.** You are going to obtain a geometric picture of real numbers.

Consider a horizontal line ***L***, and suppose that it extends indefinitely both to the left and to the right. Choose any point on ***L*** and label this O (oh). This point O represents the number 0 (zero) and is called the **origin.** Now choose another point, P, to the *right* of O. This second point represents the number 1. (See Figure 1.1.)

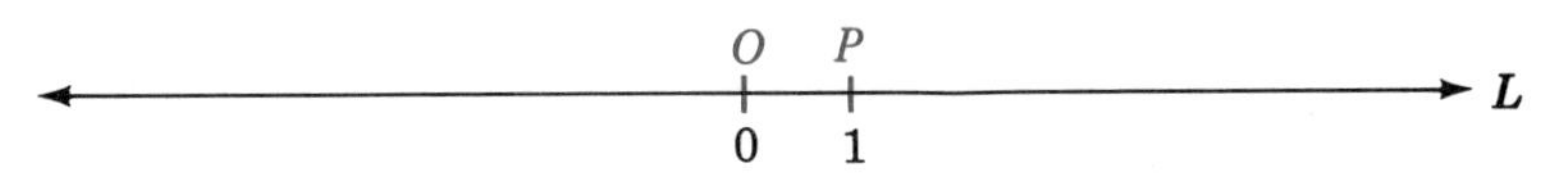

FIGURE 1.1. The line L

A portion of a line that lies between two points (including these "end points") is called a **line segment.** That part of the line ***L*** between O and P is a line segment, and is denoted by $\overline{OP}$. The *length* of $\overline{OP}$ determines the basic **unit of distance** on the line ***L***. The number 2 is represented one (distance) unit to the right of P. The number 3 is represented one unit further to the right. Continue this process to represent the numbers 4, 5, 6, etc. (See Figure 1.2.)

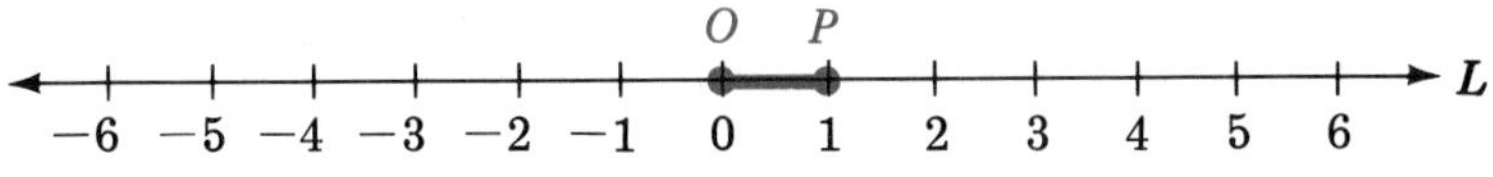

FIGURE 1.2. The line segment $\overline{OP}$ is in brown.

Now return to the origin, O, and then go one distance unit to the *left* of O. This point represents the number -1. Go one unit further to the left to represent the number -2. Continue to the left to represent the numbers $-3, -4, -5$, etc. (See Figure 1.2 on page 1)

Definition

> *INTEGERS. The real numbers represented in this fashion are called* ***integers.***
> *The integers are*
>
> $$\ldots, -3, -2, -1, 0, 1, 2, 3, \ldots$$
>
> *The three dots to the left and to the right are each read "and so forth."*

EXAMPLE 1

On the line ***L:***

(a) The integer 0 corresponds to the point O.
(b) The integer 4 corresponds to the point 4 units to the right of O.
(c) The integer 25 corresponds to the point 25 units to the right of O.
(d) The integer -4 corresponds to the point 4 units to the left of O.
(e) The integer -25 corresponds to the point 25 units to the left of O.

EXAMPLE 2

Consider the line ***L***. Label the points that represent:

(a) 5 (b) 8 (c) -4 (d) -7

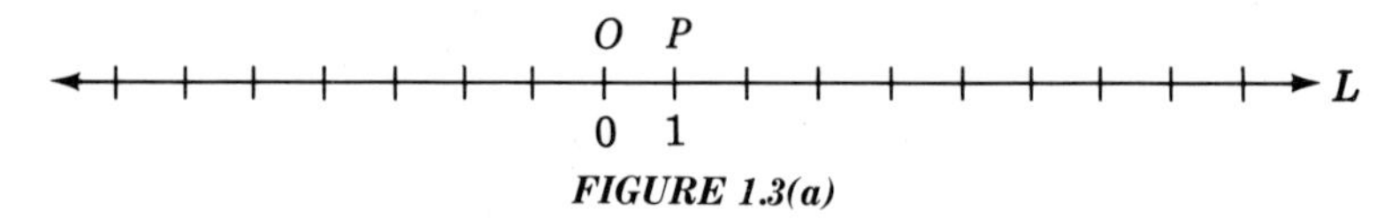

FIGURE 1.3(a)

Solution.

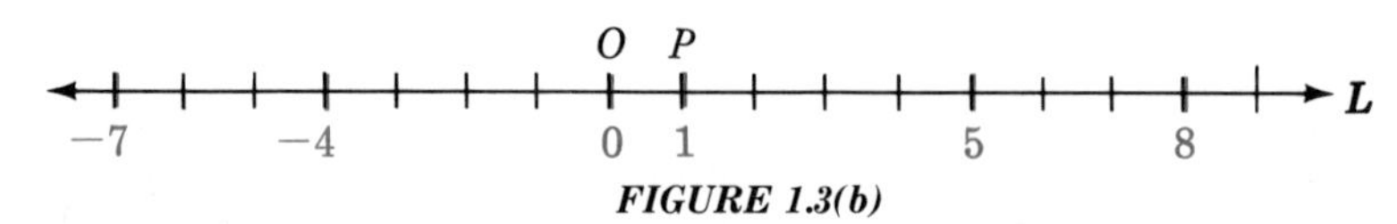

FIGURE 1.3(b)

EXAMPLE 3

(a) Which integer corresponds to the point 13 units to the right of O?
(b) Which integer corresponds to the point 13 units to the left of O?

Solution.
(a) 13 (b) -13

EXERCISES

For exercises 1–8, draw a horizontal line ***L***. Select a point O to represent

0 and another point P to the right of O to represent 1, as in Figure 1.4. Locate the following integers on L:

1. 3	2. 5	3. 9	4. −1
5. −3	6. −6	7. −9	8. −10

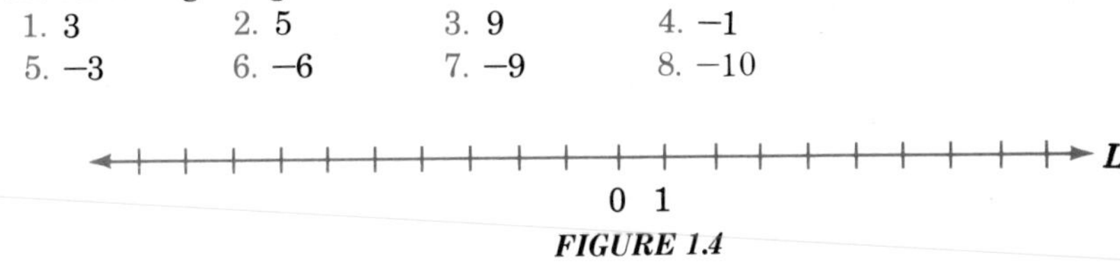

FIGURE 1.4

9. In Figure 1.5, indicate which number is represented by each of the points Q, R, S, T, U, V.

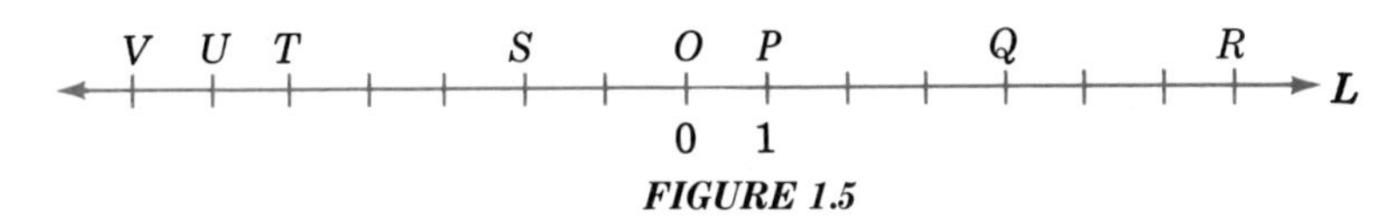

FIGURE 1.5

10. In Figure 1.6, indicate which number is represented by each of the points Q, R, S, T, U, V.

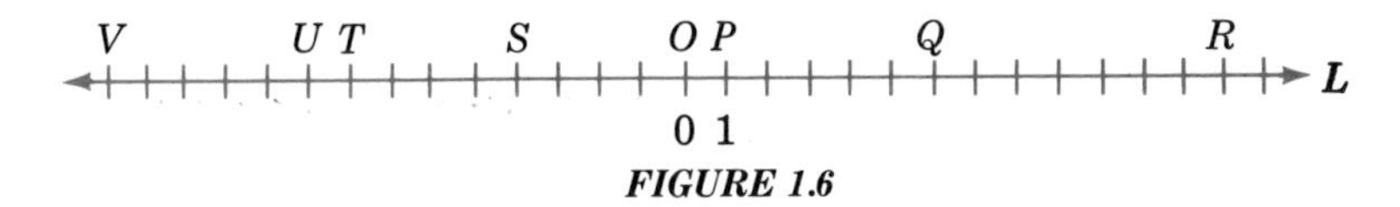

FIGURE 1.6

11. Which integer corresponds to the point 19 units to the right of O?
12. Which integer corresponds to the point 19 units to the left of O?
13. Which integer corresponds to the point 37 units to the right of O?
14. Which integer corresponds to the point 61 units to the left of O?

1.2 RATIONAL NUMBERS

What Is a Rational Number?

From now on, identify every real number with the point on L that represents it. For instance, "the point 0" means "the point O" and "the point 1" means "the point P."

Rational numbers are real numbers obtained by dividing the line segments between integer points into equal parts. For example, (with the aid of a ruler) divide the line segment between 0 and 1 into two equal parts. The midpoint obtained represents the number $\frac{1}{2}$. (See Figure 1.7 on page 4) Here

$$\frac{1}{2}$$

means that the number 1 has been divided by 2.

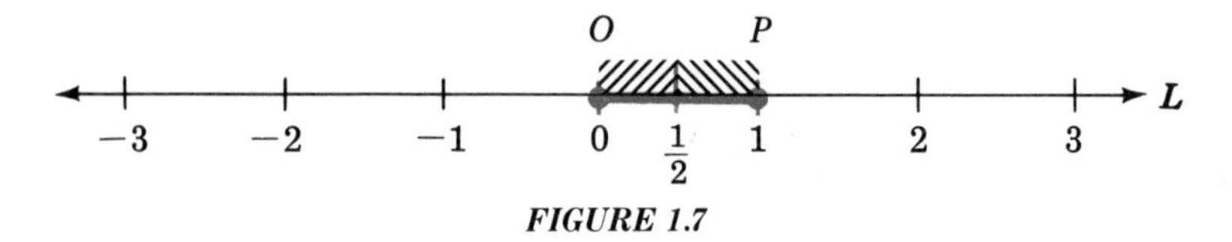

FIGURE 1.7

To obtain $\frac{2}{5}$, divide the line segment between 0 and 2 into 5 equal parts. The first point of division *to the right of* 0 represents $\frac{2}{5}$. (See Figure 1.8.) Here

$$\frac{2}{5}$$

means that the number 2 has been divided by 5.

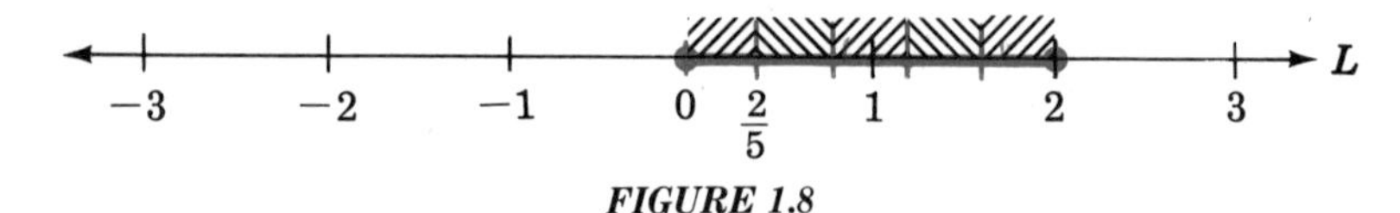

FIGURE 1.8

In general,

$$\frac{a}{b}$$

indicates that the number a has been divided by b.

To obtain $\frac{3}{2}$, divide the line segment between 0 and 3 into 2 equal parts. The point of division represents $\frac{3}{2}$. (See Figure 1.9.)

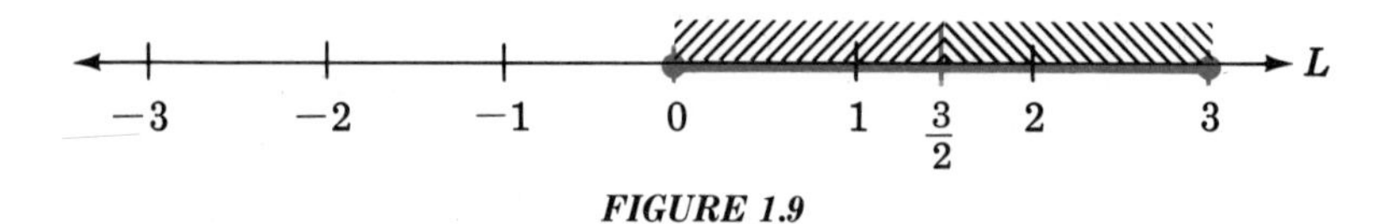

FIGURE 1.9

To obtain $\frac{-1}{4}$, divide the line segment between 0 and −1 into 4 equal parts. The first point of division *to the left of* 0 represents $\frac{-1}{4}$. (See Figure 1.10.)

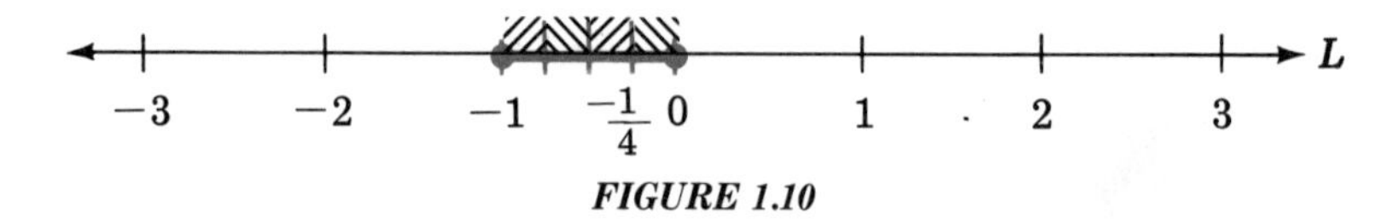

FIGURE 1.10

The definition of rational numbers that you will use is based on this geometric idea of dividing line segments between integer points.

Definition

> *A* ***rational number*** *is a real number that can be written in the form* $\frac{N}{D}$*, where N and D are integers and* $D \neq 0$*. (Read: D is not equal to 0.) Here, N is called the* ***numerator*** *and D, the* ***denominator.***

EXAMPLE 1

(a) $\frac{3}{4}$ is a rational number with numerator 3 and denominator 4.

(b) $\frac{2}{7}$ is a rational number with numerator 2 and denominator 7.

(c) $\frac{-1}{3}$ is a rational number with numerator -1 and denominator 3.

(d) $\frac{5}{2}$ is a rational number with numerator 5 and denominator 2.

Integers and Mixed Numbers

An integer, such as 5, is a rational number. For,

$$5 = \frac{5}{1}$$

(Both numbers, 5 and $\frac{5}{1}$, correspond to the same point on the line ***L***.) Therefore, 5 is a real number that can be written in the form $\frac{N}{D}$, where N and D are integers and $D \neq 0$. Here $N = 5$ and $D = 1$.

A "mixed number," such as $3\frac{1}{2}$ is a rational number. For

$$3\frac{1}{2} \text{ stands for } 3 + \frac{1}{2},$$

and as you will see,

$$3 + \frac{1}{2} = \frac{6}{2} + \frac{1}{2} = \frac{7}{2}.$$

EXAMPLE 2

Express each of the following integers in the form $\frac{N}{D}$, where N and D are integers and $D \neq 0$:

(a) 17 (b) -17 (c) -1

Solution.
(a) $17 = \frac{17}{1}$ (b) $-17 = \frac{-17}{1}$ (c) $-1 = \frac{-1}{1}$

EXAMPLE 3

Express each of the following mixed numbers in the form $\frac{N}{D}$, where N and D are integers and $D \neq 0$:

(a) $1\frac{1}{5}$ (b) $2\frac{1}{4}$ (c) $-2\frac{1}{4}$

Solution.

(a) $1\frac{1}{5} = 1 + \frac{1}{5} = \frac{5}{5} + \frac{1}{5} = \frac{6}{5}$

(b) $2\frac{1}{4} = 1 + \frac{1}{4} = \frac{8}{4} + \frac{1}{4} = \frac{9}{4}$

(c) $-2\frac{1}{4} = -2 + \frac{-1}{4} = \frac{-8}{4} + \frac{-1}{4} = \frac{-9}{4}$

Decimals

Decimals, such as

$$.3, \quad -.7, \quad .21, \quad 1.1,$$

are rational numbers. In fact,

$$.3 = \frac{3}{10}, \quad -.7 = \frac{-7}{10}, \quad .21 = \frac{21}{100}, \quad 1.1 = \frac{11}{10}$$

EXAMPLE 4

Express each of the following decimals in the form $\frac{N}{D}$, where N and D are integers and $D \neq 0$:

(a) $-.49$ (b) $.113$ (c) 4.71

Solution.

(a) $-.49 = \frac{-49}{100}$ (b) $.113 = \frac{113}{1000}$ (c) $4.71 = \frac{471}{100}$

Irrational Numbers

You may be surprised to learn that there are points on the line $\boldsymbol{L}$ that do not represent rational numbers. Such points correspond to **irrational numbers.** An example of an irrational number is π. You may know that the circumference (or length) of a circle is $2\pi r$, where r is the length of the radius. It can be shown that π *cannot* be expressed in the form $\frac{N}{D}$ for any integers N and D. Therefore π is irrational.

The *irrational* number π is *approximately* equal to the *rational* $\frac{22}{7}$ $\left(\text{or } 3\frac{1}{7}\right)$. Note, however, that

$$\pi \neq \frac{22}{7}.$$

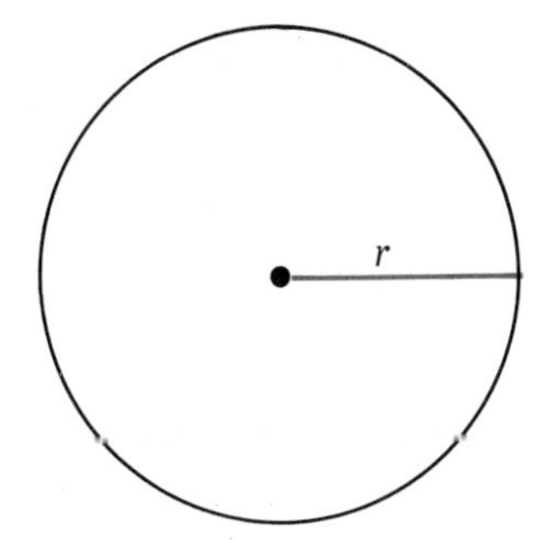

FIGURE 1.11. The circumference is $2\pi r$.

Other irrational numbers will be discussed in Section 11.2.

*Every point on **L** represents exactly one real number, rational or irrational. Also, every real number corresponds to exactly one point on **L**.* Because of this correspondence, you can picture real numbers geometrically.

From now on "number" will mean "real number."

EXERCISES

For exercises 1–8, draw a horizontal line ***L*** and label the "integer points" as in Figure 1.12. Locate the following rational numbers on ***L***.

1. $\frac{1}{2}$
2. $\frac{1}{4}$
3. $\frac{3}{4}$
4. $\frac{3}{2}$
5. $\frac{5}{4}$
6. $\frac{-1}{3}$
7. $\frac{-2}{3}$
8. $\frac{-4}{3}$

−2 −1 0 1 2 *L*

FIGURE 1.12

9. Each of the points Q, R, S, T, U, V, W in Figure 1.13 represents one of the following rational numbers:

$$\frac{1}{5}, \frac{4}{5}, \frac{7}{5}, 1\frac{1}{5}, \frac{-1}{5}, \frac{-2}{5}, \frac{-7}{5}$$

Indicate which number is represented by each point.

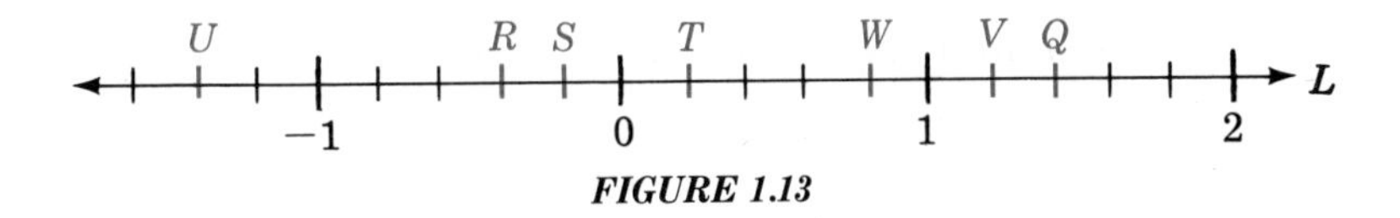

FIGURE 1.13

10. Each of the points Q, R, S, T, U, V, W in Figure 1.14 represents one of the

following rational numbers:

$$\frac{1}{7}, \frac{3}{7}, \frac{9}{7}, \frac{-1}{7}, \frac{-6}{7}, \frac{-10}{7}, -1\frac{1}{7}$$

Indicate which number is represented by each point.

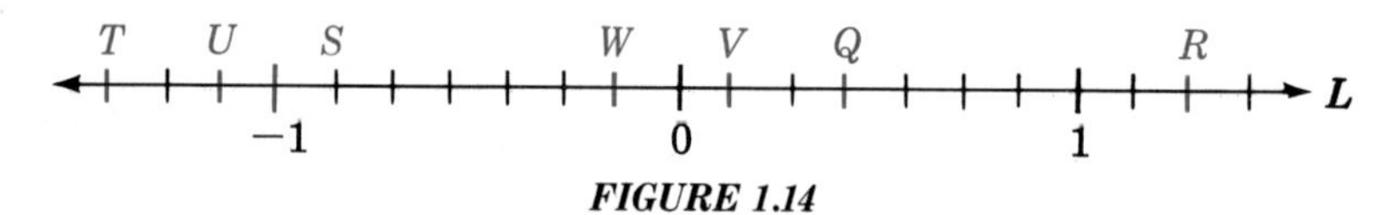

FIGURE 1.14

In exercises 11–20, express each number in the form $\frac{N}{D}$, where N and D are integers and $D \neq 0$.

> **SAMPLE.**
>
> $1\frac{1}{2} = 1 + \frac{1}{2} = \frac{2}{2} + \frac{1}{2} = \frac{3}{2}$

11. $1\frac{1}{4}$ 12. $2\frac{1}{5}$ 13. $1\frac{2}{5}$ 14. $-1\frac{1}{3}$ 15. -1

16. -2 17. $3\frac{1}{4}$ 18. $-5\frac{1}{2}$ 19. -4 20. 0

In exercises 21–30, express each decimal in the form $\frac{N}{D}$, where N and D are integers and $D \neq 0$.

21. .1 22. −.3 23. .31 24. .59 25. −.83

26. 1.7 27. 4.81 28. .101 29. −.599 30. 2.107

1.3 INEQUALITIES

Inequalities and the Real Line

Definition

> *LESS THAN. Let **a** and **b** be real numbers. If **a** lies to the left of **b** (on **L**), then you say that **a is less than b,** or **a is smaller than b,** written*
>
> $$a < b,$$
>
> *and that **b is greater than a,** or **b is larger than a,** written*
>
> $$b > a.$$
>
> *In this case b lies to the right of a.*

Note that the symbols

$$<, >$$

each point to the *smaller* number. Also,

$$\text{if } a < b, \text{ then } b > a.$$

EXAMPLE 1

(a) $3 < 5$ because 3 lies to the left of 5. Also, $5 > 3$. Note that 5 lies to the right of 3.
(b) $-2 < 0$ because -2 lies to the left of 0. Also, $0 > -2$.
(c) $-3 < 1$ because -3 lies to the left of 1. Thus -3 *is smaller than* 1.
(d) $\frac{1}{4} < \frac{1}{2}$ because $\frac{1}{4}$ lies to the left of $\frac{1}{2}$.
(e) $1.9 < 2.1$ because 1.9 lies to the left of 2.1.

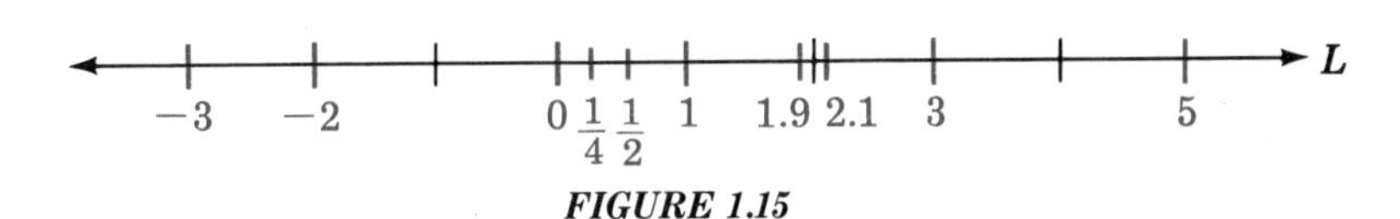

FIGURE 1.15

Positive and Negative Numbers

Definition

> *POSITIVE, NEGATIVE. A real number is called* ***positive*** *if it lies to the right of 0, and* ***negative*** *if it lies to the left of 0.*

The number 0 is neither positive nor negative.

EXAMPLE 2

(a) $3, \frac{2}{5}$, and π are each positive. Each of these lies to the right of 0.

(b) $-2, \frac{-1}{4}$, and $\frac{-5}{2}$ are each negative. Each of these lies to the left of 0.

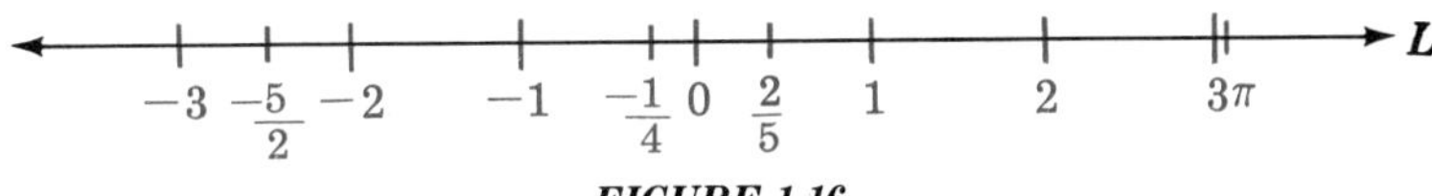

FIGURE 1.16

Positive numbers p are numbers such that

$$p > 0.$$

Negative numbers n are numbers such that

$$n < 0.$$

Thus $$7 > 0, \text{ whereas } -4 < 0$$

Every negative number is less than every positive number because a negative number lies to the left of a positive number on $\boldsymbol{L}$.

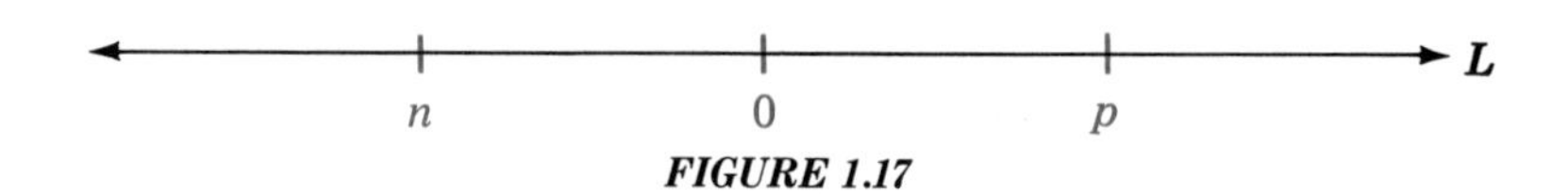

FIGURE 1.17

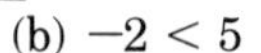

(a) $-2 < 2$ (b) $-2 < 5$ (c) $-5 < 2$

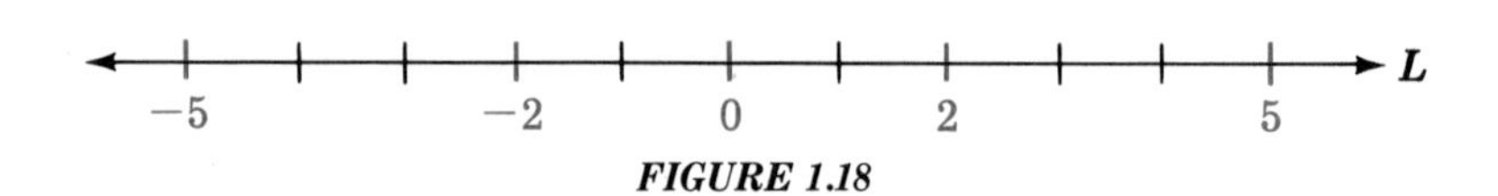

FIGURE 1.18

There are arbitrarily large positive numbers because the real line $\boldsymbol{L}$ extends indefinitely to the right. Similarly, there are arbitrarily small negative numbers because $\boldsymbol{L}$ extends indefinitely to the left.

Let a and b be real numbers. Then exactly one of the following is true:

$$a = b$$
$$a < b$$
$$a > b$$

This is because on $\boldsymbol{L}$,

	a and b are the same point	$(a = b)$
or	a lies to the left of b	$(a < b)$
or	a lies to the right of b.	$(a > b)$

EXERCISES

In exercises 1–20, fill in "$<$" or "$>$".

SAMPLE. 5 ☐ 10	***Solution.*** $<$

1. 2 ☐ 5 2. 12 ☐ 10 3. 0 ☐ 7 4. $\frac{1}{2}$ ☐ 1

5. $1 \square \frac{4}{3}$ 6. $3 \square \frac{5}{2}$ 7. $-.8 \square 0$ 8. $.5 \square -5$

9. $-4 \square 2$ 10. $-2 \square 4$ 11. $-3 \square \frac{1}{2}$ 12. $\frac{1}{4} \square \frac{1}{3}$

13. $\frac{-1}{4} \square \frac{-1}{3}$ 14. $\frac{1}{4} \square \frac{-1}{3}$ 15. $\frac{1}{10} \square \frac{1}{100}$ 16. $-9 \square -3$

17. $-10000 \square -1$ 18. $\frac{-1}{2} \square \frac{-1}{5}$

19. $-400 \square \frac{1}{400}$ 20. $\frac{5}{8} \square \frac{1}{2}$

In exercises 21–30, fill in "left" or "right".

SAMPLE. 8 lies to the $\square$ of 6 on ***L***.	***Solution.*** right

21. 4 lies to the $\square$ of 7 on ***L***.
22. 10 lies to the $\square$ of 1 on ***L***.
23. 0 lies to the $\square$ of −2 on ***L***.
24. $\frac{1}{3}$ lies to the $\square$ of $\frac{1}{5}$ on ***L***.
25. $\frac{-1}{3}$ lies to the $\square$ of $\frac{-1}{5}$ on ***L***.
26. 6 lies to the $\square$ of $\frac{1}{6}$ on ***L***.
27. −6 lies to the $\square$ of $\frac{-1}{6}$ on ***L***.
28. π lies to the $\square$ of 3 on ***L***.
29. −1 000 000 lies to the $\square$ of −1000 on ***L***.
30. 0 lies to the $\square$ of $\frac{1}{100}$ on ***L***.

In exercises 31–36, rearrange the numbers so that you can write "<" between any two numbers.

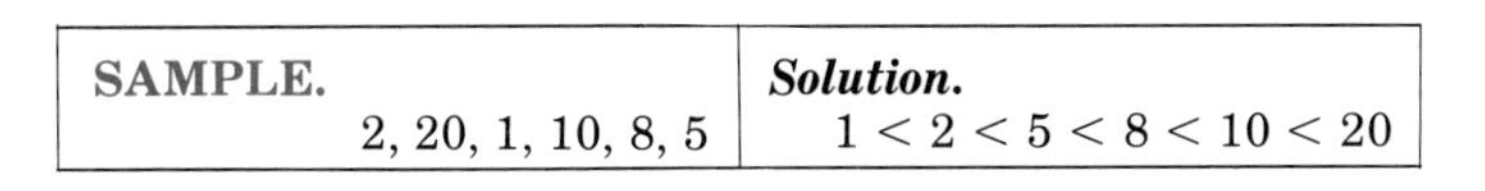

SAMPLE. 2, 20, 1, 10, 8, 5	***Solution.*** $1 < 2 < 5 < 8 < 10 < 20$

31. 4, 2, 12, 3, 18, 10
32. −5, −8, −2, −3, −7, −1
33. −7, 7, 10, −9, −12, 0
34. $\frac{1}{4}, \frac{1}{2}, \frac{1}{6}, \frac{1}{3}, \frac{1}{10}, \frac{1}{8}$
35. $\frac{1}{5}, \frac{2}{3}, 1, \frac{4}{3}, 2, \frac{9}{5}$
36. $\frac{-1}{9}, \frac{1}{5}, \frac{-1}{10}, -1, \frac{1}{2}, 1$

In exercises 37–42, rearrange the numbers so that you can write ">" between any two numbers.

SAMPLE. 15, 7, 12, 13, 18, 10	***Solution.*** $18 > 15 > 13 > 12 > 10 > 7$

37. 8, 6, 10, 2, 5, 7

38. −2, −1, −4, −10, −7, 0

39. $\frac{1}{2}, \frac{1}{10}, \frac{1}{5}, \frac{1}{7}, \frac{1}{12}, 1$

40. $\frac{-1}{6}, \frac{-1}{2}, \frac{-1}{3}, \frac{-2}{3}, -1, 0$

41. .3, .2, −2, −4, 1, −9

42. $-10, -4, \frac{1}{2}, \frac{1}{4}, \frac{-1}{2}, \frac{-1}{4}$

1.4 INVERSE AND ABSOLUTE VALUE

Inverse

The numbers 5 and −5 lie on opposite sides of the origin, 0, but are the same distance from 0. The *positive* number 5 lies 5 units to the *right* of the origin, whereas the *negative* number −5 lies 5 units to the *left* of the origin.

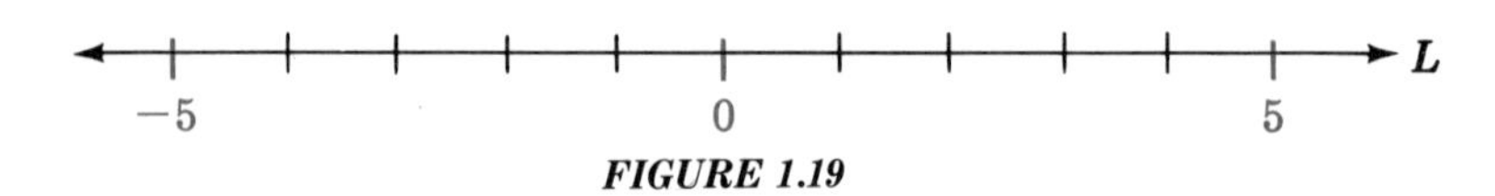

FIGURE 1.19

Definition

> *INVERSE. Let **a** be a nonzero real number. The **additive inverse of a,** or for short, the **inverse of a,** is the number that lies on the opposite side of the origin and is the same distance from the origin as is **a.** The **(additive) inverse of** 0 is 0 itself.*

For every number a, let

$$-a$$

denote the inverse of a. Thus −5 is the inverse of 5.

EXAMPLE 1

(a) The inverse of 10 is −10.
(b) The inverse of −10 is 10.

$$-(-10) = 10$$

Observe that *the inverse of a positive number is a negative number. The inverse of a negative number is a positive number.*

Absolute Value

Definition

> *ABSOLUTE VALUE. Let **a** be any real number. The **absolute value of a** is its distance from the origin.*

Let

$$|a|$$

denote the absolute value of a.

EXAMPLE 2

(a) $|7| = 7$ because 7 lies 7 units from the origin.
(b) $|-7| = 7$ because -7 lies 7 units from the origin.
(c) $|0| = 0$
(d) $\left|\frac{-1}{2}\right| = \frac{1}{2}$

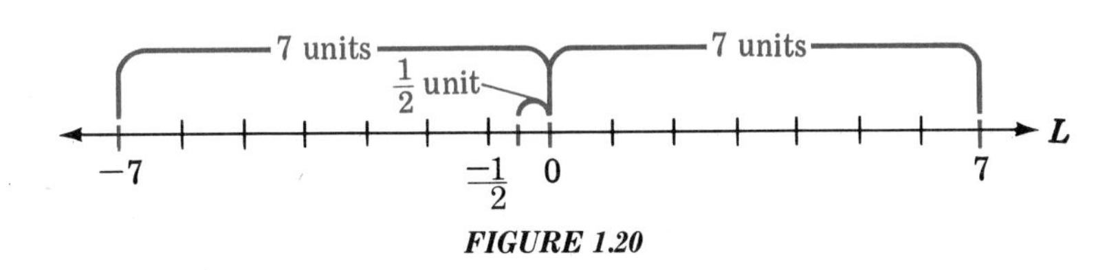

FIGURE 1.20

For every number a,

$$|a| = |-a|$$

Both numbers are the same distance from, but on opposite sides of, the origin. *Absolute value measures distance, but neglects direction, from the origin.*

Observe that

$$|a| = a, \qquad \text{if } a \text{ is positive or } 0;$$
$$|a| = -a, \qquad \text{if } a \text{ is negative.}$$

In fact, if a is *negative,* a lies to the *left* of the origin, and $-a$ lies to the *right* of the origin. Thus $-a$ which is *positive,* is the distance from a to the origin.

EXAMPLE 3

(a) $|4| = 4$
(b) $|-4| = -(-4) = 4$

EXERCISES

In exercises 1 – 12, write the inverse of each number.

1. 2 2. 20 3. 0 4. $\frac{1}{2}$ 5. π

6. $\frac{7}{5}$ 7. -3 8. $\frac{-1}{3}$ 9. -200 10. $\frac{-10}{3}$

11. $|5|$ 12. $|-5|$

In exercises 13 – 20, find each absolute value.

13. $|8|$ 14. $|-8|$ 15. $|0|$ 16. $\left|\frac{1}{4}\right|$

17. $\left|\frac{-1}{4}\right|$ 18. $|\pi|$ 19. $|-10000|$ 20. $\left|\frac{1}{4000000}\right|$

In exercises 21 – 30, fill in "=", "<", or ">".

SAMPLE. $|-4| \; \square \; -4$	*Solution.* $>$

21. $3 \; \square \; |3|$ 22. $7 \; \square \; |-7|$ 23. $|9| \; \square \; |-9|$

24. $0 \; \square \; |-0|$ 25. $|-2| \; \square \; -(-2)$ 26. $\left|\frac{-1}{3}\right| \; \square \; \frac{1}{3}$

27. $|-12| \; \square \; 1$ 28. $|-7| \; \square \; |-5|$ 29. $|-\pi| \; \square \; |-3|$

30. $-|-\pi| \; \square \; -3$

SAMPLE (for exercises 31 – 34). Which numbers are 12 units from the origin?	*Solution.* 12 and -12

31. Which numbers are 8 units from the origin?
32. Which numbers are $\frac{1}{2}$ unit from the origin?
33. Which positive number is 11 units from the origin?
34. Which negative number is 13 units from the origin?
35. Which number is further from the origin, 27 or -44?
36. Which number is further from the origin, $\frac{1}{2}$ or $\frac{-1}{5}$?

What Have You Learned in Chapter 1?

You can locate integers and rational numbers on the number line **L**.

You can express mixed numbers and decimals as rational numbers.

You know that there are irrational numbers, and one of these, π, is approximately $\frac{22}{7}$.

You know that a is less than b if a lies to the left of b on **L**.

And you can find the absolute value as well as the inverse of a number a.

Let's Review Chapter 1.

1.1 Geometric Background

1. In Figure 1.21, indicate which number is represented by each of the points Q, R, S, T, U, V.

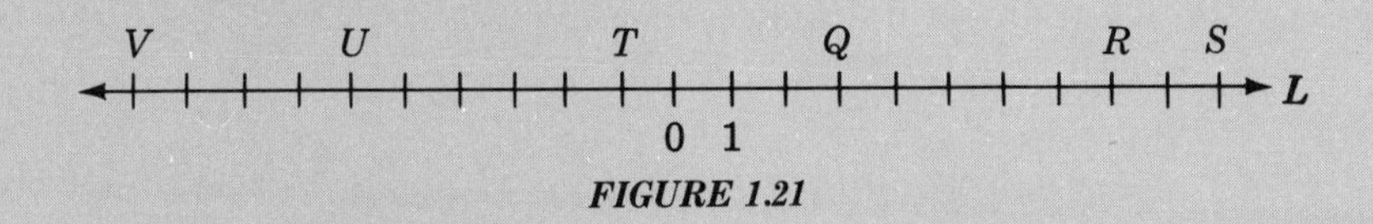

FIGURE 1.21

2. (a) Which integer corresponds to the point 12 units to the left of the origin?
   (b) Which integer corresponds to the point 12 units to the right of the origin?

1.2 Rational Numbers

For exercises 3–6, draw a horizontal line **L** and label the integer points as in Figure 1.22. Locate the following rational numbers on **L**.

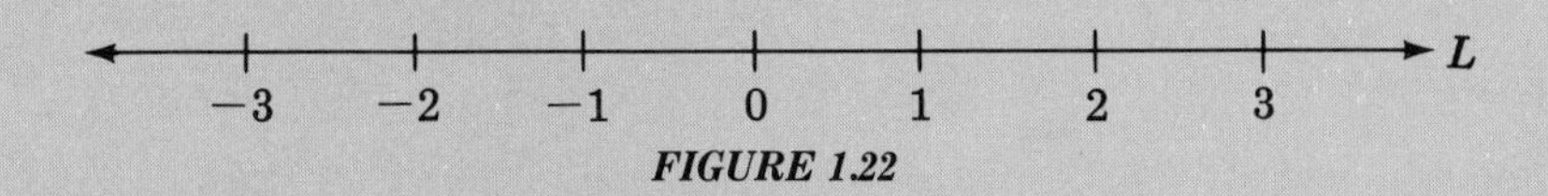

FIGURE 1.22

3. $\frac{1}{4}$ 4. $\frac{2}{5}$ 5. $\frac{-1}{2}$ 6. $\frac{-5}{2}$

7. Express $1\frac{2}{3}$ in the form $\frac{N}{D}$, where N and D are integers and $D \neq 0$.

8. Express .37 in the form $\frac{N}{D}$, where N and D are integers and $D \neq 0$.

1.3 Inequalities

9. Fill in "<" or ">":
   (a) $3 \; \square \; 8$
   (b) $-3 \; \square \; -8$
   (c) $-8 \; \square \; 3$
10. Fill in "left" or "right":
   (a) $\frac{1}{4}$ lies to the $\square$ of 4 on ***L.***
   (b) $\frac{-1}{4}$ lies to the $\square$ of -4 on ***L.***
11. Rearrange the following numbers so that you can write "<" between any two numbers.
   $1, 5, \frac{1}{5}, -1, -5, \frac{-1}{5}$
12. Rearrange the following numbers so that you can write ">" between any two numbers.
   $-7, -14, 1, 0, \frac{1}{4}, \frac{1}{2}$

1.4 Inverse and Absolute Value

13. Write the inverse of each number:
   (a) $\frac{1}{4}$ (b) -2 (c) 0
14. Find each absolute value:
   (a) $|-3|$ (b) $\left|\frac{2}{5}\right|$ (c) $|0|$
15. Fill in "=", "<", or ">":
   (a) $10 \; \square \; |-10|$
   (b) $-|7| \; \square \; |-7|$
16. Which numbers are 7 units from the origin?

Try These Exam Questions for Practice.

1. Which integer corresponds to the point 6 units to the left of the origin?
2. Express $2\frac{1}{2}$ in the form $\frac{N}{D}$, where N and D are integers and $D \neq 0$.

3. Express $-.9$ in the form $\frac{N}{D}$, where N and D are integers and $D \neq 0$.
4. Fill in "<" or ">": $-4 \ \square \ -2$
5. Fill in "left" or "right": $\frac{-1}{4}$ lies to the $\square$ of $\frac{-1}{2}$ on ***L***.
6. Rearrange the following numbers so that you can write "<" between any two numbers. $-2, \frac{1}{2}, \frac{-1}{2}, 12, -12, \frac{1}{12}$
7. Write the inverse of -19.
8. Find $|-19|$.

2

Arithmetic

2.1 ADDITION AND SUBTRACTION

Let a and b be real numbers. The **sum of *a* and *b*** is denoted by

$$a + b. \qquad \text{(read: "}a\text{ plus }b\text{")}$$

The **difference between *a* and *b*** (in this order) is denoted by

$$a - b. \qquad \text{(read: "}a\text{ minus }b\text{")}$$

Addition

First consider addition; subtraction is then defined in terms of addition.

EXAMPLE 1

(a) $4 + 2$ (or 6) lies 2 units to the *right* of 4.
(b) $4 + (-3)$ (or 1) lies $|-3|$ units (or 3 units) to the *left* of 4.

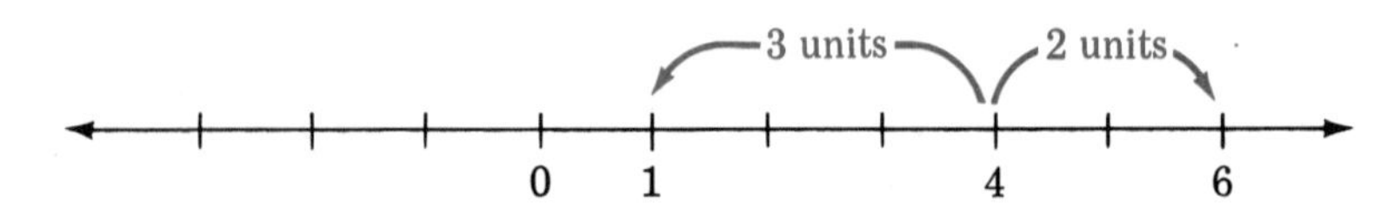

FIGURE 2.1. ***$4 + 2 = 6$ and $4 + 2$ lies 2 units to the*** *right* ***of 4. $4 + (-3) = 1$ and $4 + (-3)$ lies 3 units to the*** *left* ***of 4.***

Let p be *positive* and n *negative*. Then $a + p$ lies p units to the *right* of a. However, $a + n$ lies $|n|$ units to the *left* of a.

Subtraction

Definition

$$a - b = a + (-b)$$

Thus, *to subtract a number b, add its inverse −b.*

EXAMPLE 2

(a)
$$8 - 5 = 8 + (-5) = 3$$

Note that $8 - 5$ lies 5 units to the *left* of 8. [Figure 2.2(a).]

(b)
$$8 - (-5) = 8 + 5 = 13$$

Thus $8 - (-5)$ lies 5 units to the *right* of 8. [Figure 2.2(b).]

Suppose that a and b are both positive.

$$\text{If } a > b, \text{ then } a - b > 0$$

Thus
$$\underbrace{8 - 5}_{3} > 0$$

$$\text{If } a < b, \text{ tnen } a - b < 0$$

Thus
$$\underbrace{5 - 8}_{-3} < 0$$

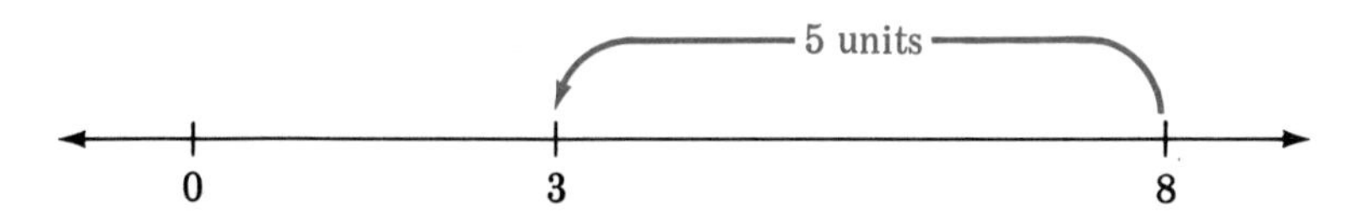

FIGURE 2.2(a). 8 − 5 > 0 because 8 > 5.

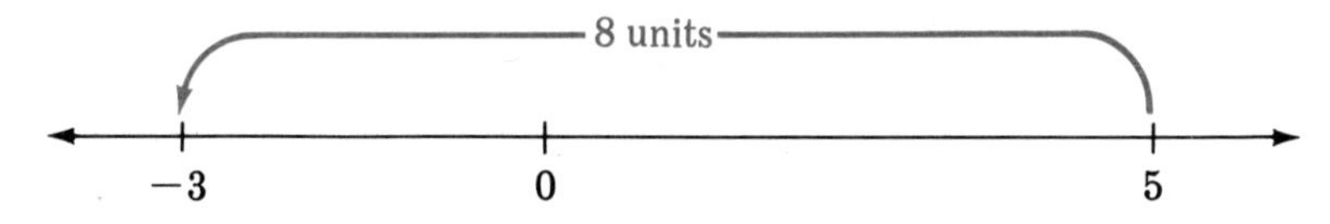

FIGURE 2.2(b). 5 − 8 < 0 because 5 < 8.

Finally, the following important rules are easily seen:

$$a + 0 = a$$
$$a - 0 = a$$
$$0 - a = 0 + (-a) = -a$$

Adding Negative Numbers

Negative numbers are added in terms of their inverses, as in the following example.

EXAMPLE 3

$$(-5) + (-4) = -(5 + 4) = -9$$

For any (positive) numbers a and b,

$$(-a) + (-b) = -(a + b)$$

Thus the *sum of the inverses* is the *inverse of the sum.*

Adding Numbers With Different Signs

The **sign** of a positive number is + and the sign of a negative number is −. Thus the sign of 8 is +; the sign of −8 is −.

EXAMPLE 4

$$8 + (-8) = 0$$

Note that −8 is the inverse of 8.

The inverse of a is the number added to a to obtain 0. This is because a and $-a$ are each the same distance from, but on opposite sides of, the origin, 0. Thus

$$a + (-a) = 0$$

Now consider numbers with *different signs* and *different absolute values.*

EXAMPLE 5

(a) $8 + (-5) = 3$
For $8 + (-5)$ lies 5 units to the *left* of 8. (See Figure 2.3.) Note that 8 is larger than −5 in absolute value. The positive sign of 8 prevails.

(b) $-8 + 5 = -3$
For $-8 + 5$ lies 5 units to the *right* of −8. (See Figure 2.3.) Note that −8 is larger than 5 in absolute value. The negative sign of −8 prevails.

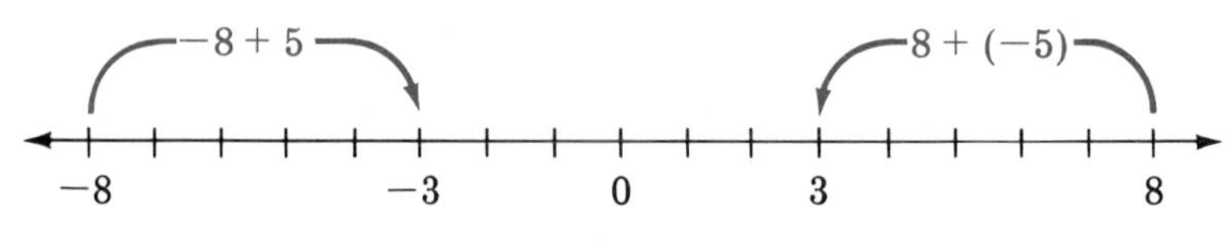

FIGURE 2.3

To add numbers with *different signs* and *different absolute values:*

1. *Subtract* the smaller absolute value from the larger absolute value.
2. The sign of the number with larger absolute value prevails.

Thus in Example 5(a),

$$|8| > |-5|$$

Therefore $$8 + (-5) = +(8 - 5) = 3$$

the sign of 8

But in Example 5(b),

$$|-8| > |5|$$

Thus $$-8 + 5 = -(8 - 5) = -3$$

the sign of -8

EXAMPLE 6

Add: $-12 + 7$

Solution. $|-12| > |7|$

Therefore $$-12 + 7 = -(12 - 7) = -5$$

the sign of -12

EXAMPLE 7

Add: $(-192) + 281$

Solution. $|281| > |-192|$

Subtract:

$$\begin{array}{r} 281 \\ \underline{192} \\ 89 \end{array}$$

Thus $$(-192) + 281 = +(281 - 192) = 89$$

the sign of 281

EXAMPLE 8

How much change do you receive if you pay for a \$27 pair of shoes with a \$50 bill?

Solution. *Subtract* 27 from 50 (or add -27 to 50).

$$\begin{array}{r} 50 \\ \underline{27} \\ 23 \end{array}$$

You receive \$23 change.

EXERCISES

In exercises 1–20, add as indicated.

1. $14 + 8$
2. $25 + 15$
3. $22 + 0$
4. $0 + (-17)$
5. $(-19) + (-17)$
6. $(-24) + (-31)$
7. $18 + (-15)$
8. $(-18) + 15$
9. $16 + (-24)$
10. $94 + (-94)$
11. $\begin{array}{r} 472 \\ \underline{176} \end{array}$
12. $\begin{array}{r} 382 \\ \underline{199} \end{array}$
13. $\begin{array}{r} -153 \\ \underline{-138} \end{array}$
14. $\begin{array}{r} -934 \\ \underline{-179} \end{array}$
15. $\begin{array}{r} 386 \\ \underline{-179} \end{array}$
16. $\begin{array}{r} -849 \\ \underline{273} \end{array}$
17. $\begin{array}{r} -1086 \\ \underline{789} \end{array}$
18. $\begin{array}{r} -8407 \\ \underline{-3089} \end{array}$
19. $\begin{array}{r} -7047 \\ \underline{8439} \end{array}$
20. $\begin{array}{r} 2085 \\ \underline{-7986} \end{array}$

In exercises 21–36, *subtract* the bottom number from the top one by adding the inverse of the bottom number.

SAMPLE. Subtract: $\begin{array}{r} 9 \\ \underline{-4} \end{array}$	***Solution.*** Add the inverse of −4 to 9. $\begin{array}{r} 9 \\ \underline{4} \\ 13 \end{array}$

21. $\begin{array}{r} 10 \\ \underline{3} \end{array}$
22. $\begin{array}{r} 10 \\ \underline{-3} \end{array}$
23. $\begin{array}{r} -10 \\ \underline{3} \end{array}$
24. $\begin{array}{r} -10 \\ \underline{-3} \end{array}$
25. $\begin{array}{r} 36 \\ \underline{17} \end{array}$
26. $\begin{array}{r} 104 \\ \underline{91} \end{array}$
27. $\begin{array}{r} -38 \\ \underline{-22} \end{array}$
28. $\begin{array}{r} -83 \\ \underline{-96} \end{array}$
29. $\begin{array}{r} 47 \\ \underline{54} \end{array}$
30. $\begin{array}{r} 51 \\ \underline{-65} \end{array}$
31. $\begin{array}{r} -214 \\ \underline{107} \end{array}$
32. $\begin{array}{r} -284 \\ \underline{392} \end{array}$
33. $\begin{array}{r} 2817 \\ \underline{-1719} \end{array}$
34. $\begin{array}{r} -1916 \\ \underline{-1827} \end{array}$
35. $\begin{array}{r} 1492 \\ \underline{9384} \end{array}$
36. $\begin{array}{r} -7105 \\ \underline{8808} \end{array}$

In exercises 37–46, subtract as indicated.

37. $9 - 4$
38. $9 - (-4)$
39. $-9 - 4$
40. $-9 - (-4)$
41. $12 - 0$
42. $0 - 12$
43. $-18 - 17$
44. $-18 - (-17)$

45. $102 - 104$

46. $-102 - 104$

47. Fill in "<" or ">": $6 + 2$ ☐ $6 - 2$

48. Find $|5 + (-7)|$.

49. Find $|(-2) - (-1)|$.

50. Subtract 17 from 24.

51. Subtract -12 from 19.

52. Subtract -30 from -60.

53. A car travels 40 miles north and then 50 miles south. How far from the starting point is the car? Is it north or south of the starting point?

54. How much change do you receive when you purchase a \$13 item with a \$20 bill?

55. If you owe Tom \$12 and you owe Bill \$9, how much do you owe?

56. A man buys a dozen eggs and breaks three of them on the way home. How many eggs does he have left?

57. On their first date, Jerry's girl orders a steak dinner priced at \$11. Jerry has only a \$20 bill in his wallet. If he must leave \$3 for taxes and a tip, how much can he spend on his own dinner?

2.2 COMMUTATIVE AND ASSOCIATIVE LAWS

Commutative Law

EXAMPLE 1

(a) $8 + 6 = 6 + 8 = 14$
(b) $8 - 6 = 2$
(c) $6 - 8 = -2$

You can add numbers in either order. But when you subtract, the order is crucial. Thus for real numbers a and b,

$$a + b = b + a$$

This is called the **Commutative Law of Addition.** However, if $a \neq b$, then

$$a - b \neq b - a$$

Associative Law

When you add three numbers a, b, and c, you get the same result if you first add a and b, and then c, or if you add the sum of b and c to a:

$$(a + b) + c = a + (b + c)$$

This is known as the **Associative Law of Addition.** Note that parentheses indicate which sum occurs first. Because of the Associative Law you can write the sum

$$a + b + c$$

without parentheses. Thus

$$a + b + c = (a + b) + c = a + (b + c)$$

EXAMPLE 2

$$\begin{aligned} (8 + 5) + 6 &= 13 + 6 = 19 \\ 8 + (5 + 6) &= 8 + 11 = 19 \end{aligned}$$

Use of These Laws

The Associative and Commutative Laws enable you to rearrange numbers when adding them. For example,

$$\begin{aligned} -5 + 6 + (-2) &= [-5 + 6] + (-2) \\ &= [6 + (-5)] + (-2), && \text{by the Commutative Law} \\ &= 6 + [(-5) + (-2)], && \text{by the Associative Law} \\ &= 6 + [-7] \\ &= -1 \end{aligned}$$

If the numbers are not all of the same sign:

1. First add the positive numbers to obtain A.
2. Then add the negative numbers to obtain $-B$.
3. The sum of all the numbers is $A + (-B)$, or

$$A - B.$$

EXAMPLE 3

Add:

$$\begin{array}{r} 12 \\ -8 \\ 19 \\ 20 \\ -9 \\ \underline{-17} \end{array}$$

Solution. First add the positive and negative numbers separately:

$$\begin{array}{rr} 12 & -8 \\ 19 & -9 \\ \underline{20} & \underline{-17} \\ 51 & -34 \end{array}$$

The sum of all the numbers is $51 - 34$, or 17.

Parentheses

When addition and subtraction are combined in the same example, parentheses and brackets must often be used to clarify the intention. First combine the numbers within each pair of parentheses.

EXAMPLE 4

Find the value of

$$8 + (5 - 3) - (3 - 4).$$

Solution.

$$\begin{aligned} 8 + \underbrace{(5 - 3)}_{2} - \underbrace{(3 - 4)}_{-1} &= 8 + 2 - (-1) \\ &= 8 + 2 + 1 \\ &= 11 \end{aligned}$$

There may be pairs of parentheses within a pair of brackets. Combine within the *inner* pairs first, as in Examples 5 and 6.

EXAMPLE 5

Find the value of

$$6 - [2 - (7 - 3)].$$

Solution.

$$\begin{aligned} 6 - [2 - \underbrace{(7 - 3)}_{4}] &= 6 - \underbrace{[2 - 4]}_{-2} \\ &= 6 - (-2) \\ &= 6 + 2 \\ &= 8 \end{aligned}$$

EXAMPLE 6

Find the value of

$$10 - [(7 - 2) - (8 - 3)].$$

Solution. Here two pair of parentheses lie within a pair of brackets. Combine within *both* inner parentheses first.

$$\begin{aligned} 10 - [\underbrace{(7 - 2)}_{5} - \underbrace{(8 - 3)}_{5}] &= 10 - [5 - 5] \\ &= 10 - 0 \\ &= 10 \end{aligned}$$

If no parentheses are given, combine in the order written, from left to right.

EXAMPLE 7

Find the value of

$$20 - 7 - 8 + 9 - 3.$$

Solution.

$$\begin{array}{l} 20 - 7 - 8 + 9 - 3 \\ \quad 13 - 8 + 9 - 3 \\ \qquad 5 + 9 - 3 \\ \qquad\quad 14 - 3 \\ \qquad\qquad 11 \end{array}$$

The value is 11.

EXERCISES

In exercises 1–51, find each value.

1. $8 + 4 + 10$
2. $15 + 7 + 20 + 5$
3. $(-9) + (-3) + (-5)$
4. $(-12) + (-7) + (-2) + (-11)$

5. $\begin{array}{r} 17 \\ 4 \\ \underline{18} \end{array}$
6. $\begin{array}{r} 93 \\ 21 \\ \underline{55} \end{array}$
7. $\begin{array}{r} 31 \\ 82 \\ \underline{17} \end{array}$
8. $\begin{array}{r} 101 \\ 27 \\ \underline{212} \end{array}$

9. $\begin{array}{r} -11 \\ -19 \\ \underline{-17} \end{array}$
10. $\begin{array}{r} -31 \\ -63 \\ \underline{-8} \end{array}$
11. $\begin{array}{r} -15 \\ -102 \\ -711 \\ \underline{-99} \end{array}$
12. $\begin{array}{r} -305 \\ -909 \\ -882 \\ \underline{-37} \end{array}$

13. $17 + (-15) + 9$
14. $12 + (-7) + (-15) + 18$
15. $7 + 6 + (-3) + (-5) + 10$
16. $13 + (-2) + (-5) + (-19) + 17 + 18$
17. $54 + 19 + (-17) + (-36) + (-22) + 47$
18. $27 + (-36) + (-79) + 45 + (-55) + (-32)$

19. $\begin{array}{r} 27 \\ -32 \\ 33 \\ \underline{-49} \end{array}$
20. $\begin{array}{r} 43 \\ -8 \\ 17 \\ \underline{92} \end{array}$
21. $\begin{array}{r} 53 \\ -19 \\ 29 \\ -34 \\ -59 \\ \underline{-85} \end{array}$

22. $\begin{array}{r} 107 \\ 393 \\ -242 \\ -116 \\ -55 \\ \underline{808} \end{array}$
23. $\begin{array}{r} 916 \\ -53 \\ -741 \\ 808 \\ -747 \\ \underline{831} \end{array}$
24. $\begin{array}{r} 1082 \\ 5019 \\ -1131 \\ -2163 \\ -1442 \\ \underline{3042} \end{array}$

25. $16 - 5 - 2$
26. $27 - 13 - 19$
27. $8 - 4 - 2 - 1$
28. $10 - 3 - 2 - 4 - 1$
29. $6 - (5 - 3)$
30. $6 - 5 - 3$
31. $8 - (3 - 4)$
32. $10 - 4 - (7 - 2)$
33. $19 - (8 - 4) - (17 - 12)$
34. $10 - (6 - 3) - (9 - 2) - (1 - 8)$
35. $(7 - 2) - (4 - 3)$
36. $6 - (3 - 2) - (1 - 6)$
37. $16 - [(4 - 5) - 1]$
38. $16 - [4 - (5 - 1)]$
39. $(10 - 4) - [6 - (3 - 1)]$
40. $[(7 - 12) - 5] - [8 - (2 - 7)]$
41. $20 - [(7 - 3) - (2 - 1)]$
42. $[(8 - 3) - (9 - 13)] - 2$
43. $6 + 8 - 4 + 7$
44. $12 - 9 + 7 - 4 - 2$
45. $8 + 10 - 6 + 7 - 2$
46. $16 - 4 - 12 - 8 + 2$
47. $12 + (4 - 3) - (4 + 5)$
48. $10 - 8 - 7 - (2 + 5 + 1)$
49. $7 + [(6 - 3 + 2) - (7 + 19)]$
50. $3 - [(8 - 5) - (4 + 7)]$
51. $|2 - (5 - 9)|$
52. Subtract 7 from the sum of 10 and 5.
53. The sum of 9 and 8 is subtracted from 17. Find the resulting value.
54. A gambler wins $20, loses $35, wins $37, and then loses $6. How much has he won or lost?
55. A traveling salesman drives 10 miles west, then 4 miles east, and then 6 miles west. Where is he in relation to his starting point?
56. How much change do you receive if you purchase $5, $6, and $8 items with a $20 dollar bill?

2.3 MULTIPLICATION

Multiplication by a Positive Integer

The product of a and b is usually written as

$$ab,$$
$$a \cdot b,$$
$$a \times b,$$

or as

$$\begin{array}{r} a \\ \times\, b \\ \hline \end{array}$$

The numbers a and b are called the **factors** of this product. Sometimes the factors of a product are enclosed in parentheses, as in

$$(-5)(-2).$$

Multiplication by a positive integer p amounts to repeated addition. Thus

$$m \cdot p$$

means $$\underbrace{m + m + m + \ldots + m}_{p \text{ times}}.$$

EXAMPLE 1

(a) $5 \cdot 3 = 5 + 5 + 5 = 15$
(b) $(-3)2 = (-3) + (-3) = -6$

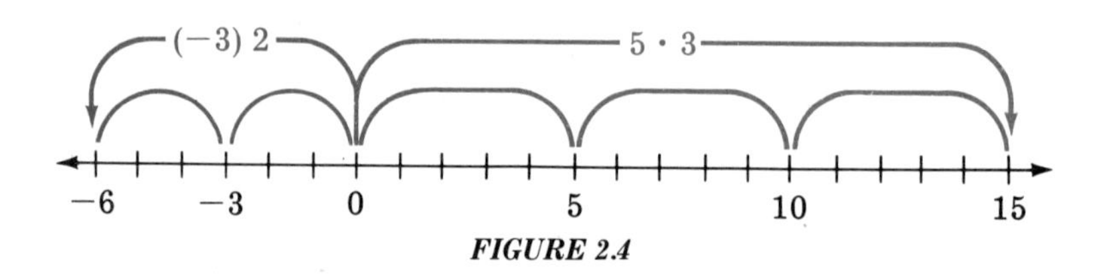

FIGURE 2.4

EXAMPLE 2

Multiply 729 by 214.

Solution.

$$\begin{array}{r} 729 \\ \times 214 \\ \hline 2916 \\ 729 \\ 1458 \\ \hline 156006 \end{array}$$

EXAMPLE 3

Each of the 23 workers in a factory receives \$135 per week. How large is the weekly payroll?

Solution.

$$\begin{array}{r} \$135 \\ \times 23 \\ \hline 405 \\ 270 \\ \hline \$3105 \end{array}$$

Commutative and Associative Laws

Multiplication is both commutative and associative, as is addition. Thus let a, b, and c be any numbers.

$ab = ba$	Commutative Law of Multiplication
$(ab)c = a(bc)$	Associative Law of Multiplication

Because of the Associative Law, omit parentheses and write

$$abc = (ab)c = a(bc).$$

EXAMPLE 4

(a) $4 \cdot 5 = 5 \cdot 4 = 20$

(b) $(3 \cdot 2)4 = 6 \cdot 4 = 24$
$3(2 \cdot 4) = 3 \cdot 8 = 24$

Multiplying By 0

When *at least one* of the factors is 0, the product is 0. If *none* of the factors is 0, the product is nonzero.

EXAMPLE 5

(a) $6 \cdot 0 = 0 \cdot 2 = 0$

(b) $3 \cdot 8 \cdot 7 \cdot 10 \cdot 0 = 0$

(c) If $a \neq 0$ and $b \neq 0$, then $ab \neq 0$

Negative Factors

When you multiply a number a by -1, you obtain $-a$, the inverse of a. *When you multiply a nonzero number a by a negative number, change the sign of a to obtain the sign of the product.* Thus, if a is positive, the product is negative; if a is negative, the product is positive.

EXAMPLE 6

(a) $3(-1) = -3$

(b) $(-3)(-1) = 3$

(c) $3(-1)(-2) = 6$. Here there are two negative factors, and thus two sign changes. The product is positive.

(d) $3(-1)(-2)(-2) = -12$. There are three negative factors. The product is negative.

(e) $3(-1)(-2)(-2)(-2) = 24$. With four negative factors, the product is positive.

1. The product of nonzero factors is positive if there is an *even number* (0, 2, 4, 6 . . .) of negative factors (or if all of the factors are positive).
2. The product of nonzero factors is negative if there is an *odd number* (1, 3, 5, . . .) of negative factors.

In particular, *ab is positive if both factors are positive or if both are negative, and ab is negative if one factor is positive and the other negative.*

EXAMPLE 7

(a) $\underbrace{2 \cdot 5}_{10}$ and $\underbrace{(-2)(-5)}_{10}$ are each positive.

(b) $\underbrace{2(-5)}_{-10}$ and $\underbrace{(-2)5}_{-10}$ are each negative.

EXERCISES

In exercises 1–32, multiply as indicated.

1. $8 \cdot 9$
2. $12 \cdot 14$
3. $62 \cdot 7$
4. $53 \cdot 9$
5. $\begin{array}{r} 67 \\ \times 31 \\ \hline \end{array}$
6. $\begin{array}{r} 88 \\ \times 92 \\ \hline \end{array}$
7. $\begin{array}{r} 385 \\ \times 79 \\ \hline \end{array}$
8. $\begin{array}{r} 207 \\ \times 110 \\ \hline \end{array}$
9. $\begin{array}{r} 582 \\ \times 916 \\ \hline \end{array}$
10. $\begin{array}{r} 1847 \\ \times 656 \\ \hline \end{array}$
11. $(-6)9$
12. $12(-8)$
13. $(-7)(-10)$
14. $(-11)(-12)$
15. $\begin{array}{r} 26 \\ \times -21 \\ \hline \end{array}$
16. $\begin{array}{r} -34 \\ \times 36 \\ \hline \end{array}$
17. $\begin{array}{r} -93 \\ \times -88 \\ \hline \end{array}$
18. $\begin{array}{r} -362 \\ \times 107 \\ \hline \end{array}$
19. $\begin{array}{r} -294 \\ \times -389 \\ \hline \end{array}$
20. $\begin{array}{r} -2067 \\ \times -3002 \\ \hline \end{array}$
21. $6 \cdot 5 \cdot 4$
22. $3 \cdot 8 \cdot 2 \cdot 2$
23. $3(-2)(-1)$
24. $5(-2)(-4)$
25. $(-7)(-8)(-2)$
26. $(-3)9(-2)(-1)$
27. $(-4)(-2)(-1)(-5)$
28. $(-2)(-3)(-2)(-1)(-2)$
29. $16 \cdot 0$
30. $4 \cdot 5 \cdot 0(-1)$
31. $12(-10)9(-1)(-2)(-1)$
32. $(-8)(-4)(-6)(-3)(-5)$

In exercises 33–36, fill in "<" or ">".

33. $2 \cdot 3$ ☐ $2 \cdot 4$ 34. $2(-3)$ ☐ $2(-4)$ 35. $(-1)0$ ☐ $(-1)1$

36. $(-2)(-4)$ ☐ $5(-5)$

37. Find the product of 15 and −3.

38. Find the product of 2, 3, and −5.

39. A box can hold 250 sheets of paper. How many sheets can 6 boxes hold?

40. A man loses $2 in each of 8 races. How much does he lose altogether?

41. Twenty-two students in a class each contribute three dollars to a political candidate. How much money is collected from these students?

42. Two thousand four hundred eight fans each pay $4 to attend a basketball game. How much money is grossed from attendance?

2.4 DIVISION

Multiplication and Division

Subtraction was defined in terms of addition. Now division will be defined in terms of multiplication.

EXAMPLE 1

$$10 = 5 \cdot 2$$

Here 10 is the *product.* 5 and 2 are each *factors.* In division, you will write

$$\frac{10}{5} = 2,$$

and call 10 the *dividend,* 5 the *divisor,* and 2 the *quotient.*

Definition

DIVISION. Let a, b, and c be real numbers, with $b \neq 0$. Let

$$\frac{a}{b} = c,$$

if

$$a = bc.$$

In this case, a is called the ***dividend,*** *b, the* ***divisor,*** *and c, the* ***quotient.***

Division of a by b is also written as $a \div b$ and as $b\overline{)a}$.

It is understood that whenever $\frac{a}{b} = c$, then $a = bc$. *Thus to check division, find the product bc (or cb, if more convenient). You should obtain the dividend, a.*

EXAMPLE 2

Find the value of:

$$32\overline{)416}$$

Check your answer.

Solution. You are asked to divide 416 by 32.

$$\begin{array}{rrl} \textbf{\textit{quotient}} \rightarrow & 13 & \\ \textbf{\textit{divisor}} \rightarrow \quad 32 & \overline{)416} & \leftarrow \textbf{\textit{dividend}} \\ & 32 & \\ \hline & 96 & \\ & 96 & \\ \hline \end{array}$$

CHECK.

$$\begin{array}{rr} \textbf{\textit{divisor}} \rightarrow & 32 \\ \textbf{\textit{quotient}} \rightarrow & \times 13 \\ \hline & 96 \\ & 32 \\ \hline \textbf{\textit{dividend}} \rightarrow & 416 \end{array}$$

EXAMPLE 3

A profit of \$1045 is divided equally among 5 business partners. How much is each partner's share?

Solution.

$$\frac{1045}{5} = 209$$

Each partner receives \$209.

Division With a Remainder

$$12 = 4 \cdot 3$$

Thus when you divide 12 by 4, you obtain 3 as the quotient.

$$\frac{12}{4} = 3$$

But if you divide 13 by 4, you obtain the *rational number* $\frac{13}{4}$. In terms of integers, you obtain a "remainder."

$$\frac{13}{4} = 3\frac{1}{4}$$

Note that

$$13 = \underbrace{4 \cdot 3}_{12} + 1.$$

Let a and b be *integers,* with $b > 0$. Either

$$a = bc$$

for some integer c, or else

$$a = bc + r$$

for some integer c and for some *positive* integer r *less than* b.

If $a = bc + r$, as above, then

$$\frac{a}{b} = c + \frac{r}{b}$$

Here c is called the **quotient** and r is called the **remainder.** For example,

$$\frac{13}{4} = 3 + \frac{1}{4} \left(\text{or } 3\frac{1}{4}\right)$$

Thus when 13 is divided by 4, the quotient is 3 and the remainder is 1.

To check division when there is a remainder, recall that

$$a = bc + r.$$

Thus, *first multiply the divisor, b, by the quotient, c. Then add the remainder, r. You should obtain the dividend, a.*

EXAMPLE 4

Find the quotient and remainder when 37 is divided by 5. Check your result.

Solution.

$$\frac{37}{5} = 7 + \frac{2}{5}$$

The quotient is 7 and the remainder is 2.
CHECK:

$$5 \cdot 7 + 2 = 35 + 2 = 37$$

EXAMPLE 5

Find the quotient and remainder when 4382 is divided by 53. Check your result.

Solution.

$$
\begin{array}{rrl}
 & \textbf{\textit{quotient}} & \\
 & \downarrow & \\
 & 82 + \frac{36}{53} & \leftarrow \textbf{\textit{remainder}} \\
 & & \leftarrow \textbf{\textit{divisor}} \\
\textbf{\textit{divisor}} \rightarrow & 53 \overline{)4382} & \leftarrow \textbf{\textit{dividend}} \\
 & \underline{424} & \\
 & 142 & \\
 & \underline{106} & \\
 & 36 & \leftarrow \textbf{\textit{remainder}}
\end{array}
$$

The quotient is 82 and the remainder is 36.
CHECK.

$$
\begin{array}{rl}
53 & \leftarrow \textbf{\textit{divisor}} \\
\underline{\times 82} & \leftarrow \textbf{\textit{quotient}} \\
106 & \\
\underline{424} & \\
4346 & \\
\underline{+36} & \leftarrow \textbf{\textit{remainder}} \\
4382 & \leftarrow \textbf{\textit{dividend}}
\end{array}
$$

Division by 0

Division by 0 *is not defined.* To understand why not, first suppose you divide a *nonzero* number, such as 6, by 0. Try to use the above definition of division in terms of multiplication:

$$\frac{6}{0} = c$$

would mean that

$$6 = 0 \cdot c$$

But

$$0 \cdot c = 0 \text{ (instead of 6)}$$

Next, try to divide 0 by 0, according to the definition. Note that

$$0 = 0 \cdot 1,$$
$$0 = 0 \cdot 5,$$
$$0 = 0 \cdot 0.$$

You would obtain

$$\frac{0}{0} = 1, \quad \frac{0}{0} = 5, \quad \frac{0}{0} = 0.$$

There would be no *single* quotient c as there is when the divisor is nonzero. Therefore, division by 0 is not allowed.

Although you cannot *divide by* 0, the number 0 can be *divided by* any nonzero number. Thus

$$\frac{0}{4} = 0 \qquad \text{because } 0 = 4 \cdot 0$$

$$\frac{0}{-2} = 0 \qquad \text{because } 0 = (-2)0$$

For all $a \neq 0$,

$$\frac{0}{a} = 0$$

Sign of the Quotient.

Suppose $b \neq 0$. Then $\frac{a}{b}$ is positive if a and b have the same sign, and $\frac{a}{b}$ is negative if a and b have different signs.

EXAMPLE 6

(a) $\frac{20}{5} = 4$

(b) $\frac{-20}{-5} = 4$

(c) $\frac{-20}{5} = -4$

(d) $\frac{20}{-5} = -4$

When you divide products of numbers, count the *total number of negative factors in both numerator and denominator.* If there is an *even* number of negative factors, the quotient is *positive.* If there is an *odd* number, the quotient is *negative.*

EXAMPLE 7

(a)

$$\frac{14(-2)(-1)}{-7} = \frac{28}{-7} = -4$$

Here there are three negative factors, two in the numerator and one in the denominator.

(b)

$$\frac{(-3)(-5)}{(-2)(-4)} = \frac{15}{8} \left(\text{or } 1\frac{7}{8}\right)$$

Here there are four negative factors. The quotient is positive.

EXERCISES

In exercises 1–26, divide:

1. $\frac{8}{4}$
2. $\frac{15}{5}$
3. $\frac{36}{6}$
4. $\frac{35}{5}$

5. $\frac{12}{3}$ 6. $\frac{-12}{3}$ 7. $\frac{12}{-3}$ 8. $\frac{-12}{-3}$

9. $\frac{20}{10}$ 10. $\frac{-50}{-10}$ 11. $\frac{24}{-6}$ 12. $\frac{-84}{7}$

13. $\frac{42}{-6}$ 14. $\frac{-132}{-11}$ 15. $\frac{-160}{10}$ 16. $\frac{0}{-9}$

17. $15\overline{)180}$ 18. $18\overline{)198}$ 19. $22\overline{)1694}$ 20. $27\overline{)1053}$

21. $24\overline{)864}$ 22. $93\overline{)-3069}$ 23. $-104\overline{)-21\,008}$

24. $-95\overline{)14\,440}$ 25. $721\overline{)218\,463}$ 26. $1062\overline{)1\,005\,714}$

In exercises 27 – 36, divide and check your answer.

27. $\frac{108}{12}$ 28. $\frac{132}{-12}$ 29. $25\overline{)675}$ 30. $51\overline{)918}$

31. $73\overline{)2774}$ 32. $65\overline{)5980}$ 33. $117\overline{)14\,157}$

34. $365\overline{)292\,730}$ 35. $272\overline{)229\,024}$ 36. $991\overline{)316\,129}$

In exercises 37 – 48, find the quotient and remainder.

37. $\frac{102}{5}$ 38. $\frac{137}{12}$ 39. $15\overline{)160}$ 40. $20\overline{)252}$

41. $18\overline{)872}$ 42. $31\overline{)973}$ 43. $26\overline{)9085}$ 44. $73\overline{)7294}$

45. $802\overline{)7946}$ 46. $254\overline{)8493}$

47. $463\overline{)27\,452}$ 48. $1243\overline{)82\,467}$

In exercises 49 – 58, find the quotient and remainder. Also, check your answer.

49. $\frac{127}{11}$ 50. $\frac{152}{12}$ 51. $43\overline{)392}$ 52. $29\overline{)983}$

53. $38\overline{)1026}$ 54. $93\overline{)1876}$ 55. $132\overline{)6894}$

56. $794\overline{)82\,463}$ 57. $389\overline{)72\,651}$ 58. $1072\overline{)84\,931}$

In exercises 59–62, find the quotient, or indicate that division is not defined.

59. $\frac{0}{7}$ 60. $\frac{0}{-7}$ 61. $\frac{0}{0}$ 62. $\frac{7}{0}$

In exercises 63–68, multiply and divide, as indicated.

63. $\frac{4 \cdot 2 \cdot 3}{12}$ 64. $\frac{5(-1)(-2)}{10}$ 65. $\frac{(-1)(-4)}{(-1)2}$

66. $\frac{(-1)(-2)(-3)(-4)}{-6}$ 67. $\frac{(-3)(-6)(-1)}{(-1)(-2)}$ 68. $\frac{(-4)(-5)0(-1)}{(-2)(-1)}$

69. Each of 25 players receives an equal share of a World Series earning of $300225. How much is each player's share?

70. There are 32 rows in an auditorium. Each row contains the same number of seats. Altogether there are 1184 seats. How many seats are there in a row?

71. A cardboard box can hold 450 sheets of paper. How many boxes are needed to hold 48600 sheets of paper?

72. A carton can hold 32 cans of pineapple. How many cartons are needed to ship 26176 cans of pineapple?

2.5 EXPONENTS

Squares and Cubes

For any number a,

$$a^2 \text{ means } a \cdot a.$$

Call a^2 the **square of *a*,** or ***a* squared.**

EXAMPLE 1

(a) $4^2 = 4 \cdot 4 = 16$
(b) $9^2 = 9 \cdot 9 = 81$
(c) $0^2 = 0 \cdot 0 = 0$

For negative numbers, parentheses are used to clarify the intended meaning.

EXAMPLE 2

(a) $(-2)^2 = (-2)(-2) = 4$
(b) $-2^2 = -4$ because, *by convention,*

$$-2^2 = -(2^2) = -4$$

Here, in part (b), first square 2, and then consider the inverse of 2^2.

For any number a,

$$a^3 \text{ means } a \cdot a \cdot a.$$

Call a^3 the **cube of *a*** or ***a* cubed.**

EXAMPLE 3

(a) $3^3 = 3 \cdot 3 \cdot 3 = 27$
(b) $1^3 = 1 \cdot 1 \cdot 1 = 1$
(c) $(-2)^3 = (-2)(-2)(-2) = -(2 \cdot 2 \cdot 2) = -8$
(d) $-2^3 = -(2^3) = -8$
(e) $10^3 = 10 \cdot 10 \cdot 10 = 1000$

Positive Integral Powers

Let a be any number.

$$a^1 \text{ means } a.$$

Let n be a positive integer, where $n > 1$. Then

$$a^n \text{ means } \underbrace{a \cdot a \cdot a \ldots a}_{n \text{ factors}}.$$

Thus when considering a^n, a appears as a factor n times. Call a^n the ***n*th power of *a*,** or ***a* to the *n*th.**

EXAMPLE 4

Find: (a) 2^4 (b) 2^5 (c) 2^6

Solution.
(a) $2^4 = 2 \cdot 2 \cdot 2 \cdot 2 = 16$
(b) $2^5 = 2 \cdot 2 \cdot 2 \cdot 2 \cdot 2 = 32$
(c) $2^6 = 2 \cdot 2 \cdot 2 \cdot 2 \cdot 2 \cdot 2 = 64$

EXAMPLE 5

Find: (a) 10^4 (b) 10^5 (c) 10^8

Solution.
(a) $10^4 = 10000$ (with 4 zeros)
(b) $10^5 = 100000$ (with 5 zeros)
(c) $10^8 = 100000000$ (with 8 zeros)

Base and Exponent

Definition

> *BASE, EXPONENT. In the expression*
>
> $$a^n,$$
>
> *a is called the* ***base*** *and n the* ***exponent.***
>
> $$a^n$$
>
> *is also called the* ***nth power of a.***

EXAMPLE 6

$$4^3 = 64$$

Here 4 is the *base* and 3 the *exponent.* 4^3, or 64, is the *third power (or cube) of* 4.

EXAMPLE 7

(a) $1^n = 1$ for all positive integers n
(b) $0^n = 0$ for all positive integers n
(c) $(-1)^1 = -1$
$(-1)^2 = (-1)(-1) = 1$
$(-1)^3 = (-1)(-1)(-1) = -1$
$(-1)^4 = (-1)(-1)(-1)(-1) = 1$

Note the pattern:

$(-1)^n = -1$ for *odd* (positive) integers n
$(-1)^n = 1$ for *even* (positive) integers n

Example 7 can be generalized. *Odd powers of negative numbers are negative. Even powers of negative numbers are positive.* To see this, count the number of negative factors, as indicated by the exponent.

EXERCISES

In exercises 1–12 find the indicated squares or their inverses.

1. 3^2 2. 5^2 3. 8^2 4. 11^2

5. $(-3)^2$ 6. $(-7)^2$ 7. $(-10)^2$ 8. -10^2

9. 12^2 10. 20^2 11. $(-25)^2$ 12. -100^2

In exercises 13–22, find the indicated cubes or their inverses.

13. 2^3 14. 4^3 15. 5^3 16. 8^3

17. $(-1)^3$　18. -1^3　19. $(-5)^3$　20. $(-10)^3$

21. -20^3　22. $(-50)^3$

In exercises 23–40, find the indicated powers.

23. 3^4　24. 3^5　25. 3^6　26. $(-3)^4$

27. $(-3)^5$　28. $(-3)^6$　29. 4^3　30. 4^4

31. $(-5)^3$　32. $(-5)^4$　33. $(-1)^{10}$　34. $(-1)^{15}$

35. $(-1)^{20}$　36. $(-1)^{27}$　37. $(-2)^4$　38. $(-2)^5$

39. 10^7　40. $(-10)^{10}$

SAMPLE. What power of 2 is 8?	***Solution.*** $2^3 = 8$. Thus 8 is the *third* power of 2.

41. What power of 10 is 100?

42. What power of 3 is 27?

43. What power of (−2) is 4?

44. What power of 2 is 32?

2.6 ORDER OF OPERATIONS

Role of Parentheses

By convention, the expression

$$2 \cdot 3^2$$

means

twice the square of 3.

Thus

$$2 \cdot 3^2 = 2 \cdot 9 = 18$$

Note that the exponent 2 refers to the base 3. If you want to indicate the square of the product of 2 and 3, use parentheses. Thus,

$$(2 \cdot 3)^2 = 6^2 = 36$$

The expression

$$3 + 5 \cdot 2$$

means

add the product of 5 and 2 to 3.

Thus

$$3 + 5 \cdot 2 = 3 + 10 = 13$$

If the sum of 3 and 5 is to be multiplied by 2, use parentheses. Thus

$$(3 + 5)2 = 8 \cdot 2 = 16$$

Rules for Ordering Operations

The order in which addition, subtraction, multiplication, division, and raising to a power are applied in an example is often crucial. *If parentheses are given, first perform the operations within parentheses.* Otherwise:

1. First raise to a power.
2. Then multiply or divide from left to right.
3. Then add or subtract from left to right.

EXAMPLE 1

Find the value of: $7(2 - 1)$

Solution. First perform operations within parentheses.

$$7(2 - 1) = 7 \cdot 1 = 7$$

EXAMPLE 2

Find the value of: $2 + 5^2 \cdot 3$

Solution. First raise to a power; then multiply; then add.

$$\begin{aligned} 2 + 5^2 \cdot 3 &= 2 + 25 \cdot 3 && \text{because } 5^2 = 25 \\ &= 2 + 75 && \text{because } 25 \cdot 3 = 75 \\ &= 77 \end{aligned}$$

EXAMPLE 3

Find the value of:

$$\frac{-20}{5} - 1$$

Solution. First divide; then subtract.

$$\frac{-20}{5} - 1 = -4 - 1 = -5$$

EXAMPLE 4

Find the value of:

$$24 \div 6 + 3 \cdot 2$$

Solution. First divide 24 by 6 and multiply 3 by 2. Then add these results.

$$24 \div 6 + 3 \cdot 2 = 4 + 6 = 10$$

EXAMPLE 5
Find the value of:

$$\left(\frac{11-5}{3}-1\right)^2$$

Solution. Think of this as

$$([(11-5)\div 3]-1)^2$$

and work from the innermost pair of parentheses outward. Thus first find $11-5$, then divide $11-5$ by 3, then subtract 1, and finally square the result. Therefore

$$\begin{aligned}\left(\frac{11-5}{3}-1\right)^2 &= \left(\frac{6}{3}-1\right)^2\\ &= (2-1)^2\\ &= 1^2\\ &= 1\end{aligned}$$

Distributive Laws

Observe that

$$4(3+2) = 4\cdot 5 = 20.$$

But you could also calculate as follows:

$$\begin{aligned}4(3+2) &= 4\cdot 3 + 4\cdot 2\\ &= 12+8\\ &= 20\end{aligned}$$

Next, observe that

$$4(3-2) = 4\cdot 1 = 4$$

As above, you can obtain this result by a second method:

$$\begin{aligned}4(3-2) &= 4\cdot 3 - 4\cdot 2\\ &= 12-8\\ &= 4\end{aligned}$$

The second method, though longer, has an important use in algebra, as you will see in the next chapter. This method illustrates the following **Distributive Laws,** which concern products of sums and of differences.

Let a, b, and c be any real numbers. Then

$$\begin{aligned}a(b+c) &= ab+ac\\ (b+c)a &= ba+ca\\ a(b-c) &= ab-ac\\ (b-c)a &= ba-ca\end{aligned}$$

EXAMPLE 6

Use the Distributive Laws to find the value of:

(a) $(2 + 8)(-5)$
(b) $(7 - 3)6$

Solution.

(a) $(2 + 8)(-5) = 2(-5) + 8(-5)$
$= -10 - 40$
$= -50$

(b) $(7 - 3)6 = 7 \cdot 6 - 3 \cdot 6$
$= 42 - 18$
$= 24$

Observe that

$$-a = (-1)a.$$

Thus

$$-2 = (-1)2$$

and

$$-(3 - 4) = (-1)(3 - 4)$$

EXAMPLE 7

Use the Distributive Laws to find the value of:

$$-(3 - 4)$$

Solution.

$$\begin{aligned} -(3 - 4) &= (-1)(3 - 4) \\ &= (-1)3 - (-1)4 \\ &= -3 + 4 \\ &= 1 \end{aligned}$$

In general,

$$\begin{aligned} -(b + c) &= (-1)(b + c) \\ &= (-1)b + (-1)c \qquad \text{by the Distributive Law} \\ &= -b - c \end{aligned}$$

Similarly,

$$\begin{aligned} -(b - c) &= (-1)(b - c) \\ &= (-1)b - (-1)c \qquad \text{by the Distributive Law} \\ &= -b + c \end{aligned}$$

Therefore

$$\begin{aligned} a - (b - c) &= a + [-(b - c)] \\ &= a - b + c \end{aligned}$$

Because

$$(a - b) - c = a - b - c,$$

it follows that when $c \neq 0$,

$$a - (b - c) \neq (a - b) - c$$

Thus *the Associative Law does not hold for subtraction.*

The Distributive Laws also apply when more than two numbers are added or subtracted. For example,

$$a(b + c + d) = ab + ac + ad$$
$$a(b + c - d - e) = ab + ac - ad - ae$$

EXAMPLE 8

Use the Distributive Laws to find the value of:

$$4(3 + 2 - 1)$$

Solution.

$$\begin{aligned} 4(3 + 2 - 1) &= 4 \cdot 3 + 4 \cdot 2 + 4(-1) \\ &= 12 + 8 - 4 \\ &= 16 \end{aligned}$$

EXERCISES

In exercises 1–24, find each value.

1. $3 + 5 \cdot 4$
2. $(3 + 5)4$
3. $4 \cdot 2^2$
4. $(4 \cdot 2)^2$
5. $\frac{4 + 2}{2}$
6. $\frac{4}{2} + 2$
7. $4 + \frac{2}{2} - 3 \cdot 7$
8. $4^2 - 2^2$
9. $4 - 2^2$
10. $(4 - 2)^2$
11. $(2 - 4)^2$
12. $2 - 4^2$
13. $(2 + 5)^2 - 1$
14. $2 + 5^2 - 1$
15. $\frac{(7 - 2)^2}{5}$
16. $\frac{7 - 2^2}{5}$
17. $5(3 - 1) + 2^2$
18. $[5(3 - 1) + 2]^2$
19. $\left(\frac{6 + 5}{11} - 2\right)^2$
20. $\frac{8 + 7}{5} - \frac{3 + 4}{-7}$
21. $\left[\frac{3 - (4 + 5)}{2}\right]^3$
22. $2\left(\frac{3 - 4 + 5}{2}\right)^3$
23. $\frac{(6 - 2)^2}{2} - 3(5 - 3)^3$
24. $\frac{(5 - 2)^2 - (-1 + 3)^2}{5}$
25. $6 \cdot 6 - 6 \div 6$
26. $6(6 - 6 \div 6)$
27. $6(6 - 6) \div 6$
28. $(6 \cdot 6 - 6) \div 6$

In exercises 29–44, use the Distributive Laws to find each value.

SAMPLE. Find the value of: $(6 + 2)(-2)$	*Solution.* $(6 + 2)(-2) = 6(-2) + 2(-2)$ $= -12 - 4$ $= -11$

29. $(8 + 1)3$
30. $2(1 + 4)$
31. $(-3)(6 + 1)$
32. $(8 + 2)(-5)$
33. $-(4 + 8)$
34. $-(4 - 1)$
35. $(3 - 8)(-2)$
36. $(-7)(7 - 4)$
37. $6(5 - 9)$
38. $7(8 - 10)$
39. $3(2 + 1 + 4)$
40. $6(8 + 5 - 1)$
41. $9(6 - 3 + 2)$
42. $2(-4 - 1 - 5)$
43. $8(2 + 1 + 3 + 4)$
44. $10(6 + 7 + 2 - 1)$
45. Find the value of:
    (a) $4 - (3 - 2)$ (b) $(4 - 3) - 2$

SAMPLE. Divide the sum of 8 and 2 by 5.	*Solution.* $\frac{8 + 2}{5} = \frac{10}{5}$ $= 2$

46. Multiply the sum of 7 and 2 by 6.
47. Add 8 to the product of 4 and 3.
48. Divide the sum of 4 and 5 by 3.
49. Subtract 4 from the cube of 2.
50. Divide the square of (-4) by the cube of (-2).

What Have You Learned in Chapter 2?

You can now add, subtract, multiply, and divide integers.

You can work with exponents.

You know in what order to apply these arithmetic operations in a given problem.

Let's Review Chapter 2.

2.1 Addition and Subtraction

1. Add: $27 + (-35)$
2. Add:
$$\begin{array}{r} -517 \\ \underline{-895} \end{array}$$
3. *Subtract* the bottom number from the top one by adding the inverse of the bottom number.
$$\begin{array}{r} -421 \\ \underline{-536} \end{array}$$
4. Subtract as indicated: $-29 - (-15)$

2.2 Commutative and Associative Laws

In exercises 5–7, find each value.

5. $(-7) + (-4) + (-6)$
6.
$$\begin{array}{r} 492 \\ -85 \\ -723 \\ 186 \\ \underline{-253} \end{array}$$
7. $8 - [5 - (2 - 3) - (1 - 4)]$
8. Subtract the sum of 5 and 10 from 36.

2.3 Multiplication

In exercises 9–11, multiply as indicated.

9.
$$\begin{array}{r} 281 \\ \underline{\times 124} \end{array}$$
10. $12(-9)$
11. $(-6)(-3)(-1)(-4)$
12. Find the product of 7, 3 and -10.

2.4 Division

13. Divide: (a) $\frac{-144}{12}$ (b) $\frac{-144}{-12}$
14. Divide and check your answer: $61\overline{)427}$
15. Find the quotient and remainder. Also, check your answer: $27\overline{)147}$
16. Multiply and divide, as indicated: $\frac{(-5)(-9)}{3(-1)}$

2.5 Exponents

17. Find: 9^2
18. Find: $(-3)^3$
19. Find: $(-1)^5$
20. What power of 2 is 16?

2.6 Order of Operations

In exercises 21–25, find each value.

21. $5 + 2 \cdot 3$
22. $(5 + 2)3$
23. $\frac{6}{2} + 7 + \frac{-8}{4}$
24. $\frac{(2-4)^3}{2} - \left(\frac{4}{2}\right)^2$
25. $5(7 + 4 - 3)$
26. Subtract 5 from the square of 10.

And these from Chapter 1:

27. Express $-.9$ in the form $\frac{N}{D}$, where N and D are integers and $D \neq 0$.
28. Fill in "$<$" or "$>$":
    (a) $4 + 1$ ☐ $4 - 1$
    (b) $6 - (2 - 1)$ ☐ $6 - 2 - 1$
29. Find: $|4 - 7|$
30. Which numbers are 20 units from the origin?

Try These Exam Questions for Practice.

1. Add:

$$\begin{array}{r} 4087 \\ 245 \\ -1159 \\ -2721 \\ 5003 \\ \hline \end{array}$$

2. Multiply: $(-4)2(-5)3(-1)$

3. Find the quotient and remainder. Also, check your answer:
$34\overline{)989}$

4. Multiply and divide, as indicated: $\dfrac{(-2)(-8)(-4)}{16}$

5. Find: $(-5)^2$

6. Find: $\dfrac{(5-2)^2}{3} - [2-(5-7)]$

7. What power of 3 is 27?

8. Multiply the sum of 5 and 2 by -3.

3

Algebraic Expressions

3.1 TERMS

Variables

In arithmetic you add, subtract, multiply, and divide individual numbers. Sometimes the same arithmetic process can apply to all numbers under discussion. It is important to have symbols available that can stand for any one of several numbers. For example, if you want to consider adding 10 to a number, you can express this by

$$x + 10.$$

If you want to express twice a number, write

$$2x.$$

Definition

> *A **variable** is a symbol that designates any one of the numbers being discussed.*

Generally, the letters x, y, and z will be used as variables. Other letters, such as a, b, c, m, n, r, s, t can also be used as variables.

A variable can be used to designate *any* real number. However, a variable can also be used to stand for only *some* of the real numbers. For example, a variable can stand only for integers or only for positive integers.

EXAMPLE 1

When you say,

Let x be any real number,

the symbol x is a variable. It represents every number.

EXAMPLE 2

When you say,

Let n be a nonzero integer,

the symbol n is a variable. It can be replaced by each integer, other than 0. Thus n represents

$$\ldots, -3, -2, -1, \quad 1, 2, 3, \ldots$$

Whatever you say about n applies to each of these integers.

Variables make it possible to apply arithmetic concepts to many numbers at one time. For example, if the variables represent real numbers, you can add, subtract, multiply, and divide these variables just as you can real numbers. Powers of variables can also be considered. The arithmetic rules of Chapter 2 can be adapted to variables. Algebraic expressions containing variables, such as

$$x + 1$$
$$x - y$$
$$xy$$
$$x^2$$
$$x^2 + 3x,$$

will be used throughout the book. Most of this chapter will be a preparation for using variables in later applications.

What Is a Term?

Definition

> *TERM. A* ***term*** *is a product of numbers and variables. A number, by itself, is also considered to be a term, as is a variable, by itself.*

EXAMPLE 3

Show that each of the following expressions is a term:

(a) $2x$
(b) xyz
(c) 5
(d) t

Solution.
(a) $2x$ is the product of 2 and the variable x.
(b) xyz is the product of the variables x, y, and z.
(c) 5 is a number. As such, it is a term.
(d) A variable, such as t, is also a term.

EXAMPLE 4

Show that each of the following is a term:
(a) x^2
(b) $3x^2y^3$
(c) $(5x)(2y)$

Solution.
(a) x^2 stands for $x \cdot x$. Thus it is the product of variables.
(b) $3x^2y^3 = 3 \cdot x \cdot x \cdot y \cdot y \cdot y$
Thus $3x^2y^3$ is the product of a number and variables.
(c) $(5x)(2y)$ is the product of the numbers 5 and 2 and the variables x and y. The Commutative Law of Multiplication will enable you to rewrite this expression as $5 \cdot 2 \cdot x \cdot y$ or $10xy$.

Coefficients

Definition

> *COEFFICIENT. The numerical factor (or the product of numerical factors) in a term is known as the* ***numerical coefficient,*** *or simply, the* ***coefficient,*** *of the term. If only variables appear, the coefficient of the term is understood to be 1 or −1. Thus the coefficient of xy is 1, and of $-xy$ is −1.*

If the term is a number, then it is its own coefficient. For example, the coefficient of the term 2 is 2, itself.

EXAMPLE 5

Find the coefficient of each term.
(a) $3x$ (b) $-4x^2y$ (c) $(5x)(-3z)$ (d) $\frac{1}{3}$ (e) $\frac{t^2}{3}$

Solution.
(a) 3
(b) −4
(c) −15 [Note that $-15 = 5(-3)$.]
(d) $\frac{1}{3}$ [The numerical term $\frac{1}{3}$ is its own coefficient.]
(e) $\frac{1}{3}$ $\left[\text{Note that } \frac{t^2}{3} = \frac{1t^2}{3} = \frac{1}{3}\,t^2.\right]$

When you multiply by 0, the product is 0. Thus *any term,* such as $0xy$, *that has* 0 *as its coefficient reduces to* 0.

Like Terms

Definition

> ***Like terms** are terms that differ only in their coefficients or in the order of their variables (or in both of these).*

EXAMPLE 6

(a) $5xy$ and $-3xy$ are like terms. They differ only in their coefficients, 5 and -3.
(b) $2xz$ and $3zx$ are like terms. They differ only in their coefficients and in the order of their variables.
(c) $2xz$ and $2zx$ are like terms. They differ only in the order of their variables.
(d) $2x^2yz$ and $\frac{zx^2y}{2}$ are like terms. They differ only in their coefficients, 2 and $\frac{1}{2}$, and in the order of their variables.
(e) s^2t^2 and $stst$ are like terms. Note that

$$s^2t^2 = stst.$$

EXAMPLE 7

(a) 3 and $2x$ are unlike terms. Only one of these contains a variable.
(b) x^2 and x^3 are unlike terms. They differ in the powers of their variables.
(c) x^2y and xy are unlike terms.

Any two numbers are like terms because neither contains any variables. Thus 5 and $\frac{-3}{4}$ are like terms.

EXERCISES

In exercises 1–10, write each term as a product of a number and variables.

SAMPLE.	*Solution.*
$\frac{3x^2y^3}{4}$	$\frac{3}{4} \cdot x \cdot x \cdot y \cdot y \cdot y$

1. t^2
2. x^4
3. $5xy^2$
4. $10x^2yz^2$
5. $\frac{a^2x^2}{2}$
6. a^3b^4
7. $\frac{3u^2v^3}{5}$
8. $\frac{-5x^2y^3z}{6}$
9. $\frac{-u^2v^2w}{2}$
10. $-a^2b^3c^5$

In exercises 11–24, find the coefficient of each term.

11. $7x$
12. $-3yz$
13. $5x^2$
14. $-3x^2uv$
15. $\frac{x^2}{2}$

16. $\frac{-t}{5}$ 17. $\frac{3xy}{4}$ 18. $\frac{z^2}{-2}$ 19. $0xyz^2$ 20. $(3x)(2y)$

21. 2 22. -7 23. πr^2 24. $2\pi r$

In exercises 25–38, indicate which pairs are like terms.

25. $2x$ and $4x$
26. $3y$ and $-3y$
27. $5xy$ and $5yx$
28. $13xy$ and $13xz$
29. $-2x^2$ and $2x^3$
30. 15 and 30
31. xyz and $-yxz$
32. $2u^2v$ and $4uvu$
33. x^3y and xy^3
34. 1 and -1
35. $7x^2yz^3$ and $-2xyzxz^2$
36. x^2 and $0x^2$
37. $3x^2$ and $2x^3$
38. $\frac{x^2y^3}{5}$ and $6yxyxy$

3.2 ADDITION OF TERMS

Adding Like Terms

To add or subtract *like* terms, extend the *Distributive Laws.* For example,

$$3x + 2x = (3 + 2)x$$
$$= 5x$$

and

$$3x - 2x = (3 - 2)x$$
$$= 1x$$
$$= x$$

Let a and b be any fixed numbers, and let x be a variable. Define

$$ax + bx = (a + b)x$$

and

$$ax - bx = (a - b)x.$$

The Distributive Laws also apply to like terms that contain 2 or more variables.

EXAMPLE 1

$$5xy + 4xy = (5 + 4)xy$$
$$= 9xy$$

EXAMPLE 2

$$3x^2z^3 - x^2z^3 = (3 - 1)x^2z^3$$
$$= 2x^2z^3$$

As with numbers, the Distributive Laws apply to adding or subtracting more than two like terms.

EXAMPLE 3

$$\begin{aligned} xyz + 4xyz - 2xyz &= (1 + 4 - 2)xyz \\ &= 3xyz \end{aligned}$$

Adding Unlike Terms

The Distributive Laws do not apply to unlike terms. Simply combine unlike terms with + or − signs.

EXAMPLE 4

Add xy and $3z$.

Solution. xy and $3z$ are unlike terms. Write

$$xy + 3z.$$

This expression cannot be simplified.

Note that the expression you obtain in Example 4,

$$xy + 3z,$$

is *not* a term. (It is not a *product* of numbers and variables.) It is, however, the sum of two (unlike) terms.

EXAMPLE 5

Subtract $2t$ from the sum of r and $3s$.

Solution. $2t$, r, and $3s$ are unlike terms. You obtain the expression,

$$(r + 3s) - 2t.$$

This can be simplified to

$$r + 3s - 2t.$$

Adding Several Terms

To add several terms, arrange them so that like terms are grouped together. You may either group like terms in columns, as in Example 6, or you may group like terms by means of parentheses, as in Example 7. In rearranging the terms, you are extending the Commutative Law

$$a + b = b + a$$

and the Associative Law

$$(a + b) + c = a + (b + c)$$

of addition.

EXAMPLE 6

Add and subtract, as indicated:

$$2x + 3y - z + 4x - 4y + z$$

Solution. Arrange like terms in columns.

$$\begin{array}{l} 2x + 3y - z \\ \underline{4x - 4y + z} \\ 6x - y \end{array}$$

EXAMPLE 7

Add:

$$5x + y - t + u - v + 2x - 3t$$

Solution. Group like terms together within parentheses.

$$(5x + 2x) + y + (-t - 3t) + u - v$$

Insert this sign

$$= 7x + y - 4t + u - v$$

EXERCISES

In exercises 1–40, add or subtract, as indicated. In some cases, you will not be able to simplify the given expression.

1. $2x + x$
2. $5y + 2y$
3. $2x - x$
4. $5y + (-2)y$
5. $xy + 3xy$
6. $uv - 2uv$
7. $abc + 4abc$
8. $x^2y + 5x^2y$
9. $10xy - 4yx$
10. $2x^2yz + zyx^2$
11. $3x + x + x$
12. $5t - 2t + t$
13. $2s - s - s$
14. $5r - 3r + 6r$
15. $4w + w + 2w + w$
16. $3x - x + 2x - x$
17. $2xyz - xyz + 5xyz$
18. $5r^2st^2 + 3r^2st^2$
19. $2x + 2y$
20. $xy + 3x^2$
21. $6u - 3v$
22. $5x + 3$
23. $2x - y + 3x + y$
24. $4s - 2t + s$
25. $2a + 6b + 5a + 6b - a$
26. $2m - 4n + 7n - 3m$
27. $2ab - 3cd + 5ab - cd$
28. $3xy + 5xy - 2z + 6z^2$

29. $4a + b - c + 2a - b - 2c$

30. $3x - 2y + 5x - z + 6x - 2z$

31. $3m - 2m + n - 3m + 2n$

32. $3x^2y + 5z^2 + 3x^2y + 5z^2$

33. $4xyz - 3abc + 2xyz - 5abc + 7xyz$

34. $2a + 5b - c + d + 6a - d + 7a$

35. $5x^2 + 3y^2 - z^2 + 6x^2 - z^2 - 11x^2$

36. $6x + y + 1 - 2x - 3 + y + 10$

37. $9 - 4u^2 - v^2 + 5 + 4u^2 + v^2$

38. $7 - 2u + 5v - 3 + 2u + 5v$

39. $9m + 2n - 6p + 2n - 4p - 9m$

40. $2x - 3y + 5z - 3y - 5z + 6y - 2x$

41. Add $3x$, $5x$, and $6x$.

42. Subtract $2yz$ from the sum of $7yz$ and $-3yz$.

43. Add $2a$, $5b$, $3a$, and $-5b$.

44. Subtract $3y$ from the sum of $4y$, $2x$, and $-y$.

3.3 POLYNOMIALS

What Is a Polynomial?

Definition

> *A* nonzero *term is also called a* ***monomial.*** *The sum (or difference) of two unlike (nonzero) terms is called a* ***binomial,*** *and of three unlike (nonzero) terms a* ***trinomial.***

EXAMPLE 1

(a) $3xy$ is a nonzero term, and hence a monomial.
(b) $2x + 5y$ is the sum of two unlike terms. It is a binomial.
(c) $5x - y + z$ is the sum of three unlike terms. It is a trinomial.

EXAMPLE 2

$2x + 5x + x$ is a monomial. For, $2x$, $5x$, and x are like terms. Thus

$$2x + 5x + x = 8x$$

Definition

> *A **polynomial** is either a term or a sum of terms.*

Monomials, binomials, and trinomials are all polynomials.

EXAMPLE 3

Each of the following is a polynomial:

$$4x - y$$
$$2x^2 + 3x - 1$$
$$2a + b - c + d$$
$$7$$
$$0$$
$$x^5 + 3x^4 - x^3 + x^2 - 2x + 1$$

Adding Polynomials

To add or subtract polynomials, group like terms together, as in the preceding section.

EXAMPLE 4

Add the polynomials $2x + y$, $3x + y - z$, and $5y + z$.

Solution. Rearrange the terms so that like terms are in the same column.

$$\begin{array}{r} 2x + y \phantom{{}+z} \\ 3x + y - z \\ 5y + z \\ \hline 5x + 7y \phantom{{}+z} \end{array}$$

Subtracting Polynomials

Definition

> *The **inverse of a polynomial P** is the polynomial obtained by changing the sign of each term of P.*

Let $-P$ denote the inverse of P.

EXAMPLE 5

The inverse of $x - 2y + 3z$ is $-x + 2y - 3z$. Note the addition:

$$\begin{array}{r} x - 2y + 3z \\ + \; -x + 2y - 3z \\ \hline 0 \phantom{{}- 2y + 3z} \end{array}$$

In general, *the inverse of P is the polynomial added to P to obtain* 0:

$$P + (-P) = 0$$

Recall that subtraction was defined in terms of addition. To subtract the polynomial Q from P, add $-Q$.

$$P - Q = P + (-Q)$$

EXAMPLE 6

Subtract the bottom polynomial from the top one:

$$\begin{array}{r} 4a - 3b + c \\ \underline{2a - 3b - c} \end{array}$$

Solution. Change the sign of each term of the bottom polynomial and add.

$$\begin{array}{rrr} 4a & -3b & +c \\ - & + & + \\ +2a & -3b & -c \\ \hline 2a & & +2c \end{array}$$

EXAMPLE 7

Simplify:

$$10a - [4b - (3a - b)]$$

Solution. Begin with the innermost parentheses and work outward, as you do when adding and subtracting real numbers.

$$\begin{aligned} 10a - [4b - (3a - b)] &= 10a - [4b - 3a + b] \\ &= 10a - 4b + 3a - b \\ &= (10a + 3a) + (-4b - b) \\ &= 13a - 5b \end{aligned}$$

EXERCISES

In exercises 1–12, classify each polynomial as *(i)* a monomial, *(ii)* a binomial, *(iii)* a trinomial, or *(iv)* none of these.

1. $2x$
2. $3x + 5y$
3. 5
4. -5
5. 0
6. $-x$
7. $4x^2 + 3x + 1$
8. $x^3 + x^2 + x + 1$
9. $4x + 3x + 2x + x$
10. $2x + 3y + 2x + y$
11. $2x + y + 2x - y$
12. $2x + y - 2x - y$

In exercises 13–30, add the polynomials:

13. $\begin{array}{r} 2x + y \\ \underline{x + 4y} \end{array}$
14. $\begin{array}{r} 3a + b \\ \underline{5a + 2b} \end{array}$
15. $\begin{array}{r} 6m + 9n \\ \underline{10m + 7n} \end{array}$
16. $\begin{array}{r} 2r - s \\ \underline{3r - 2s} \end{array}$

17. $\begin{array}{r} 8a - 5b + 2 \\ \underline{-7a + 5b + 1} \end{array}$

18. $\begin{array}{r} 6x - 3y + 7 \\ \underline{2x + 3y - 5} \end{array}$

19. $\begin{array}{r} a + b + 2c \\ \underline{2a + b + c} \end{array}$

20. $\begin{array}{l} 5r + s - t \\ \underline{2r \quad\ \ + t} \end{array}$

21. $\begin{array}{r} 3x + 5y - 2z \\ \underline{2x - 5y + 2z} \end{array}$

22. $\begin{array}{r} 6r^2 + s^2 - 3t^2 \\ \underline{r^2 - s^2 + t^2} \end{array}$

23. $\begin{array}{r} a + b - c \\ 2a + b - 2c \\ \underline{4a - 2b + c} \end{array}$

24. $\begin{array}{r} 5x - 2y + z \\ 6x - 3y + z \\ \underline{x - y - z} \end{array}$

25. $\begin{array}{l} 3r + 2s - t + u \\ 2r - s + t \\ \underline{\qquad\ \ s \qquad + u} \end{array}$

26. $\begin{array}{r} w + x - y + z \\ w - x + y - z \\ 2w + x - y \\ \underline{x + y + z} \end{array}$

27. $\begin{array}{r} 10a + 7b - 3c + d \\ 5a - 2b + 3c + 9d \\ 7a - 2b + 5c \\ \underline{2b \qquad + d} \end{array}$

28. $\begin{array}{r} 9t + 5u + 6v + 12w \\ 3t + 8u - 2v + 7w \\ 2t \qquad - 7v + 2w \\ \underline{3t - 6u \qquad + 12w} \end{array}$

29. $\begin{array}{r} a + b \qquad - d + 2e \\ -b + 4c \qquad + e \\ 3a - b \qquad + 7d \\ 5a \qquad - 2c + d \\ \underline{-2a \qquad\qquad - d - e} \end{array}$

30. $\begin{array}{r} 16w + 22x - 17y + 9z \\ 12w - 17x + 9y - 8z \\ 32w - 25x \qquad + 10z \\ 28w + 7x - 13y \\ \underline{12w - 5x + 7y} \end{array}$

In exercises 31–42, subtract the bottom polynomial from the top one.

31. $\begin{array}{r} 4a + 2b \\ \underline{2a + b} \end{array}$

32. $\begin{array}{r} 9x + 7y \\ \underline{8x + 7y} \end{array}$

33. $\begin{array}{r} 7c - 6d \\ \underline{2c + 5d} \end{array}$

34. $\begin{array}{r} 2m + 5n \\ \underline{-m + 5n} \end{array}$

35. $\begin{array}{r} -7s + 8t \\ \underline{-7s - 8t} \end{array}$

36. $\begin{array}{r} 12y - 15z \\ \underline{-9y + 18z} \end{array}$

37. $\begin{array}{r} 6a - 3b + 2c \\ \underline{a \qquad - c} \end{array}$

38. $\begin{array}{r} 9x - 12y + 7z \\ \underline{2x + y - z} \end{array}$

39. $\begin{array}{r} 2a - b + c - d \\ \underline{-a + 2b - c + d} \end{array}$

40. $\begin{array}{r} 3r - s + t - u \\ \underline{s \qquad - u} \end{array}$

41. $\begin{array}{r} r - s + t - 2u + v \\ \underline{3r - s - t + 2u - v} \end{array}$

42. $\begin{array}{r} 16w - 7x \qquad + z \\ \underline{9w - 8x + y \quad\ } \end{array}$

In exercises 43–54, simplify each polynomial.

43. $3a - (4b + 2a)$

44. $5x - (4y + 2z)$

45. $5m - (2m - n)$

46. $3a - (2a - b + c)$

47. $(4a - 2b) - (3a - 2b)$

48. $(5x - 2y + z) - (2x - y - z) + (4x - y)$

49. $(r - s + t) - (3r + s - t) - (2r - 3s)$

50. $3x - y - [2x - y - (x + y)]$

51. $6x - 2y - [-x - (y - x)]$

52. $5a - 2b + c - [(3a - b) - (2a - b + c)]$

53. $7 - 2a - 3b + [5a - (3b - c)]$

54. $2x - 3y - [3x - (z - 2y) - (x - y + z)]$

55. Add $3a + b$, $5a - 2b$, and $2a$.

56. Add $2x + y - z$, $3y - z$, and $x + z$.

57. Subtract $2x + 9y$ from $6x + 10y$.

58. Subtract $x - y$ from $x + y$.

59. Subtract $a - 2b$ from the sum of $3a + b$ and $5a - b$.

60. Subtract $2x^2$ from 0.

3.4 EVALUATING POLYNOMIALS

Substituting Numbers for Variables

Recall that a variable designates any number under discussion. Polynomials are constructed from numbers and variables by addition and multiplication. Thus polynomials pertain to numbers. When you are given numbers that the variables represent, you can evaluate the polynomial by substituting the numbers for the variables.

EXAMPLE 1

Evaluate $5x$ when $x = 6$.

Solution. Substitute 6 for x in the expression $5x$, and obtain $5 \cdot 6$, or 30.

EXAMPLE 2

Evaluate $3y - 1$ when $y = -2$.

Solution. Replace y by -2:

$$\begin{aligned} 3(-2) - 1 &= -6 - 1 \\ &= -7 \end{aligned}$$

When a variable occurs more than once in a polynomial, *each time it occurs replace it by the same number.*

EXAMPLE 3

Evaluate

$$t^2 + 4t + 3$$

when $t = 5$.

Solution. Replace *each occurrence of* t by 5:

$$\begin{aligned} 5^2 + 4 \cdot 5 + 3 &= 25 + 20 + 3 \\ &= 48 \end{aligned}$$

If a polynomial contains 2 or more variables, replace each variable by a specified number.

EXAMPLE 4

Evaluate

$$x^2 - 2y$$

when $x = 3$ and $y = -4$.

Solution. Substitute 3 for x and **−4** for $\boldsymbol{y}$, and obtain

$$\begin{aligned} 3^2 - 2(-4) &= 9 + 8 \\ &= 17. \end{aligned}$$

EXAMPLE 5

Evaluate

$$x^2 + 3xy + yz$$

when $x = 2$, $y = 1$, $z = 4$.

Solution. Replace *each occurrence of* x by 2, *of* $\boldsymbol{y}$ by **1**, *and of* z by 4:

$$\begin{aligned} 2^2 + 3 \cdot 2 \cdot 1 + 1 \cdot 4 &= 4 + 6 + 4 \\ &= 14 \end{aligned}$$

Geometric Formulas

Formulas are algebraic expressions that indicate the relationship between various quantities. Geometric formulas are often expressed in terms of polynomials.

EXAMPLE 6

The area, A, of a square is given by the formula

$$A = s^2,$$

where s is the length of a side. Find the area if a side is of length (a) 10 inches, (b) 20 inches. (See Figure 3.1 on page 62.)

Solution.

(a) $A = 10^2 = 100$
The area is 100 square inches.
(b) $A = 20^2 = 400$
The area is 400 square inches.

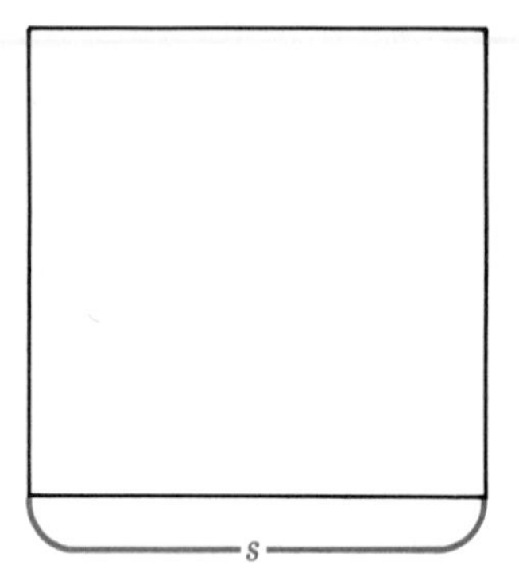

FIGURE 3.1. The area of a square is s^2.

EXAMPLE 7

The area, A, of a triangle is given by the formula

$$A = \frac{bh}{2},$$

where b is the length of the base and h is the length of the altitude. Find the area if
(a) $b = 6$ feet, $h = 10$ feet;
(b) $b = 12$ feet, $h = 9$ feet.

Solution.

(a) Substitute 6 for b and **10** for $\boldsymbol{h}$ in the formula $A = \frac{b\boldsymbol{h}}{2}$, and obtain

$$A = \frac{6 \cdot \mathbf{10}}{2} = 30$$

Thus the area is 30 square feet.

(b) Substitute 12 for b and **9** for $\boldsymbol{h}$:

$$A = \frac{12 \cdot \mathbf{9}}{2} = 54$$

Thus the area is 54 square feet.

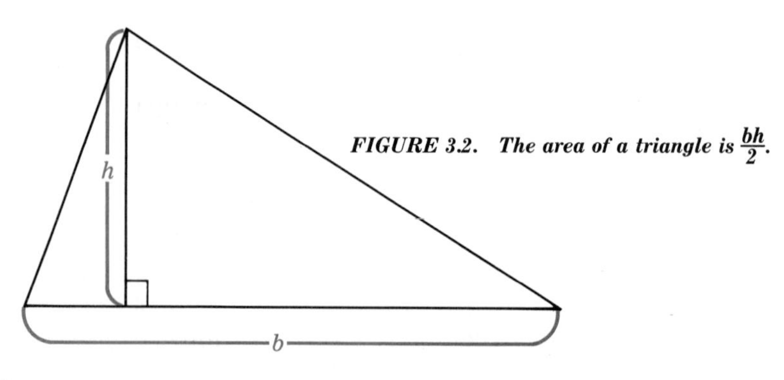

FIGURE 3.2. The area of a triangle is $\frac{bh}{2}$.

EXERCISES

In exercises 1–16, evaluate each polynomial for the specified number.

1. $3x$; $\boxed{x = 4}$
2. $t + 5$; $\boxed{t = 2}$
3. $6x + 1$; $\boxed{x = 1}$
4. $2x - 3$; $\boxed{x = 1}$
5. $7y + 9$; $\boxed{y = 0}$
6. $5z - 3$; $\boxed{z = -1}$
7. z^2; $\boxed{z = 6}$
8. u^3; $\boxed{u = -3}$
9. $x^2 - 2$; $\boxed{x = 4}$
10. $x^2 + 4x$; $\boxed{x = -1}$
11. $x^2 + x + 1$; $\boxed{x = 3}$
12. $x^2 + 5x + 1$; $\boxed{x = -5}$
13. $t^4 - t^2 + 1$; $\boxed{t = 1}$
14. $t^5 - 4t^2 - 2t + 3$; $\boxed{t = -1}$
15. $x^3 - 7x^2 + 14x - 3$; $\boxed{x = 3}$
16. $z^4 - z^2 + 2z + 5$; $\boxed{z = 10}$

In exercises 17–30, evaluate each polynomial for the specified numbers.

17. xy; $\boxed{x = 4, y = -3}$
18. $2xy + 1$; $\boxed{x = 10, y = 4}$
19. $x + 2y$; $\boxed{x = -6, y = 8}$
20. $x^2 - y^2$; $\boxed{x = 1, y = -1}$
21. $s^2 - 3t^2$; $\boxed{s = 10, t = 5}$
22. $5mn + 2m - n$; $\boxed{m = 8, n = -2}$
23. $3a^2 - 2b^2 + c$; $\boxed{a = 4, b = 3, c = 2}$
24. $x^2 + y^2 - z^2 + xyz$; $\boxed{x = 4, y = -3, z = -2}$
25. $2a - 7b + 9c - 2$; $\boxed{a = b = c = 3}$
26. $a + 2b - c + 5d$; $\boxed{a = 10, b = 6, c = -1, d = 2}$
27. $a^3 - bc^2 + 2ab$; $\boxed{a = 2, b = 5, c = -2}$
28. $x^2y - yx^2 + x - y$; $\boxed{x = 10, y = -1}$
29. $x^3 + x^2 - 2y + 3xy$; $\boxed{x = 100, y = -1}$
30. $2u^2 - 3v^2 + w^2$; $\boxed{u = 1, v = 2, w = 3}$

In exercises 31–40, evaluate each polynomial for each specified number.

SAMPLE. Evaluate	***Solution.***
$x^2 - 2x + 5$ when (a) $x = 2$, (b) $x = -3$, (c) $x = 10$.	(a) $2^2 - 2 \cdot 2 + 5 = 4 - 4 + 5$ $= 5$ (b) $(-3)^2 - 2(-3) + 5$ $= 9 + 6 + 5$ $= 20$ (c) $10^2 - 2 \cdot 10 + 5$ $= 100 - 20 + 5$ $= 85$

31. $x + 10$; (a) $x = 2$, (b) $x = -2$, (c) $x = -10$

32. $2x - 8$; (a) $x = 6$, (b) $x = 0$, (c) $x = 4$

33. $5y + 9$; (a) $y = 0$, (b) $y = 2$, (c) $y = 10$

34. $z^2 + 1$; (a) $z = 0$, (b) $z = 1$, (c) $z = -1$

35. $x^2 + 2x$; (a) $x = 0$, (b) $x = 4$, (c) $x = -4$

36. $x^2 + 5x + 2$; (a) $x = 2$, (b) $x = 3$, (c) $x = 4$

37. $2x^2 - 3x + 1$; (a) $x = 4$, (b) $x = -4$, (c) $x = 6$

38. $u^3 - 2u + 6$; (a) $u = 0$, (b) $u = -1$, (c) $u = 2$

39. $b^4 - b^2 + 1$; (a) $b = 1$, (b) $b = -2$, (c) $b = 3$

40. $m^3 - 2m^2 + m - 4$; (a) $m = -1$, (b) $m = 2$, (c) $m = -3$

41. Find the area of a square if a side is of length (a) 11 inches, (b) 15 inches, (c) 30 inches.

42. Find the area of a triangle if
(a) $b = 20$ inches, $h = 30$ inches;
(b) $b = 18$ inches, $h = 32$ inches.

43. The area of a rectangle is given by the formula $A = lw$, where l is the length and w the width. Find the area if $l = 40$ feet, $w = 22$ feet. (See Figure 3.3.)

44. The perimeter of (distance around) a rectangle is given by the formula $P = 2l + 2w$, where l and w are as in Exercise 43. Find the perimeter if $l = 40$ inches, $w = 22$ inches. (See Figure 3.3.)

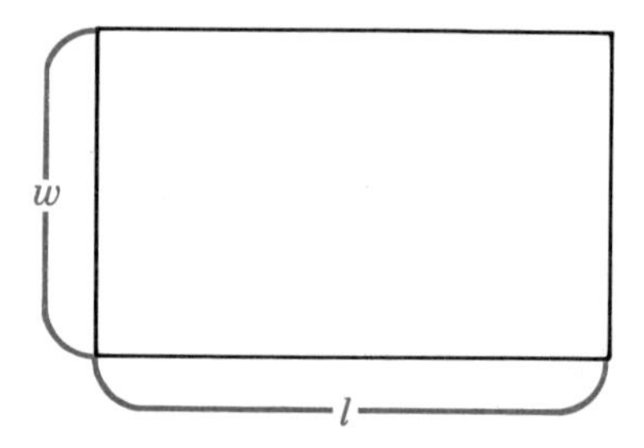

FIGURE 3.3. The area of a rectangle is lw. The perimeter is 2l + 2w.

45. The area of a circle is given by the formula $A = \pi r^2$, where r is the length of the radius. Find the area if (a) $r = 10$ inches, (b) $r = 12$ inches, (c) $r = 13$ inches. (See Figure 3.4.)

46. The circumference of a circle is given by the formula $C = 2\pi r$, where r is the length of the radius. Find the circumference if (a) $r = 10$ inches, (b) $r = 12$ inches, (c) $r = 13$ inches. (See Figure 3.4.)

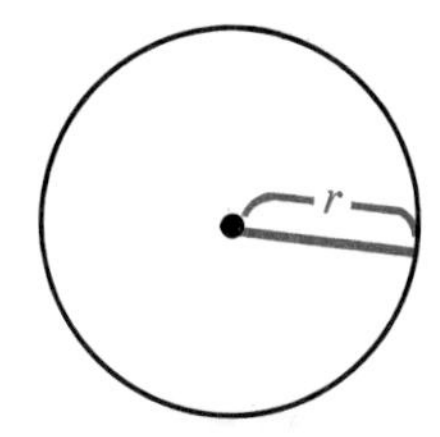

FIGURE 3.4. The area of a circle is πr^2. The circumference is $2\pi r$.

47. The volume of a rectangular box is given by the formula $V = lwh$, where l is the length, w the width, and h the height. Find the volume if:
(a) $l = 10$ inches, $w = 6$ inches, $h = 8$ inches;
(b) $l = 20$ inches, $w = 10$ inches, $h = 12$ inches. (See Figure 3.5.)

48. The surface area of a rectangular box is given by the formula $S = 2\,lw + 2\,lh + 2\,wh$, where l, w, and h are as in exercise 47. Find the surface area if
(a) $l = 10$ feet, $w = 6$ feet, $h = 8$ feet;
(b) $l = 20$ feet, $w = 10$ feet, $h = 12$ feet. (See Figure 3.5.)

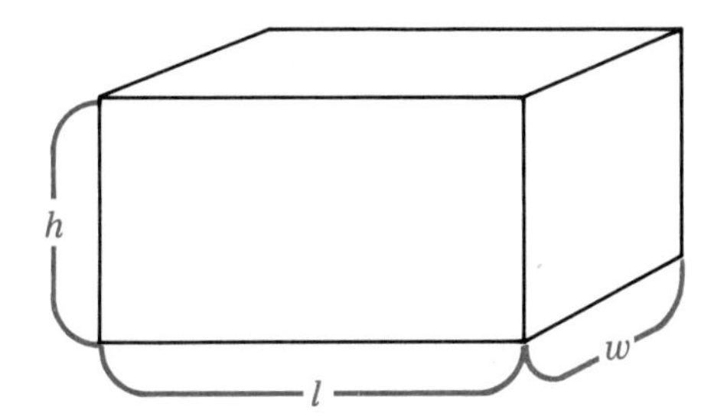

FIGURE 3.5. The volume of a rectangular box is lwh. The surface area is $2lw + 2lh + 2wh$.

What Have You Learned in Chapter 3?

You have learned that a variable designates any one of the numbers being discussed, that a term is a number or a variable or a product of numbers and variables, and that a polynomial is a term or a sum of terms.

You can add and subtract polynomials.

And you can evaluate polynomials for specified numbers.

Let's Review Chapter 3.

3.1 *Terms*

1. Write each term as a product of a number and variables.

   (a) $5xy^3$ (b) $\dfrac{uv^2w^3}{2}$

2. Find the coefficient of $-12x^2y$.
3. Are $3x^2y$ and $6xyx$ like terms?
4. Are $10xy^3$ and $10x^3y$ like terms?

3.2 *Addition of Terms*

In exercises 5–8, add and subtract, as indicated.

5. $7x + 3x$
6. $2uv + 3uv - uv$
7. $3m + 5n - m + 10n$
8. $4x + 3y - z + x - y - 2z$

3.3 *Polynomials*

In exercises 9 and 10, add the polynomials.

9. $$\begin{array}{r} 3a + b \\ \underline{5a + 6b} \end{array}$$

10. $$\begin{array}{rrrr} w & +\, x & +\, 2y & -\, z \\ w & -\, x & +\, 4y & \\ & x & -\, y & +\, z \\ \hline 2w & & -\, y & -\, z \end{array}$$

11. Subtract the bottom polynomial from the top one:

    $$\begin{array}{rrr} 3a & +\, b & -\, c \\ \hline 2a & & -\, 2c \end{array}$$

12. Simplify: $(5x + y - 2z) - (3x - z) + (x - y + 4z)$
13. Simplify: $15a + b - [4a - (2a - b)]$
14. Subtract $2m + 5n$ from $5m - 2n$.

3.4 Evaluating Polynomials

15. Evaluate $x^2 + 3x + 10$ when $x = 5$.
16. Evaluate $x^2 - y^2$ when $x = 9$, $y = 6$.
17. Evaluate $x^3 - 2x^2 + 4$ when (a) $x = 2$, (b) $x = -2$ inches.
18. Find the area of a square if a side is of length 20 inches.

And these from Chapters 1 and 2:

19. Write the inverse of (a) 2, (b) -2.
20. Find the value of: $8 - [(5 - 3) - (4 - 9)]$
21. Find the indicated product: $4(-7)0(-1)$
22. Use the Distributive Laws to find the value of: $3(6 + 2)$

Try These Exam Questions for Practice.

1. Find the coefficient of $\frac{x^3}{2}$.
2. Add: $2m + 5m + 3m$
3. Add and subtract, as indicated: $2a - 3b + 5a + 3b$
4. Add:

$$\begin{array}{rrrr} a + & 7b - & 10c + & d \\ 5a - & b - & 3c + & 9d \\ 6a & & - & c \\ \hline & 5b + & c + & d \end{array}$$

5. Subtract the bottom polynomial from the top one:

$$\begin{array}{r} 5x - 3y + 9z \\ \underline{2x - 6y + 10z} \end{array}$$

6. Simplify: $4m - (3n + 2p) - [6m - (2n - p)]$
7. Evaluate $x^2 + 8x - 3$ when $x = 4$.
8. Evaluate $2x^2 + 5y^2$ when $x = 10$, $y = 5$.

Products and Factors

4.1 MONOMIAL PRODUCTS

Products of Powers

What happens when you multiply two powers of the same number? Observe that

$$2^2 \cdot 2^1 = 4 \cdot 2 = 8 = 2^3.$$

Thus

$$2^2 \cdot 2^1 = 2^{2+1}$$

Next, consider what happens when you multiply two powers of the same variable.

EXAMPLE 1

Multiply: $x^3 \cdot x^2$

Solution.

$$x^3 \cdot x^2 = (xxx)(xx) = xxxxx = x^5$$

Thus $$x^3 \cdot x^2 = x^{3+2}$$

Let m and n be positive integers. To find the product

$$a^m \cdot a^n,$$

observe that

$$a^m = \underbrace{a \cdot a \cdot a \ldots a}_{m \text{ factors}},$$

$$a^n = \underbrace{a \cdot a \cdot a \ldots a}_{n \text{ factors}}.$$

Therefore,
$$a^m \cdot a^n = (\underbrace{a \cdot a \cdot a \ldots a}_{m \text{ factors}})(\underbrace{a \cdot a \cdot a \ldots a}_{n \text{ factors}})$$

$$= \underbrace{a \cdot a \cdot a \ldots a \cdot a \cdot a \cdot a \ldots a}_{m+n \text{ factors}}$$

$$= a^{m+n}$$

Thus
$$a^m \cdot a^n = a^{m+n}$$

In other words, to multiply two powers of the same number (or variable) a, write down the base a and add the exponents.

EXAMPLE 2

Use the above rule to find $y^8 \cdot y^5$.

Solution.

$$y^8 \cdot y^5 = y^{8+5}$$
$$= y^{13}$$

Three or more powers of the same number (or variable) are multiplied by writing down the base and adding the exponents.

EXAMPLE 3

Multiply: $t^4 \cdot t^2 \cdot t$

Solution.

$$t^4 \cdot t^2 \cdot t = t^{4+2+1}$$
$$= t^7$$

Associative and Commutative Laws

According to the *Associative Law of Multiplication,*

$$(ab)c = a(bc)$$

According to the *Commutative Law of Multiplication,*

$$ab = ba$$

Observe how these laws are used in the following example.

EXAMPLE 4

Multiply: $(2x)3$

Solution.

$$\begin{aligned}(2x)3 &= 3(2x) && \text{by the Commutative Law}\\ &= (3 \cdot 2)x && \text{by the Associative Law}\\ &= 6x\end{aligned}$$

When multiplying monomials, use the Associative and Commutative Laws to rearrange the factors.

1. *Group all coefficients at the beginning, and multiply them.*
2. *Group powers of the same variable together and multiply them, as previously explained.*

EXAMPLE 5

Multiply: $(3ab)(4ab)$

Solution.

$$(3ab)(4ab) = \underbrace{3 \cdot 4}\ \underbrace{a\ a}\ \underbrace{b\ b}$$

group all coefficients at the beginning — group powers of the same variable together

$$= \underbrace{12}\ \underbrace{a^2b^2}$$

multiply coefficients and powers of the same variable

EXAMPLE 6

Multiply: $(4xy^2)(7x^2y^4)$

Solution.

$$\begin{aligned}(4xy^2)(7x^2y^4) &= (4 \cdot 7)(x\,x^2)(y^2y^4)\\ &= 28\,x^3y^6\end{aligned}$$

EXAMPLE 7

Multiply: $(4ax^2y)(-2x^2y^3)(-3a^2bx)$

Solution.

$$\begin{aligned}(4ax^2y)(-2x^2y^3)(-3a^2bx) &= [4(-2)(-3)](aa^2)\,b(x^2x^2x)(yy^3)\\ &= 24a^3bx^5y^4\end{aligned}$$

Simple Division

Recall that

$$\frac{6}{2} = 3$$

because $6 = 2 \cdot 3.$

Similarly, division of polynomials can be defined in terms of multiplication. Thus

$$\frac{4x}{2} = 2x$$

because $4x = 2 \cdot 2x,$

and

$$\frac{x^5}{x^2} = x^3$$

because $x^5 = x^2 \cdot x^3.$

Note that

$$\begin{aligned}\frac{x^5}{x^2} &= \frac{xxxxx}{xx}\\ &= xxx\\ &= x^3.\end{aligned}$$

Thus

$$\frac{x^5}{x^2} = x^{5-2}$$

To divide powers of the same number (or variable) a, write down the base a and subtract the exponents:

$$\frac{a^m}{a^n} = a^{m-n}$$

where m and n are positive integers, and $m > n$.

EXAMPLE 8

Divide: $\dfrac{y^8}{y^3}$

Solution.

$$\begin{aligned}\frac{y^8}{y^3} &= y^{8-3}\\ &= y^5\end{aligned}$$

Powers of several variables may have to be divided. Thus

$$\frac{x^6y^2}{x^2y} = x^4y$$

because

$$x^6y^2 = (x^2y)(x^4y)$$

Note that

$$\frac{x^6y^2}{x^2y} = x^{6-2}y^{2-1}.$$

EXAMPLE 9

Divide: $\dfrac{a^7b^4c^3}{a^4b^3c^2}$

Solution.

$$\begin{aligned}\frac{a^7b^4c^3}{a^4b^3c^2} &= a^{7-4}b^{4-3}c^{3-2}\\ &= a^3bc\end{aligned}$$

In the final example, first divide the coefficients. Also, note that the powers of y "cancel." In fact,

$$\frac{y^3}{y^3} = 1$$

because

$$y^3 = y^3(1)$$

EXAMPLE 10

Divide: $\dfrac{10x^6y^3}{5x^4y^3}$

Solution.

$$\begin{aligned}\frac{10x^6y^3}{5x^4y^3} &= 2x^{6-4}\\ &= 2x^2\end{aligned}$$

EXERCISES

In exercises 1–50, multiply, as indicated.

1. $a \cdot a$
2. $b \cdot b \cdot b$
3. $c^2 \cdot c$
4. $x^3 \cdot x$
5. $y \cdot y^6$
6. $z \cdot z^9$
7. $x^2 \cdot x^2$
8. $a^2 \cdot a^3$
9. $b^3 \cdot b^2$
10. $c^4 \cdot c^6$
11. $m^{10} \cdot m^8$
12. $n^9 \cdot n^{20}$
13. $4a^2 \cdot a^4$
14. $7b^5 \cdot b^3$
15. $m^3 \cdot m^3 \cdot 4$
16. $z^2 \cdot z^7 \cdot 5$
17. $2 \cdot x \cdot 3 \cdot x^2$
18. $(-1)\, y^2 \cdot 5 \cdot y^2$
19. $a^2 \cdot a \cdot a$
20. $b \cdot b^2 \cdot b^3$
21. $c^4 \cdot c^2 \cdot c^3$
22. $d^4 \cdot d^7 \cdot d$
23. $a^2 \cdot a \cdot a^3 \cdot a$
24. $b^2 \cdot b^2 \cdot b \cdot b^3$
25. $x^2 \cdot x^3 \cdot x \cdot x^2 \cdot x^3$
26. $y^4 \cdot y^3 \cdot y^7 \cdot y \cdot y^5$
27. $2a^2\,(-1)\, a^3 \cdot a^4$
28. $(-3)\, b^2\,(-2)\, b^4 \cdot b^5$
29. $(x^2y)y$
30. $(xy)x$
31. $(x^2y)(xy^2)$
32. $(xy^2)(x^2y^2)$
33. $(a^2b)(a^3b^2)$
34. $(c^2d^4)(cd^3)$
35. $(5xy)(6xy)$
36. $(2c^2d)(5cd^2)$
37. $(-xy)(-2x^2y^2)$
38. $(-4mn)(2m^2n^3)$
39. $(3ab)(-2a^2b)$
40. $(-u^2v)(-3uv^2)$
41. $(a^2x)(-3x^3)$
42. $(3r^2s^5)(-2rs^7)$

43. $(5xy)(3x^2y)(2x^2)$

44. $(-4a^2b)(-ab)(abc)$

45. $(50a)(2b)(10ab)$

46. $(-xy)(2xy)(3y)$

47. $(5mn)(2mn)(3amn)$

48. $(2xy^2z)(xyz^2)(3xz)(yz^2)$

49. $(-2abc)(bcd)(3abc^2)(4a^2d)$

50. $(rs)(-tu)(r^2st)(-su^2)$

51. Find the product of $2xy$ and $9x^2y$.

52. Find the product of $-2a^2bc$, $-5ac$, and abc^2.

53. Evaluate $a^2(ab^2)$ when $a = 2$ and $b = -2$.

54. Evaluate $(2ab^2)(-3a^2b)$ when $a = 1$ and $b = -1$.

In exercises 55–76, divide, as indicated.

55. $\frac{8x}{2}$

56. $\frac{-30x}{5}$

57. $\frac{28x^2}{-7}$

58. $\frac{x^4}{x^2}$

59. $\frac{y^7}{y^3}$

60. $\frac{a^{10}}{a^7}$

61. $\frac{x^2y^2}{xy}$

62. $\frac{a^3b^2}{ab}$

63. $\frac{m^4n^6}{m^2n^2}$

64. $\frac{s^5t^8}{s^4t^5}$

65. $\frac{x^4y^3z^2}{xy^2z}$

66. $\frac{a^4b^5c^6}{a^2b}$

67. $\frac{4x^2}{2x}$

68. $\frac{20x^5}{-5x^4}$

69. $\frac{-9a^8}{-3a^4}$

70. $\frac{15b^{10}}{5b^8}$

71. $\frac{16a^4b}{4a}$

72. $\frac{30ab^2}{-10ab}$

73. $\frac{21x^3y^6}{7x^2y^4}$

74. $\frac{35s^8t^7}{-5s^7t}$

75. $\frac{64a^{12}b^9c}{16a^3b^7}$

76. $\frac{48x^3y^3z^7}{-12x^2yz^5}$

4.2 DISTRIBUTIVE LAWS FOR POLYNOMIALS

Multiplication

According to the *Distributive Laws,*

$$a(b + c) = ab + ac$$
$$(b + c)a = ba + ca$$

Similarly,

$$a(b - c) = ab - ac$$
$$(b - c)a = ba - ca$$

EXAMPLE 1

$$6(x + y) = 6x + 6y$$

EXAMPLE 2

$$\begin{aligned}-2(x - y) &= -2x - (-2)y \\ &= -2x + 2y\end{aligned}$$

Here are some further examples.

EXAMPLE 3

Use the Distributive Laws to find:

$$5a(a + b)$$

Solution.

$$\begin{aligned} 5a(a + b) - (5a)a + (5a)b \\ &= 5(aa) + 5(ab) \\ &= 5a^2 + 5ab \end{aligned}$$

EXAMPLE 4

Multiply: $x^2(x + y)$

Solution.

$$\begin{aligned} x^2(x + y) &= x^2 \cdot x + x^2 \cdot y \\ &= x^3 + x^2y \end{aligned}$$

EXAMPLE 5

Multiply: $5abc(a^2b - 2b^2c$

Solution.

$$\begin{aligned} 5abc(a^2b - 2b^2c) &= 5abc(a^2b) - 5abc(2b^2c) \\ &= 5a^3b^2c - 10ab^3c^2 \end{aligned}$$

EXAMPLE 6

Multiply: $4xyz^2(xy - 2z + 3x^2z)$

Solution.

$$\begin{aligned} 4xyz^2(xy - 2z + 3x^2z) &= 4xyz^2(xy) - 4xyz^2(2z) + 4xyz^2(3x^2z) \\ &= 4x^2y^2z^2 - 8xyz^3 + 12x^3yz^3 \end{aligned}$$

Division

Observe that

$$\frac{ab + ac}{a} = b + c$$

because, by the Distributive Laws,

$$ab + ac = a(b + c).$$

Thus

$$\frac{3x + 3y}{3} = x + y$$

because

$$3x + 3y = 3(x + y)$$

Note that *both* terms of the polynomial $3x + 3y$ are divided by 3. In fact,

$$\begin{aligned} \frac{3x + 3y}{3} &= \frac{3x}{3} + \frac{3y}{3} \\ &= x + y \end{aligned}$$

Also,
$$\frac{x^3 + 5x}{x} = x^2 + 5$$

because
$$x^3 + 5x = x(x^2 + 5)$$

Both terms of $x^3 + 5x$ are divided by x:

$$\frac{x^3 + 5x}{x} = \frac{x^3}{x} + \frac{5x}{x}$$
$$= x^2 + 5$$

EXAMPLE 7

Divide: $\frac{2x + 4}{2}$

Solution.
$$\frac{2x + 4}{2} = \frac{2x}{2} + \frac{4}{2}$$
$$= x + 2$$

EXAMPLE 8

Divide: $\frac{2x^4 - 7x^3}{x^2}$

Solution.
$$\frac{2x^4 - 7x^3}{x^2} = \frac{2x^4}{x^2} - \frac{7x^3}{x^2}$$
$$= 2x^2 - 7x$$

EXAMPLE 9

Divide: $\frac{5x^2y + 10xy^2 - 15xy^3}{5xy}$

Solution.
$$\frac{5x^2y + 10xy^2 - 15xy^3}{5xy} = \frac{5x^2y}{5xy} + \frac{10xy^2}{5xy} - \frac{15xy^3}{5xy}$$
$$= x + 2y - 3y^2$$

EXERCISES

In exercises 1–42, multiply, as indicated.

1. $4(x + y)$
2. $3(y + z)$
3. $-10(a + b)$
4. $9(x^2 + y)$
5. $2(a - b)$
6. $12(a - b)$
7. $-3(m - n)$
8. $-2(x^2 - y^2)$
9. $5(x + y)$
10. $2(x - y)$
11. $2(ax + 3y)$
12. $-3(3x + 2y^2)$
13. $-5(2a^2 - 4b)$
14. $3\pi(2x - 3y)$
15. $a(a + 4)$
16. $x(x^2y + x^3)$
17. $a^2(a + b)$
18. $c(2c + 5d^2)$
19. $2a(a^2 + a^3)$
20. $-4y^2(y + yz)$
21. $5xyz(x^2 + y^2)$
22. $4xy(xyz + y^2)$
23. $2ab(3a^2 - 5b)$
24. $3xy(2x^2 - 7xy)$
25. $-mn(2m^2 - 3n^2)$
26. $3xyz(xz - y^2z)$
27. $2uv(5u^2 - 3u^2v^2)$

28. $5s^2t(2s^3 - 4t^2)$
29. $10x^2yz^3(5x^4yz^3 + 8xy^2z^2)$
30. $7ab^2c(4a^2bc^4 - 3a^7bc^2)$
31. $8(x + y + z)$
32. $-2(a + b + c)$
33. $5(a - b - c)$
34. $-6(a + 7b - c)$
35. $10(x^2 + 2x + 3)$
36. $-3(y^2 - 4y + 9)$
37. $3x(x + y + z)$
38. $2ab(a^2 + ab + b^2)$
39. $-2ab(a^2 + 5ab - b^2)$
40. $4xyz^2(x^2z - 4xy + 3xz)$
41. $-x^2yz^2(x + 2xy - 3xyz)$
42. $-2a^2bc^2(-ab + 5bc - 2a^2bc^7)$
43. Evaluate $3(x + 5y)$ when $x = 5$ and $y = 2$.
44. Evaluate $-2(a - 3b)$ when $a = 6$ and $b = 4$.

In exercises 45–52, simplify each expression.

SAMPLE. Simplify:	***Solution.***
$4a - 7(2 - 3a)$	$4a - 7(2 - 3a) = 4a - [7(2) - 7(3a)]$ $= 4a - [14 - 21a]$ $= 4a - 14 + 21a$ $= 25a - 14$

45. $3x - 2(y + 5z)$
46. $3(4 - 2a) + 2(a + 1)$
47. $3x(y - 2z) - 4x$
48. $4x^2(y + z^2) - 2(x - y)$
49. $a(a - b) + b(b - a)$
50. $x(x^2 - 2) + x^2(x + 2)$
51. $2(a + 2b) - 3(a - b) + 5(a + 4b)$
52. $6(x + 3y) - 4(x - y) - 2(3x + 2y)$

In exercises 53–68, divide, as indicated.

53. $\frac{5x + 5y}{5}$
54. $\frac{7a + 7b}{7}$
55. $\frac{9x - 18z}{9}$
56. $\frac{4m - 16n}{-4}$
57. $\frac{a^2 + a}{a}$
58. $\frac{b^5 - b^2}{b}$
59. $\frac{m^7 + m^4}{m^3}$
60. $\frac{x^3 - x^2}{x^2}$
61. $\frac{x^2y + x}{x}$
62. $\frac{a^3b^2 + ab}{ab}$
63. $\frac{ax^2 - a^2x}{ax}$
64. $\frac{3x^3y^2 + 9xy}{3}$
65. $\frac{4ab - 12a}{4a}$
66. $\frac{20x^4y^3 + 30x^3y^3}{5x^2y^3}$
67. $\frac{12a^2b + 18ab + 9ab^2}{3ab}$
68. $\frac{16x^2y^2z^2 + 12xy^2z^2 - 4xyz}{4xyz}$

4.3 PRIME FACTORS

Primes

In addition to multiplying polynomials, you must sometimes reverse this process and write a polynomial as a product of other polynomials. This reverse process is known as ***factoring.*** First you will learn how to factor integers, that is, how to write integers as products of other integers.

EXAMPLE 1

$$6 = 2 \cdot 3$$

Here 2 and 3 are each called "factors" of 6. Also, 6 is called a "multiple" of 2 and of 3.

Definition

> *FACTOR. Let a and b be integers, where $b \neq 0$. Then b is called a* ***factor of a*** *(or a* ***divisor of a****) if $a = bc$ for some integer c. In this case, c is also called a factor of a. And a is said to be a* ***multiple of b and of c.***

EXAMPLE 2

Find all *positive* factors of 10.

Solution.

$$10 = 10 \cdot 1$$

and

$$10 = 5 \cdot 2$$

The *positive* factors of 10 are 1, 2, 5, and 10.

Note that the only way you can write 5 as the product of positive integers is

$$5 = 5 \cdot 1 \qquad [\text{or } 5 = 1 \cdot 5].$$

Thus the only *positive* factors of 5 are 5 and 1.

Definition

> *PRIME. Let p be an integer, where $p > 1$. Then p is said to be a* ***prime*** *if the only positive factors of p are p, itself, and 1.*

EXAMPLE 3

2, 3, 5, and 7 are each primes.

Definition

> *COMPOSITE. An integer n, $n > 1$, that is not a prime is called a* ***composite.***

EXAMPLE 4

(a) 6 and 10 are each composites.
(b) 4 is also a composite because

$$4 = 2 \cdot 2.$$

EXAMPLE 5

Which of the following are primes? Which are composites?

(a) 11 (b) 15 (c) 29 (d) 39

Solution.
(a) 11 is a prime.
(b) 15 is a composite because

$$15 = 3 \cdot 5.$$

(c) 29 is a prime. No smaller positive integer other than 1 is a factor of 29.
(d) 39 is a composite because $39 = 3 \cdot 13$.

Prime Factors

Observe that

$$12 = 4 \cdot 3.$$

But the factor 4(which equals $2 \cdot 2$) can be further simplified. Thus

$$12 = 2 \cdot 2 \cdot 3$$

Every composite can be expressed as the product of primes. Thus

$$\begin{aligned} 40 &= 8 \cdot 5 \\ &= (2 \cdot 2 \cdot 2)5 \\ &= 2^3 \cdot 5 \end{aligned}$$

Use exponents to express prime factorization. Except for the order of the factors, there is only one prime factorization. No matter how you begin factoring, the *prime* factors are the same. Thus

$$\begin{aligned} 40 &= 4 \cdot 10 \\ &= (2 \cdot 2)(2 \cdot 5) \\ &= 2^3 \cdot 5 \end{aligned}$$

Also, you can write

$$-40 = -(2^3 \cdot 5).$$

EXAMPLE 6

Find the prime factorization of:
(a) 72 (b) 96

Solution.
(a) $72 = 8 \cdot 9 = 2^3 \cdot 3^2$
(b) $96 = 3 \cdot 32 = 3 \cdot 2^5$

Common Factors

Clearly 5 is a factor of both 10 and 15.

Definition

> *COMMON FACTOR. Let m and n be integers. Then the positive integer c is called a* ***common factor of*** *m and n if c is a factor of both m and n. The largest common factor of m and n is called the* ***greatest common divisor of*** *m and n, and is written*
>
> $$\gcd(m, n).$$

EXAMPLE 7

(a) 1 and 5 are the common factors of 10 and 15. The larger of these is 5. Thus,

$$\gcd(10, 15) = 5$$

(b) 1 is the only common factor of 8 and 9. Thus,

$$\gcd(8, 9) = 1$$

(c) The common factors of 6 and 12 are 1, 2, 3, and 6. Thus,

$$\gcd(6, 12) = 6$$

Example 7(c) illustrates the fact that *if m is a factor of n, then* $\gcd(m, n) = m$.
As a second example,

$$\gcd(10, 100) = 10$$

You can also speak of common factors and greatest common divisors of three or more integers. The gcd notation is again used.

EXAMPLE 8

(a) What are the common factors of 8, 12, and 20?
(b) Find gcd (8, 12, 20).

Solution.
(a) The common factors of 8, 12, and 20 are 1, 2, and 4.
(b) $\gcd(8, 12, 20) = 4$

When the integers are relatively simple, their gcd can be found at sight, as in Examples 7 and 8. When difficulties arise, use the prime factorizations of each integer.

EXAMPLE 9

Find: gcd (36, 54, 72)

Solution.

$$\begin{aligned}36 &= 4 \cdot 9 = 2^2 \cdot 3^2\\ 54 &= 2 \cdot 27 = 2 \cdot 3^3\\ 72 &= 8 \cdot 9 = 2^3 \cdot 3^2\end{aligned}$$

Consider the *smallest* power of each *common* prime factor. *The product of these smallest powers* is then a common factor of the integers 36, 54, and 72. In fact, this product is the largest common factor of these integers, and hence, the gcd.

Thus here the common prime factors are 2 and 3.

The *smallest* power of 2 that occurs is 2 (in the factorization of 54).

The smallest power of 3 is 3^2.

$$\begin{aligned}\text{gcd }(36, 54, 72) &= 2 \cdot 3^2\\ &= 18\end{aligned}$$

Coefficients of Polynomials

An important application of these notions occurs in finding the gcd of the coefficients of a polynomial. This will be used in "factoring" polynomials.

EXAMPLE 10

Find the gcd of the coefficients of $32x - 80$.

Solution. Factor both coefficients:

$$\begin{aligned}32 &= 4 \cdot 8\\ &= 2^2 \cdot 2^3\\ &= 2^5\end{aligned}$$

$$\begin{aligned}-80 &= -(16 \cdot 5)\\ &= -2^4 \cdot 5\end{aligned}$$

The only common prime factor is 2, and the smallest power of 2 that occurs is 2^4. Thus

$$\begin{aligned}\text{gcd }(32, -80) &= 2^4\\ &= 16,\end{aligned}$$

and the gcd of the coefficients of $32x - 80$ is 16.

EXERCISES

In exercises 1–10, determine which integers are primes and which are composites.

1. 9	2. 13	3. 17	4. 22	5. 23
6. 27	7. 31	8. 33	9. 37	10. 41

In exercises 11–30, find the prime factorization of each integer.

11. 10	12. 16	13. 18	14. 20	15. −24
16. 25	17. 28	18. 30	19. 36	20. −40
21. 42	22. −44	23. 65	24. 66	25. 70
26. −100	27. 108	28. 144	29. 169	30. 221

In exercises 31–50, find the gcd of the given integers.

31. 6 and 8
32. 6 and 9
33. 9 and 15
34. 9 and 18
35. 25 and 35
36. 12 and 24
37. 12 and −36
38. −24 and −36
39. 48 and 72
40. 60 and 84
41. 4, 6, and 8
42. 5, 10, and 15
43. 25, 30, and 50
44. 30, 40, and 50
45. −8, 16, and 32
46. −40, −60, and −80
47. 27, 36, and 45
48. 64, 72, and 144
49. 12, 18, 24, and 36
50. 30, 42, 60, and 96

In exercises 51–60, find the gcd of the coefficients of each polynomial.

51. $3x + 6$
52. $4x - 8$
53. $7x + 14$
54. $12x + 20$
55. $-x^2 + 5$
56. $48x^2 + 108x$
57. $12x^2 - 18x + 6$
58. $5x^2 - 10x + 25$
59. $24x^2 + 36x - 42$
60. $27x^3 - 18x^2 + 54x$
61. Find the smallest (positive) number that is the product of two different primes.
62. Find the smallest (positive) number that is the product of three different primes.
63. List all primes less than 50.
64. List all primes between 50 and 100.

4.4 COMMON FACTORS

Factors of Polynomials

Recall that a nonzero integer b is a factor of (the integer) a if

$$a = bc$$

for some integer c. The same notion applies to polynomials. You will want to call

$$2 \text{ and } x$$

factors of the polynomial

$$2x.$$

Definition

> *Let P and Q (≠ 0) be polynomials. Then Q is called a* ***factor of*** *P if*
>
> $$P = Q \cdot R$$
>
> *for some polynomial R.*

EXAMPLE 1

Show that 3 is a factor of $6x + 9$.

Solution. First note that 3 is the gcd of the coefficients of $6x + 9$. Also,

$$\frac{6x+9}{3} = \frac{6x}{3} + \frac{9}{3} = 2x + 3$$

Thus $$6x + 9 = 3(2x + 3)$$

Here $$P = 6x + 9, Q = 3, R = 2x + 3$$

EXAMPLE 2

Show that x^2 is a factor of $x^5 + 2x^2$.

Solution.

$$\frac{x^5 + 2x^2}{x^2} = \frac{x^5}{x^2} + \frac{2x^2}{x^2} = x^3 + 2$$

Thus $$x^5 + 2x^2 = x^2(x^3 + 2)$$

Here $$P = x^5 + 2x^2, Q = x^2, R = x^3 + 2$$

Numbers, such as 3 (in Example 1), as well as powers of variables, such as x^2 (in Example 2), can be factors of a polynomial. In the next example, *the product of a number and of powers of several different variables is a factor of the given polynomial.*

EXAMPLE 3

Let $P = 6x^2yz^3 + 12x^3yz$. Then

$$6,\ x^2,\ y, \text{ and } z$$

are each factors of P. Also, their *product*,

$$6x^2yz,$$

is a factor of P. In fact,

$$\frac{6x^2yz^3 + 12x^3yz}{6x^2yz} = z^2 + 2x$$

Thus $$6x^2yz^3 + 12x^3yz = 6x^2yz(z^2 + 2x)$$

EXAMPLE 4

Let $P = 8x^3y + 12x^2y^2 + 4x^2yz$. Then

$$4 = \gcd(8, 12, 4)$$

Thus 4 is a factor of P. Similarly,

$$x^2 \text{ and } y$$

are each factors of *all three terms of P*, and are therefore factors of P. The *product* of these,

$$4x^2y$$

is a factor of P because

$$8x^3y + 12x^2y^2 + 4x^2yz = 4x^2y(2x + 3y + z).$$

Definition

> *Let P be a polynomial. The* ***(greatest) common (monomial) factor of*** *P, or for short, the* ***common factor of*** *P, is defined to be the product of*
>
> **1.** *the greatest common divisor of the coefficients of P and*
>
> **2.** *the smallest power of each variable that occurs in each term of P.*

Thus in Example 4, $4x^2y$ is the common factor of P.

Isolating the Common Factor

Definition

> *To* ***isolate the common factor,*** *Q, of a polynomial, P, write*
>
> $$P = Q \cdot R,$$
>
> *where Q is the common factor of P and where R is another polynomial.*

Apply the *Distributive Laws* to isolate the common factor of a polynomial.

EXAMPLE 5

Isolate the common factor of $4a + 8b$.

Solution. Let

$$P = 4a + 8b.$$

Then the common factor of P is

$$\underbrace{4.}_{Q}$$

Isolate the common factor Q of P by writing

$$\underbrace{4a + 8b}_{P} = \underbrace{4}_{Q} \underbrace{(a + 2b)}_{R}.$$

EXAMPLE 6

Isolate the common factor of $5x^2y^2 - 10xy^3$.

Solution. Let

$$P = 5x^2y^2 - 10xy^3.$$

Then the common factor of P is

$$\underbrace{5xy^2.}_{Q}$$

Isolate the common factor Q by writing

$$\underbrace{5x^2y^2 - 10xy^3}_{P} = \underbrace{5xy^2}_{Q} \underbrace{(x - 2y)}_{R}.$$

EXAMPLE 7

Isolate the common factor of

$$2a^2bc + 4a^2c + 5a^3b.$$

Solution.

$$\gcd(2, 4, 5) = 1$$

The only variable that occurs in all three terms is a; the smallest power to which it occurs is the second power, a^2. Thus

$$a^2$$

is the common factor. Isolate the common factor by writing

$$\underbrace{2a^2bc + 4a^2c + 5a^3b}_{P} = \underbrace{a^2}_{Q} \underbrace{(2bc + 4c + 5ab)}_{R}.$$

EXAMPLE 2

$$\begin{aligned}(x+5)(x+3) &= x^2 + (5+3)x + 5 \cdot 3 \\ &= x^2 + 8x + 15\end{aligned}$$

It is probably easier to multiply by writing the factors one above the other. Write *like terms* in the *same column.*

EXAMPLE 3

Multiply: $(x-4)(x-2)$

Solution. First multiply by x:

$$\begin{array}{l} x - 4 \\ \uparrow \nearrow \\ \underline{x \qquad\quad} \\ x^2 - 4x \end{array}$$

then by -2:

$$\begin{array}{r} x - 4 \\ \nwarrow \uparrow \\ \underline{x - 2} \\ x^2 - 4x \qquad \\ \underline{\quad - 2x + 8} \\ x^2 - 6x + 8 \end{array}$$

EXAMPLE 4

Multiply: $(x+6)(x-3)$

Solution.

$$\begin{array}{rrr} x & + 6 & \\ \underline{x} & \underline{- 3} & \\ x^2 & + 6x & \\ \underline{} & \underline{- 3x} & \underline{- 18} \\ x^2 & + 3x & - 18 \end{array}$$

EXAMPLE 5

Multiply: $(x+5)^2$

Solution. $(x+5)^2 = (x+5)(x+5)$

$$\begin{array}{rrr} x & + 5 & \\ \underline{x} & \underline{+ 5} & \\ x^2 & + 5x & \\ \underline{} & \underline{5x} & \underline{+ 25} \\ x^2 & + 10x & + 25 \end{array}$$

In general,

$$(x+a)^2 = (x+a)(x+a) = x^2 + 2ax + a^2$$

EXAMPLE 6

Multiply: $(x + 4)(x - 4)$

Solution.

$$\begin{array}{r}
x + 4 \\
\underline{x - 4} \\
x^2 + 4x \\
\underline{- 4x - 16}\\
x^2 - 16
\end{array}$$

Note that the "cross-terms" $4x$ and $-4x$ are inverses. Their sum is 0.

In general,

$$(x + a)(x - a) = x^2 + \underbrace{(a - a)}_{0}x - a^2$$

Thus $\quad (x + a)(x - a) = x^2 - a^2$

This factoring technique is known as the "difference of squares."

EXAMPLE 7

Multiply: $(2x + 1)(2x + 3)$

Solution.

$$\begin{array}{r}
2x + 1 \\
\underline{2x + 3} \\
4x^2 + 2x \\
\underline{6x + 3}\\
4x^2 + 8x + 3
\end{array}$$

EXAMPLE 8

$$\begin{array}{l}
2a^2 + 3\\
\underline{\;\;a \;\; + 2}\\
2a^3 \qquad\quad + 3a\\
\underline{\qquad\; 4a^2 \qquad\;\; + 6}\\
2a^3 + 4a^2 + 3a + 6
\end{array}$$

EXAMPLE 9

$$\begin{array}{r}
5a + b \\
\underline{a - b} \\
5a^2 + ab \\
\underline{- 5ab - b^2}\\
5a^2 - 4ab - b^2
\end{array}$$

EXERCISES

In exercises 1–60, multiply as indicated.

1. $(x+1)(x+2)$
2. $(y+3)(y+1)$
3. $(x+1)(x+6)$
4. $(z+8)(z+1)$
5. $(y+2)(y+3)$
6. $(x+4)(x+2)$
7. $(z+4)(z+3)$
8. $(a+7)(a+4)$
9. $(a+5)(a-1)$
10. $(a+3)(a-2)$
11. $(b+3)(b-7)$
12. $(b-2)(b-4)$
13. $(x-9)(x+3)$
14. $(y-5)(y-2)$
15. $(m+6)(m+2)$
16. $(n-10)(n+3)$
17. $(c-2)(c+5)$
18. $(n-2)(n-7)$
19. $(x+2)(x-9)$
20. $(x-2)(x-5)$
21. $(y+1)(y+8)$
22. $5(z+2)(z-1)$
23. $-2(a-9)(a+5)$
24. $-3(b-1)(b-2)$
25. $-(c-3)(c+6)$
26. $-2(d+7)(d+2)$
27. $(x+1)^2$
28. $(y+2)^2$
29. $(a-2)^2$
30. $(b-7)^2$
31. $(c+6)^2$
32. $(v-8)^2$
33. $(x+10)^2$
34. $(z-10)^2$
35. $(x+8)(x-8)$
36. $(y-4)(y+4)$
37. $(v+1)(v-1)$
38. $(x+10)(x-10)$
39. $(x+12)(x-12)$
40. $(x-12)(x-12)$
41. $(x^2+4)^2$
42. $(x^2+3)(x^2-3)$
43. $(a^4+1)(a^4-1)$
44. $(u^4+1)(u^4+2)$
45. $(x^3+1)^2$
46. $(x^3-3)^2$
47. $(x^2+3)(x^2+2)$
48. $(b^2+5)(b^2-2)$
49. $(c^3-1)(c^3-4)$
50. $(c^3-1)(c^2+1)$
51. $(x+y)(x+y)$
52. $(x+a)(x-a)$
53. $(m+2n)(m+n)$
54. $(x+3a)(x+2a)$
55. $(m-4n)(m+2n)$
56. $(x-2y)(x+y)$
57. $(5a-3b)(-2a+6b)$
58. $(2x+2y)(x-4y)$
59. $(20m-3n)(m-2n)$
60. $(3a+b)(2a-2b)$

In exercises 61–64, multiply and simplify.

61. $(x+2)(x+3)+(x+1)(x+4)$
62. $(x+4)(x-2)-(x+3)(x+5)$
63. $(y+7)(y-2)+(y+5)^2$
64. $(y+3)^2-(y-1)^2$
65. Multiply the sum of $3x+1$ and $-2x$ by $2x+5$.
66. Evaluate $(x+8)(x-3)$ when $x=6$.
67. Evaluate $(2x+3y)(x-5y)$ when $x=4$ and $y=6$.
68. Evaluate $(2x^2-y)(x^2+xy-y^2)$ when $x=1$ and $y=-1$.

4.6 DIFFERENCE OF SQUARES

Recall that you factor a polynomial P by writing P as a product of other polynomials. In Section 4.4 you learned how to isolate the common factor of a polynomial. In the remainder of the chapter you will learn other techniques for factoring polynomials.

One of the easiest types of factoring to recognize is known as the **difference of squares.**

Recall that

$$(x + a)(x - a) = x^2 - a^2.$$

The "difference of squares" method applies to a binomial, such as

$$x^2 - a^2,$$

whose two terms are *squares* that are separated by a *minus* sign.

EXAMPLE 1

Factor: $x^2 - 100$

Solution. $100 = 10^2$

Thus

$$\begin{aligned} x^2 - 100 &= x^2 - 10^2 \\ &= (x + 10)(x - 10) \end{aligned}$$

EXAMPLE 2

Factor: $4 - a^2$

Solution. Although the number 4 is written first, the difference of squares method again applies.

$$4 = 2^2$$

Thus

$$\begin{aligned} 4 - a^2 &= 2^2 - a^2 \\ &= (2 + a)(2 - a) \end{aligned}$$

EXAMPLE 3

Factor: $s^2 - t^2$

Solution.

$$s^2 - t^2 = (s + t)(s - t)$$

A common error in applying this method is to factor the individual terms, but not the polynomial as a whole. For instance, in Example 3, do **not** write

$$s^2 - t^2 = s \cdot s - t \cdot t,$$

but rather,

$$s^2 - t^2 = (s + t)(s - t),$$

as above. (The first equality is correct, but is not the desired factored result.)

You may be given a binomial such as

$$x^2 - a^2b^2.$$

By the Associative and Commutative Laws of Multiplication,

$$a^2b^2 = (aa)(bb)$$
$$= (ab)(ab)$$
$$= (ab)^2$$

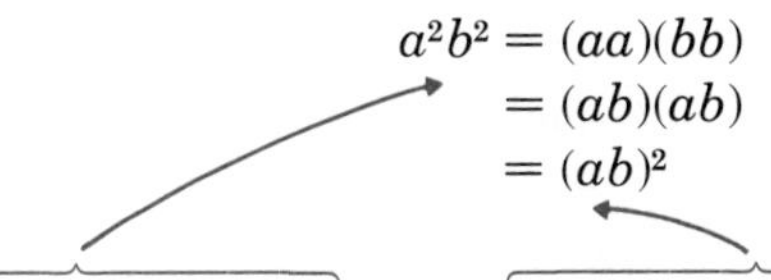

The product of the squares is the square of the product. Thus

$$x^2 - a^2b^2 = (x + ab)(x - ab)$$

EXAMPLE 4
Factor: $x^2 - 100y^2$

Solution.

$$100y^2 = 10^2y^2$$
$$= (10y)^2$$

Thus
$$x^2 - 100y^2 = x^2 - (10y)^2$$
$$= (x + 10y)(x - 10y)$$

EXAMPLE 5
Factor: $16a^2 - 25b^2$

Solution.

$$16a^2 = (4a)^2; \quad 25b^2 = (5b)^2$$

Thus
$$16a^2 - 25b^2 = (4a)^2 - (5b)^2$$
$$= (4a + 5b)(4a - 5b)$$

Sometimes you save much effort by first isolating the common factor before applying the difference of squares method.

EXAMPLE 6
Factor: $5a^2 - 20b^2$

Solution.

$$5a^2 - 20b^2 = 5(a^2 - 4b^2)$$
$$= 5(a + 2b)(a - 2b)$$

EXAMPLE 7
Factor: $ax^2 - a^3$

Solution.

$$ax^2 - a^3 = a(x^2 - a^2)$$
$$= a(x + a)(x - a)$$

The above method applies to the difference of squares, but *not to the sum of squares.*

EXAMPLE 8

Factor: $a^2 + b^2$

Solution. *This binomial cannot be factored by the present method.* Notice that the cross-terms do not cancel in the following product:

$$\begin{array}{r} a + b \\ a + b \\ \hline a^2 + ab \\ ab + b^2 \\ \hline a^2 + 2ab + b^2 \end{array}$$

EXERCISES

In exercises 1–60, factor each binomial.

1. $x^2 - 9$
2. $y^2 - 4$
3. $x^2 - 49$
4. $a^2 - 25$
5. $m^2 - 100$
6. $s^2 - 36$
7. $y^2 - 81$
8. $z^2 - 144$
9. $1 - x^2$
10. $9 - y^2$
11. $25 - b^2$
12. $121 - c^2$
13. $x^2 - y^2$
14. $z^2 - y^2$
15. $a^2 - x^2$
16. $m^2 - n^2$
17. $4x^2 - 1$
18. $9a^2 - 25$
19. $36x^2 - 49$
20. $100z^2 - 121$
21. $1 - 9u^2$
22. $25 - 81v^2$
23. $4x^2 - y^2$
24. $9x^2 - 16a^2$
25. $9m^2 - 49n^2$
26. $100z^2 - 81a^2$
27. $144y^2 - 25x^2$
28. $64r^2 - 169s^2$
29. $400t^2 - u^2$
30. $49a^2 - 225b^2$
31. $2x^2 - 2$
32. $5y^2 - 45$
33. $3s^2 - 75$
34. $7 - 7t^2$
35. $300 - 3x^2$
36. $180 - 5y^2$
37. $2x^2 - 32y^2$
38. $3y^2 - 27z^2$
39. $8a^2 - 200b^2$
40. $7x^2 - 175y^2$
41. $6m^2 - 54n^2$
42. $44a^2 - 99b^2$
43. $24x^2 - 294y^2$
44. $288c^2 - 242d^2$
45. $a^3 - a$
46. $x^3 - 4x$
47. $b^3 - 49b$
48. $25y - y^3$
49. $a^4 - a^2$
50. $x^3 - x^5$
51. $x^3 - xy^2$
52. $a^3 - 4ab^2$
53. $m^4 - 49m^2n^2$
54. $4s^4 - 9s^2t^2$
55. $2x^3 - 2x$
56. $5a^3 - 20a$
57. $3x^3 - 3xy^2$
58. $7m^2n - 7n^3$
59. $8x^3 - 50xy^2$
60. $12a^2b - 27b^3$

In exercises 61–64, indicate which one binomial cannot be factored (by the methods you have learned).

61. (a) $x^2 - 9$ (b) $x^2 - 1$ (c) $x^2 + 1$ (d) $x^2 - a^2$
62. (a) $a^2 - x^2$ (b) $x^2 - a^2$ (c) $x^2 + a^2$ (d) $4x^2 + 4a^2$
63. (a) $y^2 - 16$ (b) $3y^2 - 48$ (c) $16y^2 + 16$ (d) $y^2 + 16$
64. (a) $ax^2 + ab^2$ (b) $a^2c + ad^2$ (c) $25a^2 + 49b^2$ (d) $ax^2 + a^2y$

4.7 FACTORING TRINOMIALS $x^2 + Mx + N$

In this section you will learn how to factor trinomials such as

$$x^2 + 3x + 2$$

and

$$x^2 - 5x + 4.$$

In general, such trinomials are of the form

$$x^2 + Mx + N,$$

where M and N are integers.

EXAMPLE 1

Factor: $x^2 + 3x + 2$

Solution. In order to obtain x^2, try to multiply:

$$\begin{array}{l} x + \_\_\_ \\ \underline{x + \_\_\_} \end{array}$$

Because the numerical term of $x^2 + 3x + \mathbf{2}$ is 2, fill in the blanks with integers whose product is 2. Try 1 and 2.

$$\begin{array}{r} x + 1 \\ \underline{x + 2} \\ x^2 + x \\ \underline{ 2x + 2} \\ x^2 + 3x + 2 \end{array}$$

Note that

$$1 \cdot 2 = 2.$$

(2 is the numerical term of

$$x^2 + 3x + \mathbf{2}$$

and

$$1 + 2 = 3.)$$

(3 is the coefficient of the term $3x$ in

$$x^2 + \mathbf{3x} + 2.)$$

Thus

$$x^2 + 3x + 2 = (x + 1)(x + 2)$$

EXAMPLE 2

Factor: $x^2 + 5x + 4$

Solution. Try

$$\begin{array}{l} x + \_\_\_ \\ \underline{x + \_\_\_} \end{array}$$

Fill in the blanks with integers whose product is 4. You could try $2 \cdot 2$ or $1 \cdot 4$:

$$\begin{array}{r} x+2 \\ \underline{x+2} \\ x^2+2x \qquad \\ \underline{\qquad 2x+4} \\ x^2+\mathbf{4}x+4 \end{array}$$

The coefficient of the *middle* term is wrong! It should be 5, not 4.

$$\begin{array}{r} x+1 \\ \underline{x+4} \\ x^2+\ x \qquad \\ \underline{\qquad 4x+4} \\ x^2+\mathbf{5}x+4 \end{array}$$

This is the right coefficient of $5x$.

Thus $$x^2+5x+4=(x+1)(x+4)$$

Here $$1\cdot 4=4$$

(4 is the numerical term of

$$x^2+5x+\mathbf{4});$$

and $$1+4=5$$

(5 is the coefficient of $5x$ in

$$x^2+\mathbf{5}x+4.)$$

EXAMPLE 3

Factor: x^2-5x+4

Solution. *The coefficient of* the term $-5x$ *is negative:*

$$x^2-\mathbf{5}x+4$$

Modify the factors of Example 2 by taking -1 and -4 (instead of 1 and 4).

$$\begin{array}{r} x-1 \\ \underline{x-4} \\ x^2-\ x \qquad \\ \underline{\qquad -4x+4} \\ x^2-\mathbf{5}x+4 \end{array}$$

Thus $$x^2-5x+4=(x-1)(x-4)$$

EXAMPLE 4

Factor: y^2+2y-3.

Solution. Here, *the numerical term is negative:*

$$y^2+2y\mathbf{-3}$$

One of the factors of -3 must be positive, the other negative (because the product, -3, is negative). Try $1(-3)$ and $3(-1)$.

$$\begin{array}{r} y + 1 \\ \underline{y - 3} \\ y^2 + y \phantom{{}-3} \\ \underline{-3y - 3} \\ y^2 - \mathbf{2}y - 3 \end{array}$$

Wrong! The coefficient of the middle term should be 2, not −2.

$$\begin{array}{r} y + 3 \\ \underline{y - 1} \\ y^2 + 3y \phantom{{}-3} \\ \underline{-y - 3} \\ y^2 + \mathbf{2}y - 3 \end{array}$$

Right coefficient of $2y$.

Thus $$y^2 + 2y - 3 = (y + 3)(y - 1)$$

As you see, you may have to try more than one pair of integers. To factor

$$x^2 + Mx + N,$$

try to find integers a and b whose product is N and whose sum is M. In symbols:

$$ab = N$$

and $$a + b = M$$

Then
$$\begin{array}{l} x + a \\ \underline{x + b} \\ x^2 + ax \\ \underline{\phantom{x^2 + {}} bx + ab \phantom{{} = x^2 + Mx + N}} \\ x^2 + \underbrace{(a + b)}_{M}x + \underbrace{ab}_{N} = x^2 + Mx + N \end{array}$$

EXAMPLE 5

Factor: $x^2 + 8x + 12$

Solution. Here $M = 8$ and $N = 12$. *Both* M and N are *positive* for the polynomial

$$x^2 + 8x + 12.$$

Try *positive* integers a and b such that

$$ab = 12$$

and $$a + b = 8.$$

The possibilities are

$$1 \cdot 12,\ 2 \cdot 6, \text{ and } 3 \cdot 4.$$

Try these:

$$1 + 12 = 13 \quad \mathbf{2 + 6 = 8} \quad 3 + 4 = 7$$
(Wrong) (Right) (Wrong)

(In practice, as soon as you find the right integral factors, immediately factor the given polynomial.)

Thus $$x^2 + 8x + 12 = (x + 2)(x + 6)$$

EXAMPLE 6

Factor: $x^2 + x + 3$

Solution. The only positive factors of 3,

$$1 \text{ and } 3,$$

do not work because

$$1 + 3 = 4.$$

Thus

$$x^2 + x + 3$$

cannot be factored by this method.

In general,

$$x^2 + 2ax + a^2 = (x + a)^2.$$

Thus $$x^2 + 10x + 25 = (x + 5)^2$$ Here $a = 5$

and $$x^2 - 10x + 25 = (x - 5)^2$$ Here $a = -5$

It is often best to isolate the common factor before applying the present method.

EXAMPLE 7

Factor: $3x^3 - 21x^2 + 30x$

Solution.

$$\gcd(3, -21, 30) = 3$$

Thus $$3x^3 - 21x^2 + 30x = 3x(x^2 \mathbf{- 7}x + \mathbf{10})$$
↑ negative ↑ positive

Try $$(-1)(-10) \text{ and } (-2)(-5).$$

$$(-1) + (-10) = -11 \quad \text{(Wrong)}$$
$$(-2) + (-5) = -7 \quad \text{(Right)}$$

Therefore

$$3x^3 - 21x^2 + 30x = 3x(x - 2)(x - 5)$$

This method applies with very little modification to trinomials in 2 variables.

EXAMPLE 8

Factor:

(a) $x^2 - 5x - 14$
(b) $x^2 - 5xy - 14y^2$

Solution.

(a) $x^2 \mathbf{- 5}x \mathbf{- 14}$

$$\begin{array}{cc} \uparrow & \uparrow \\ \text{negative} & \text{negative} \end{array}$$

The possible factors of -14 are

$$\begin{array}{ll} 1 \text{ and } -14, & 2 \text{ and } -7, \\ -1 \text{ and } 14, & -2 \text{ and } 7. \end{array}$$

Because the coefficient of $-5x$ is negative, the negative factor must have larger absolute value than the positive factor. Thus try only: $1(-14)$ and $2(-7)$

$$\begin{array}{ll} 1 + (-14) = -13 & \text{(Wrong)} \\ 2 + (-7) = -5 & \text{(Right)} \end{array}$$

Therefore

$$x^2 - 5x - 14 = (x + 2)(x - 7)$$

(b) The considerations of part (a) lead you to try:

$$\begin{array}{l} x + 2y \\ \underline{x - 7y} \\ x^2 + 2xy \\ \underline{\quad\; - 7xy - 14y^2} \\ x^2 - 5xy - 14y^2 \end{array}$$

Thus

$$x^2 - 5xy - 14y^2 = (x + 2y)(x - 7y)$$

EXERCISES

In exercises 1–58, factor each trinomial.

1. $x^2 + 4x + 3$
2. $y^2 + 6y + 5$
3. $z^2 + 7z + 6$

4. $a^2 + 8a + 7$
5. $b^2 + 2b + 1$
6. $c^2 + 4c + 4$
7. $x^2 + 6x + 9$
8. $x^2 + 10x + 25$
9. $a^2 + a - 2$
10. $m^2 + 3m - 4$
11. $b^2 - b - 2$
12. $n^2 - 4n - 5$
13. $s^2 - s - 12$
14. $x^2 - x - 12$
15. $y^2 + 8y + 12$
16. $t^2 - 11t - 12$
17. $a^2 - 12a + 36$
18. $b^2 + 10b + 25$
19. $x^2 - 9x + 14$
20. $y^2 + 13y - 14$
21. $a^2 + 5a - 14$
22. $b^2 - 5b - 14$
23. $c^2 + 9c + 20$
24. $d^2 - 12d + 20$
25. $x^2 + 13x + 40$
26. $x^2 + 11x + 30$
27. $y^2 - 7y - 30$
28. $y^2 - 10y + 16$
29. $x^2 + 9x - 36$
30. $x^2 - 17x + 66$
31. $2x^2 + 6x + 4$
32. $2x^2 + 16x + 14$
33. $3x^2 + 6x - 24$
34. $4y^2 - 28y + 48$
35. $-y^2 + 9y - 20$
36. $30 + z - z^2$
37. $a^3 + 3a^2 - 10a$
38. $b^3 + 12b^2 + 32b$
39. $4m^2 + 40m + 100$
40. $-2n^2 + 24n - 72$
41. $72 + 6x - 3x^2$
42. $y^4 + 10y^2 + 24$
43. $u^2v^2 + 18uv^2 + 81v^2$
44. $uv^2 - 16uv + 64u$
45. $2t^3 - 24t^2 + 70t$
46. $63x + 12x^2 - 3x^3$
47. $x^2 + 2ax + a^2$
48. $y^2 - 2ay + a^2$
49. $x^2 + 5ax + 4a^2$
50. $y^2 - 3ay - 4a^2$
51. $m^2 + 4mn + 4n^2$
52. $m^2 + 3mn - 4n^2$
53. $x^2 + 6xy + 9y^2$
54. $u^2 - 8uv + 16v^2$
55. $2x^2 + 24xy + 72y^2$
56. $3x^2 + 3xy - 36y^2$
57. $ax^2 + 6axy + 8ay^2$
58. $x^3 + 6x^2y + 5xy^2$

In exercises 59–60, indicate which one trinomial cannot be factored (by the methods of this section).

59. (a) $x^2 + 5x + 6$ (b) $x^2 - 5x + 6$ (c) $x^2 - 6x + 5$ (d) $x^2 + 6x - 5$
60. (a) $y^2 + 7y + 12$ (b) $y^2 + 8y + 12$ (c) $y^2 + 9y + 12$ (d) $y^2 + 13y + 12$

4.8 FACTORING TRINOMIALS $Lx^2 + Mx + N$

In this section you will factor trinomials, such as

$$2x^2 + 3x + 1$$

and

$$3x^2 + 8x + 4,$$

in which the coefficient of the x^2 term is not 1. In general, these trinomials are of the form

$$Lx^2 + Mx + N.$$

EXAMPLE 1

Factor: $2x^2 + 3x + 1$

Solution. In order to obtain $2x^2$, try:

$$\begin{array}{r} 2x + \_\_\_ \\ x + \_\_\_ \end{array}$$

Because the numerical term of

$$2x^2 + \mathbf{3}x + \mathbf{1}$$

is 1, and because the coefficient of the term $3x$ is *positive,* try $1 \cdot 1$.

$$\begin{array}{r} 2x + 1 \\ x + 1 \\ \hline 2x^2 + x \\ 2x + 1 \\ \hline 2x^2 + 3x + 1 \end{array}$$

This works. Therefore

$$2x^2 + 3x + 1 = (2x + 1)(x + 1)$$

EXAMPLE 2

Factor: $2x^2 - 5x + 2$

Solution. Again try:

$$\begin{array}{r} 2x + \_\_\_ \\ x + \_\_\_ \end{array}$$

Because of the *negative* coefficient of $-5x$ in

$$2x^2 - \mathbf{5}x + 2,$$

try $(-1)(-2)$.

Now there are two combinations to try:

$$\begin{array}{r} 2x - 2 \\ x - 1 \\ \hline 2x^2 - 2x \\ - 2x + 2 \\ \hline 2x^2 - \mathbf{4}x + 2 \end{array} \text{ (Wrong)} \qquad \text{and} \qquad \begin{array}{r} 2x - 1 \\ x - 2 \\ \hline 2x^2 - x \\ - 4x + 2 \\ \hline 2x^2 - \mathbf{5}x + 2 \end{array} \text{ (Right)}$$

Thus

$$2x^2 - 5x + 2 = (2x - 1)(x - 2)$$

EXAMPLE 3

Factor: $3x^2 + 8x + 4$

Solution.

$$\begin{array}{r} 3x + \_\_\_ \\ x + \_\_\_ \end{array}$$

Here try:

$$1 \cdot 4 \text{ and } 2 \cdot 2$$

You will have to try both combinations

$$\begin{array}{r} 3x + 1 \\ x + 4 \\ \hline 3x^2 + x \\ 12x + 4 \\ \hline 3x^2 + 13x + 4 \end{array} \quad \text{(Wrong)} \qquad \text{and} \qquad \begin{array}{r} 3x + 4 \\ x + 1 \\ \hline 3x^2 + 4x \\ 3x + 4 \\ \hline 3x^2 + 7x + 4 \end{array} \quad \text{(Wrong)}$$

as well as

$$\begin{array}{r} 3x + 2 \\ x + 2 \\ \hline 3x^2 + 2x \\ 6x + 4 \\ \hline 3x^2 + 8x + 4 \end{array} \quad \text{(Right)}$$

Therefore

$$3x^2 + 8x + 4 = (3x + 2)(x + 2)$$

You may have to isolate the common factor before applying this method.

EXAMPLE 4

Factor: $-8x^2 + 6x - 2$

Solution. Because you want to consider

$$Lx^2 + Mx + N, \qquad \text{where } L > 0,$$

isolate the common factor -2 (instead of 2):

$$-8x^2 + 6x - 2 = (-2)(4x^2 - 3x + 1)$$

Consider

$$4x^2 - \mathbf{3}x + 1.$$
$$\uparrow$$
negative

Now there are two possibilities for $4x^2$:

$$\begin{array}{l} 2x + \underline{\quad} \\ 2x + \underline{\quad} \end{array} \quad \text{and} \quad \begin{array}{r} 4x + \underline{\quad} \\ x + \underline{\quad} \end{array}$$

Because the coefficient of the term $-3x$ is negative in $4x^2 - \mathbf{3}x + 1$, try

$$1(-1).$$

Remember to try the integral factors *both* ways with $4x$ and x, if the first way doesn't work.

$$\begin{array}{r} 2x + 1 \\ 2x - 1 \\ \hline 4x^2 + 2x \quad\quad \\ -2x - 1 \\ \hline 4x^2 \quad\quad - 1 \end{array} \text{ (Wrong)} \qquad \begin{array}{r} 4x - 1 \\ x + 1 \\ \hline 4x^2 - x \quad\quad \\ 4x - 1 \\ \hline 4x^2 + 3x - 1 \end{array} \text{ (Wrong)} \qquad \begin{array}{r} 4x + 1 \\ x - 1 \\ \hline 4x^2 + x \quad\quad \\ -4x - 1 \\ \hline 4x^2 - 3x - 1 \end{array} \text{ (Right)}$$

Therefore

$$4x^2 - 3x - 1 = (4x + 1)(x - 1)$$

and

$$-8x^2 + 6x - 2 = -2(4x + 1)(x - 1)$$

Now consider a polynomial in two variables, as in part (b) of the following example.

EXAMPLE 5

Factor:

(a) $6x^2 + 17x + 5$
(b) $6x^2 + 17xy + 5y^2$

Solution.

(a) Try all combinations for

$$\begin{array}{r} 6x + \underline{\quad} \\ x + \underline{\quad} \end{array} \quad \text{and for} \quad \begin{array}{r} 3x + \underline{\quad} \\ 2x + \underline{\quad} \end{array}$$

along with

$$5 \cdot 1$$

$$\begin{array}{r} 6x + 5 \\ x + 1 \\ \hline 6x^2 + 5x \quad\quad \\ 6x + 5 \\ \hline 6x^2 + 11x + 5 \end{array} \text{ (Wrong)} \qquad \begin{array}{r} 6x + 1 \\ x + 5 \\ \hline 6x^2 + x \quad\quad \\ 30x + 5 \\ \hline 6x^2 + 31x + 5 \end{array} \text{ (Wrong)}$$

$$\begin{array}{r} 3x + 5 \\ 2x + 1 \\ \hline 6x^2 + 10x \quad\quad \\ 3x + 5 \\ \hline 6x^2 + 13x + 5 \end{array} \text{ (Wrong)} \qquad \begin{array}{r} 3x + 1 \\ 2x + 5 \\ \hline 6x^2 + 2x \quad\quad \\ 15x + 5 \\ \hline 6x^2 + 17x + 5 \end{array} \text{ (Right)}$$

Thus $\quad 6x^2 + 17x + 5 = (3x + 1)(2x + 5)$

(b) Modify the above factors:

$$\begin{array}{r} 3x + y \\ 2x + 5y \\ \hline 6x^2 + 2xy \quad\quad\quad \\ 15xy + 5y^2 \\ \hline 6x^2 + 17xy + 5y^2 \end{array}$$

Thus $$6x^2 + 17xy + 5y^2 = (3x + y)(2x + 5y)$$

EXERCISES

Factor each trinomial.

1. $3x^2 + 4x + 1$
2. $5y^2 + 6y + 1$
3. $7a^2 + 8a + 1$
4. $7b^2 - 8b + 1$
5. $3x^2 + 2x - 1$
6. $3x^2 - 2x - 1$
7. $5y^2 - 6y + 1$
8. $5y^2 - 4y - 1$
9. $2x^2 + 7x + 3$
10. $2x^2 + 5x + 2$
11. $2y^2 + 5y + 3$
12. $2y^2 + 7y + 3$
13. $2a^2 + 7a + 5$
14. $2b^2 - 9b + 9$
15. $2m^2 - 5m + 2$
16. $2m^2 - 5m - 3$
17. $2n^2 - 7n + 3$
18. $2a^2 - 7a - 4$
19. $2a^2 + 7a - 4$
20. $2a^2 - 9a + 4$
21. $2a^2 + 15a + 25$
22. $4b^2 + 20b + 25$
23. $9y^2 - 6y + 1$
24. $9z^2 - 10z + 1$
25. $4a^2 + 4a + 1$
26. $4b^2 + 5b + 1$
27. $3y^2 - 4y - 7$
28. $3z^2 - 20z - 7$
29. $4m^2 + 8m + 3$
30. $4n^2 - 8n + 3$
31. $6x^2 + 5x - 6$
32. $6z^2 + 13z + 6$
33. $7a^2 + 9a + 2$
34. $7b^2 + 15b + 2$
35. $10x^2 + 3x - 4$
36. $10x^2 + 13x + 4$
37. $9x^2 - 36x - 13$
38. $9y^2 - 42y + 13$
39. $5a^2 + 17a - 12$
40. $5b^2 + 16b + 12$
41. $4x^2 + 6x + 2$
42. $6y^2 + 8y + 2$
43. $2b^2 + 23b - 12$
44. $8x^2 + 16x + 6$
45. $6b^2 + 9b + 3$
46. $8c^2 + 20c + 12$
47. $4a^2 - 6a + 2$
48. $5b^2 - 5b - 30$
49. $4x^3 + 10x^2 + 6x$
50. $10y^4 + 21y^3 + 2y^2$
51. $4x^2 + 4xy + y^2$
52. $4s^2 - 4st + t^2$
53. $2y^2 + 3yz + z^2$
54. $6x^2 + 5ax + a^2$
55. $2x^2 + 5xy + 2y^2$
56. $u^2 - 5uv + 4v^2$
57. $9a^2 - 24ab - 20b^2$
58. $16x^2 + 2xy - 3y^2$
59. $6x^2 + 14xy + 4y^2$
60. $4u^3 - 4u^2v - 3uv^2$

What Have You Learned in Chapter 4?

You can find products such as $4x^2 \cdot y^3 \cdot 2xy^2$ and quotients such as $\frac{4x^2y^3}{2xy^2}$.

Also, you can find products such as $3ab(a + 5b)$ and quotients such as $\frac{12a^2b - 6ab}{6ab}$

You know how to express an integer, such as 24, as the product of primes.

You know how to isolate the common factor of a polynomial.

You are able to multiply binomials.

And you can factor certain binomials and trinomials.

Let's Review Chapter 4.

4.1 Monomial Products

In exercises 1 and 2, multiply:

1. $(3x^2y)(2xy^3)$
2. $(4m^2n)(-2mn^2)(-mn)$

In exercises 3 and 4, divide:

3. $\frac{10x^3y^2}{5xy}$
4. $\frac{-32a^4b^3}{-4ab^3}$

4.2 Distributive Laws for Polynomials

In exercises 5 and 6, multiply:

5. $x^2(x - 5y)$
6. $4a^2bc(2a^2b + 3abc)$

In exercises 7 and 8, divide:

7. $\frac{6a + 6b}{3}$
8. $\frac{20x^2y - 15xy^2}{5xy}$

4.3 Prime Factors

9. Find all primes less than 20.
10. Find the prime factorization of:
    (a) 36 (b) 132
11. Find gcd (24, 60).
12. Find the gcd of the coefficients of: $8x^2 + 12x - 16$.

4.4 Common Factors

In exercises 13–16, isolate the common factor of each polynomial.

13. $4x + 6$
14. $x^3 + x^2$
15. $9a^6 - 15a^4$
16. $x^5y^4z^3 + x^3y^3z^3 + x^2y^3z$

4.5 Binomial Products

In exercises 17–20, multiply.

17. $(x + 1)(x + 5)$
18. $(a + 2)(a - 7)$
19. $(z^2 + 1)(z^2 - 4)$
20. $(m + 2n)(m - 3n)$

4.6 Difference of Squares

In exercises 21–24, factor each binomial.

21. $a^2 - 25$
22. $4y^2 - 49z^2$
23. $a^3 - a$
24. $3xy^2 - 12x$

4.7 Factoring Trinomials $x^2 + Mx + N$

In exercises 25–28, factor each trinomial.

25. $c^2 + 7c + 12$
26. $m^2 - m - 20$
27. $t^3 + 6t^2 + 9t$
28. $x^2 - 5x + 6$

4.8 Factoring Trinomials $Lx^2 + Mx + N$

In exercises 29–32, factor each trinomial.

29. $4a^2 + 5a + 1$
30. $2x^2 + 7x + 5$
31. $2y^2 - 7y + 3$
32. $4x^2 + xy - 3y^2$

Next, the different types of factoring are mixed up. In exercises 33–44, factor each polynomial.

33. $5m - 10n$
34. $36 - a^2$
35. $x^2 + 9x + 14$
36. $7a^2 - 28b^2$
37. $x^3 - 4x$
38. $3x^3 - 5x^2 - 2x$
39. $m^2 + 8mn + 15n^2$
40. $4m^2 - 16n^4$
41. $s^3 + 6s^2 - 7s$
42. $t^4 - 16t^2$
43. $a^2 - 10a + 16$
44. $5x^2 - 8x - 4$

And these from Chapters 1–3:

45. Rearrange the following primes so that you can write "<" between any two of them: 7, 17, 41, 29, 23, 2.
46. Evaluate $(2t + 7)(t - 3)$ when $t = 4$.
47. Evaluate $\dfrac{6m^2n + 9mn^2}{mn}$ when $m = 2$ and $n = 3$.
48. Evaluate $(4xy + 3)(xy - 2)$ when $x = -1$ and $y = 4$.

Try These Exam Questions for Practice.

In questions 1–2, multiply:

1. $(4x^2y^5)(-3xy^3)$
2. $(4x - 3)(2x + 5)$
3. Divide: $\dfrac{15x^2y^2 - 25xy^4}{5xy^2}$
4. Express 108 as a product of primes.

In questions 5–8, factor each polynomial.

5. $6x^2 - 8x$
6. $9a^2 - 16b^2$
7. $x^2 + 9x + 18$
8. $2y^2 - y - 1$

5

Division of Polynomials

5.1 FRACTIONS

Equality of Fractions

Before considering division of polynomials, let us see what happens when you divide real numbers.

Definition

> *A **fraction** is an expression of the form $\frac{a}{b}$, where a and b are real numbers and $b \neq 0$. Here, a is called the **numerator** and b the **denominator** of the fraction $\frac{a}{b}$.*

EXAMPLE 1

(a) $\frac{3}{5}$ is a fraction with numerator 3 and denominator 5.

(b) $\frac{\pi}{2}$ is a fraction with numerator π and denominator 2.

Observe that $\frac{3}{5}$ is a real number of the form $\frac{N}{D}$, where N and D are integers and $D \neq 0$. Thus the fraction $\frac{3}{5}$ represents a rational number. But $\frac{\pi}{2}$ cannot be written in the above form. The fraction $\frac{\pi}{2}$ represents an irrational number.

Consider the fractions

$$\frac{1}{2} \text{ and } \frac{2}{4}.$$

Observe that the "cross products"

$$\frac{1}{2} \boxtimes \frac{2}{4}$$

are equal, that is, $1 \cdot 4 = 2 \cdot 2.$

These *fractions* are *equal* according to the following definition.

Definition

EQUALITY OF FRACTIONS. Two fractions

$$\frac{a}{b} \text{ and } \frac{c}{d}$$

are said to be **equal** *if their cross-products*

$$\frac{a}{b} \boxtimes \frac{c}{d}$$

are equal, that is, if

$$ad = bc.$$

Write

$$\frac{a}{b} = \frac{c}{d}$$

when these fractions are equal. Thus

$$\frac{a}{b} = \frac{c}{d} \quad \text{if } ad = bc.$$

There is a geometric interpretation of equality of fractions. Consider the equal fractions $\frac{1}{2}$ and $\frac{2}{4}$.

First, $\frac{1}{2}$ is obtained by dividing the line segment between 0 and 1 into 2 equal parts. The point of division represents $\frac{1}{2}$.

To represent $\frac{2}{4}$, divide the line segment between 0 and 2 into 4 equal parts. The first point of division represents $\frac{2}{4}$, as you see in Figure 5.1. But this is the same point that corresponds to $\frac{1}{2}$. Thus the *equal fractions* $\frac{1}{2}$ *and* $\frac{2}{4}$ *correspond to the same point on the number line.*

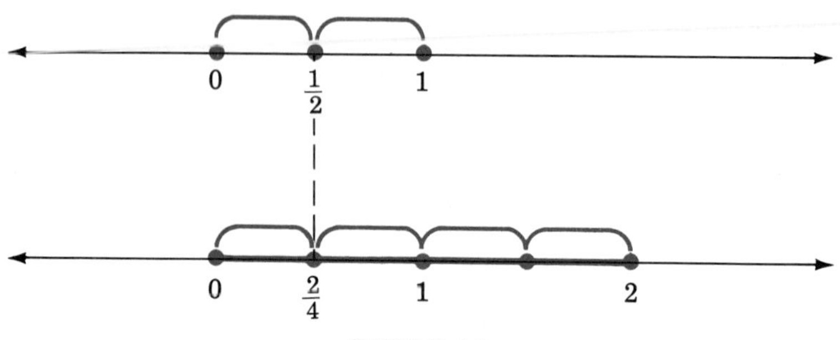

FIGURE 5.1

EXAMPLE 2

Show that $\frac{9}{12} = \frac{3}{4}$.

Solution.

$$\frac{9}{12} \bowtie \frac{3}{4}$$

Does $\quad 9 \cdot 4 = 12 \cdot 3?$

Yes. $\quad 36 = 36$

Thus $\quad \frac{9}{12} = \frac{3}{4}$

Lowest Terms

In Example 2, you found that

$$\frac{9}{12} = \frac{3}{4}.$$

Definition

> *Let N and D be integers and let $D > 0$. The rational number $\frac{N}{D}$ is expressed in **lowest terms** when N and D have no factors in common other than 1 and -1. You **simplify a fraction** of the form $\frac{N}{D}$ by expressing it in lowest terms.*

Thus $\frac{3}{4}$ is in lowest terms because 3 and 4 have no common factors and because the denominator, 4, is positive.

In practice, how do you simplify a fraction? Observe that

$$\frac{ak}{bk} = \frac{a}{b}$$

because the cross-products are equal:

$$akb = bka$$

Thus *if k is a factor common to numerator and denominator, divide each of these by k.*

To simplify a fraction, divide by all factors common to numerator and denominator. Also, if the denominator is negative, multiply numerator and denominator by -1.

EXAMPLE 3

Simplify:

(a) $\frac{15}{25}$ (b) $\frac{15}{-25}$ (c) $\frac{-15}{-25}$

Solution.

(a) gcd (15, 25) = 5. Divide numerator and denominator by 5:

$$\frac{\overset{3}{\cancel{15}}}{\underset{5}{\cancel{25}}} = \frac{3}{5}$$

(b) $\frac{15}{-25} = \frac{3}{-5}$

In order to obtain a *positive denominator,* multiply numerator and denominator by -1:

$$\frac{3}{-5} = \frac{3(-1)}{5(-1)} = \frac{-3}{5}$$

(c) $\frac{-15}{-25} = \frac{-3}{-5} = \frac{3}{5}$

EXAMPLE 4

Simplify:

(a) $\frac{12}{20}$ (b) $\frac{300}{400}$ (c) $\frac{-25}{-40}$

Solution.

(a) Divide numerator and denominator by 4:

$$\frac{12}{20} = \frac{3}{5}$$

(b) Divide numerator and denominator by 100:

$$\frac{300}{400} = \frac{3}{4}$$

(c) $\frac{-25}{-40} = \frac{-5}{-8} = \frac{5}{8}$

Note that if you divide numerator and denominator by -5, you simplify the fraction in one step.

There are three ways of expressing a *negative fraction*. Suppose $b \neq 0$. Then

$$\frac{-a}{b} = \frac{a}{-b} = -\frac{a}{b}$$

The minus sign can be in the numerator, in the denominator, or before the entire fraction.

EXAMPLE 5

$$\frac{-3}{5} = \frac{3}{-5} = -\frac{3}{5}$$

EXERCISES

In exercises 1–10, cross-multiply to show that the given fractions are equal.

SAMPLE.	***Solution.*** Cross-multiply:
$\frac{4}{10}$ and $\frac{2}{5}$	Does $4 \cdot 5 = 10 \cdot 2$?
	Yes. $20 = 20$
	Thus $\frac{4}{10} = \frac{2}{5}$

1. $\frac{4}{16}$ and $\frac{1}{4}$
2. $\frac{6}{12}$ and $\frac{1}{2}$
3. $\frac{-4}{6}$ and $\frac{-2}{3}$
4. $\frac{-5}{8}$ and $\frac{10}{-16}$
5. $\frac{14}{2}$ and $\frac{7}{1}$
6. $\frac{-36}{48}$ and $\frac{-3}{4}$
7. $\frac{-8}{14}$ and $\frac{4}{-7}$
8. $\frac{20}{25}$ and $\frac{8}{10}$
9. $\frac{13}{39}$ and $\frac{1}{3}$
10. $\frac{100}{16}$ and $\frac{25}{4}$

In exercises 11–38, simplify each fraction by dividing by factors common to both numerator and denominator. Express the simplified fraction with positive denominator.

SAMPLE.	***Solution.*** Divide numerator and denominator by -2 to obtain
$\frac{4}{-6}$	$\frac{4}{-6} = \frac{-2}{3}$

11. $\frac{4}{8}$ 12. $\frac{6}{10}$ 13. $\frac{-3}{9}$ 14. $\frac{6}{2}$ 15. $\frac{-3}{-12}$

16. $\frac{10}{4}$ 17. $\frac{4}{-3}$ 18. $\frac{7}{-14}$ 19. $\frac{-16}{20}$ 20. $\frac{32}{40}$

21. $\frac{12}{-18}$ 22. $\frac{14}{35}$ 23. $\frac{36}{72}$ 24. $\frac{40}{50}$ 25. $\frac{50}{300}$

26. $\frac{24}{48}$ 27. $\frac{15}{20}$ 28. $\frac{9}{-3}$ 29. $\frac{18}{12}$ 30. $\frac{-30}{18}$

31. $\frac{-9}{-24}$ 32. $\frac{15}{-50}$ 33. $\frac{13}{26}$ 34. $\frac{11}{121}$ 35. $\frac{24}{-72}$

36. $\frac{28}{35}$ 37. $\frac{36}{42}$ 38. $\frac{25}{60}$

39. Which of the following are equal to $-\frac{2}{5}$?

(a) $\frac{-2}{5}$ (b) $\frac{-2}{-5}$ (c) $\frac{2}{-5}$

40. Which of the following are equal to $-\frac{5}{-8}$?

(a) $\frac{-5}{8}$ (b) $\frac{5}{8}$ (c) $\frac{-5}{-8}$

5.2 EVALUATING RATIONAL EXPRESSIONS

What is a Rational Expression?

Recall that a *rational number* is a real number that can be written in the form $\frac{N}{D}$, where N and D are integers and $D \neq 0$. In this expression, N is the *numerator* and D the *denominator*. Thus $\frac{3}{4}, \frac{5}{2}$, and $\frac{-7}{9}$ are each rational numbers.

Definition

> ***A rational expression** is an algebraic expression that can be written in the form $\frac{P}{Q}$, where P and Q are polynomials and $Q \neq 0$. Here P is called the **numerator** and Q the **denominator.***

Rational numbers express division of integers. Rational expressions express division of polynomials.

EXAMPLE 1

Explain why each of the following is a rational expression:

(a) $\frac{2x}{x+2}$ (b) $\frac{y^2}{y+10}$ (c) $\frac{t^2+2t-1}{t^3-t^2+t+4}$

Solution.

(a) The numerator is the polynomial $2x$. The denominator is the polynomial $x + 2$.

(b) The numerator is the polynomial y^2. The denominator is the polynomial $y + 10$.

(c) The numerator is the polynomial $t^2 + 2t - 1$. The denominator is the polynomial $t^3 - t^2 + t + 4$.

Recall that numbers are terms, and therefore (numerical) polynomials. Thus a rational expression can have a number as its numerator or denominator.

EXAMPLE 2

Explain why each of the following is a rational expression:

(a) $\frac{5}{x^2 + 1}$ (b) $\frac{x + 4}{2}$ (c) $\frac{-7}{3}$

Solution.

(a) The numerator is the (numerical) polynomial 5. The denominator is the polynomial $x^2 + 1$.

(b) The numerator is the polynomial $x + 4$. The denominator is the polynomial 2.

(c) The numerator is the polynomial -7. The denominator is the polynomial 3.

An integer N can be written as the rational number $\frac{N}{1}$. So too, a polynomial P can be written as the rational expression $\frac{P}{1}$.

EXAMPLE 3

(a) $t^2 + 2t + 5$ is a rational expression with numerator $t^2 + 2t + 5$ and denominator 1.

(b) $\frac{1}{t^2 + 2t + 5}$ is a rational expression with numerator 1 and denominator $t^2 + 2t + 5$.

Substituting Numbers for Variables

You can evaluate a rational expression, just as you evaluated a polynomial, by substituting numbers for variables. Again, *substitute the same number for each occurrence of a variable.*

EXAMPLE 4

Evaluate $\frac{3x}{x + 5}$ when $x = 10$.

Solution. Replace x by 10 in both numerator and denominator, and obtain

$$\frac{3 \cdot 10}{10 + 5}, \quad \text{or} \quad \frac{30}{15},$$

each of which equals 2.

There may be two or more variables present in a rational expression. *For each variable, substitute the same number each time the variable occurs.*

EXAMPLE 5

Evaluate

$$\frac{xy - 4}{x^2 + y - 1}$$

when $x = 10$ and $y = 1$.

Solution. Substitute 10 for *each occurrence of* x, and **1** for *each occurrence of* $\boldsymbol{y}$:

$$\frac{10 \cdot \mathbf{1} - 4}{10^2 + \mathbf{1} - 1} = \frac{6}{100} = \frac{3}{50}$$

EXAMPLE 6

Evaluate

$$\frac{t^2 + 6t - 1}{t + 7}$$

three separate times:
(a) when $t = 2$, (b) when $t = -1$, (c) when $t = 0$.

Solution.

(a) for $t = 2$: $\dfrac{2^2 + 6 \cdot 2 - 1}{2 + 7} = \dfrac{15}{9} = \dfrac{5}{3}$

(b) for $t = -1$: $\dfrac{(-1)^2 + 6(-1) - 1}{-1 + 7} = \dfrac{-6}{6} = -1$

(c) for $t = 0$: $\dfrac{0^2 + 6 \cdot 0 - 1}{0 + 7} = \dfrac{-1}{7}$

EXERCISES

In exercises 1–10, find (a) the numerator and (b) the denominator of each rational expression.

1. $\dfrac{x}{x + 1}$
2. $\dfrac{2x + 7}{x}$
3. $\dfrac{y + 3}{y - 4}$
4. $\dfrac{x^2 - 3}{y^2 + 1}$
5. $\dfrac{2x^2 - 1}{7}$
6. $\dfrac{5}{y^2 + 2y}$
7. $\dfrac{8}{9}$
8. $\dfrac{t^4 - 3t^2 + 1}{t^3 + t + 2}$
9. $\dfrac{x^2 + 2xy + 3y}{2x + 5y}$
10. $\dfrac{uvw}{u + v + w}$

In exercises 11–26, evaluate each rational expression for the specified number.

11. $\frac{2x}{x+1}$ when $x = 10$

12. $\frac{5}{y-3}$ when $y = 5$

13. $\frac{x+3}{2}$ when $x = 2$

14. $\frac{x+7}{x+6}$ when $x = 1$

15. $\frac{2x-1}{x+1}$ when $x = 0$

16. $\frac{u^2}{u+3}$ when $u = 2$

17. $\frac{z^2+1}{z^2-1}$ when $z = 4$

18. $\frac{2m-3}{m-4}$ when $m = -1$

19. $\frac{7r+6}{r+8}$ when $r = 2$

20. $\frac{y-3}{y+3}$ when $y = -1$

21. $\frac{x^2+x}{6}$ when $x = 3$

22. $\frac{1}{x^3-2}$ when $x = 2$

23. $\frac{t^4-t^2+1}{t+1}$ when $t = 1$

24. $\frac{x^7-1}{x^4+1}$ when $x = 1$

25. $\frac{3y^2+2y+1}{y+4}$ when $y = -1$

26. $\frac{a^9-3a^4}{a+2}$ when $a = 0$

In exercises 27–40, evaluate each rational expression for the specified numbers.

27. $\frac{x+y}{2}$ when $x = 2$ and $y = 4$

28. $\frac{2s-1}{t+3}$ when $s = 1$ and $t = 2$

29. $\frac{x+4}{xy}$ when $x = 2$ and $y = 1$

30. $\frac{x+y}{x-y}$ when $x = 6$ and $y = 5$

31. $\frac{2x+3y}{y+5}$ when $x = 0$ and $y = -1$

32. $\frac{m^2-1}{n^2+1}$ when $m = 2$ and $n = 3$

33. $\frac{x^2+x}{y^2}$ when $x = 5$ and $y = -1$

34. $\frac{x^2+x+1}{y^2-3y}$ when $x = 1$ and $y = 5$

35. $\frac{ab^2}{20}$ when $a = 3$ and $b = 5$

36. $\frac{x+y+1}{2x-3}$ when $x = y = 2$

37. $\frac{xyz}{x+y+z}$ when $x = 2$, $y = z = 1$

38. $\frac{2a-b}{c+5}$ when $a = 2$, $b = c = 1$

39. $\frac{a+b+c+d}{b+7}$ when $a = 1$, $b = 2$, $c = 3$, $d = 4$

40. $\frac{x^2+y^2+z^2}{w^2}$ when $x = 1$, $y = 2$, $z = 3$, $w = 5$

In exercises 41–48, evaluate each rational expression three separate times.

41. $\frac{2x+7}{3}$ when (a) $x = 1$, (b) $x = 4$, (c) $x = 7$

42. $\frac{1}{3t-4}$ when (a) $t = 2$, (b) $t = 4$, (c) $t = 5$

43. $\frac{y+5}{y+2}$ when (a) $y=2$, (b) $y=-1$, (c) $y=0$

44. $\frac{y^2+1}{3}$ when (a) $y=1$, (b) $y=-1$, (c) $y=5$

45. $\frac{z+6}{z^2}$ when (a) $z=-1$, (b) $z=1$, (c) $z=6$

46. $\frac{x^9-1}{x^6+2}$ when (a) $x=0$, (b) $x=1$, (c) $x=-1$

47. $\frac{t^2+2t-3}{5t}$ when (a) $t=1$, (b) $t=-1$, (c) $t=10$

48. $\frac{x^3-4}{x^2+x+7}$ when (a) $x=1$, (b) $x=3$, (c) $x=10$

5.3 DIVISION OF MONOMIALS

Multiplication and Division

Division of real numbers was defined in terms of multiplication. Thus for numbers a, $b(\neq 0)$, and c,

$$\frac{a}{b}=c,$$

if

$$a=bc$$

Division of polynomials can be defined similarly.

Definition

> *DIVISION OF POLYNOMIALS. Let P, Q, S be polynomials, where $S \neq 0$. Then*
>
> $$\frac{P}{S}=Q,$$
>
> if
>
> $$P=S \cdot Q$$

You have already considered examples such as the following.

EXAMPLE 1

$$\frac{8x^2}{4x}=2x$$

because

$$\underbrace{8x^2}_{P}=\underbrace{4x}_{Q}\cdot\underbrace{2x}_{S}$$

Note that $8=2^3$ and $4=2^2$.

Thus $\frac{8x^2}{4x}=\frac{2^3x^2}{2^2x^1}=2^{3-2}x^{2-1}=2x$

Dividing Powers

Assume a is any nonzero number or variable. Let m and n be positive integers. *To divide a^m by a^n, divide by each factor a, common to numerator and denominator.*

$$\frac{a^m}{a^n} = \frac{\overbrace{a \cdot a \cdot a \ . \ . \ . \ a}^{m \text{ factors}}}{\underbrace{a \cdot a \cdot a \ . \ . \ . \ a}_{n \text{ factors}}}$$

Three cases occur, depending on the relative sizes of m and n.

1. $m > n$**:** Every factor a divides out from the denominator. But $m - n$ factors a remain in the numerator.

$$\frac{a^m}{a^n} = \frac{\overbrace{a \cdot a \cdot a \ . \ . \ . \ a}^{m \text{ factors}}}{\underbrace{a \cdot a \cdot a \ . \ . \ . \ a}_{n \text{ factors}}}$$

$$= \frac{\overbrace{\overbrace{\overset{1}{\cancel{a}} \cdot \overset{1}{\cancel{a}} \cdot \overset{1}{\cancel{a}} \ . \ . \ . \ \overset{1}{\cancel{a}}}^{n \text{ factors}} \cdot \overbrace{a \cdot a \cdot a \ . \ . \ . \ a}^{m - n \text{ factors}}}^{m \text{ factors}}}{\underbrace{\underset{1}{\cancel{a}} \cdot \underset{1}{\cancel{a}} \cdot \underset{1}{\cancel{a}} \ . \ . \ . \ \underset{1}{\cancel{a}}}_{n \text{ factors}}}$$

Note that
$n + (m - n) = n + m - n = m.$

$$= \frac{\overbrace{a \cdot a \cdot a \ . \ . \ . \ a}^{m - n \text{ factors}}}{1}$$

$$= a^{m-n}$$

Thus if $m > n$, then

$$\frac{a^m}{a^n} = a^{m-n}$$

2. $m = n$**:** Every factor a divides out from both numerator and denominator. The resulting fraction equals 1.

$$\frac{a^m}{a^n} = \frac{a^m}{a^m} = \frac{\overbrace{\overset{1}{\cancel{a}} \cdot \overset{1}{\cancel{a}} \cdot \overset{1}{\cancel{a}} \ . \ . \ . \ \overset{1}{\cancel{a}}}^{m \text{ factors}}}{\underbrace{\underset{1}{\cancel{a}} \cdot \underset{1}{\cancel{a}} \cdot \underset{1}{\cancel{a}} \ . \ . \ . \ \underset{1}{\cancel{a}}}_{m \text{ factors}}}$$

$$= 1$$

Thus $$\frac{a^m}{a^m} = 1$$

3. $m < n$**:** Every factor a divides out from the numerator. But $n - m$ factors a remain in the denominator.

$$\frac{a^m}{a^n} = \frac{\overbrace{a \cdot a \cdot a \ldots a}^{m \text{ factors}}}{\underbrace{a \cdot a \cdot a \ldots a \cdot a \cdot a \cdot a \ldots a}_{n \text{ factors}}}$$

$$= \frac{\overbrace{\overset{1}{\not{a}} \cdot \overset{1}{\not{a}} \cdot \overset{1}{\not{a}} \ldots \overset{1}{\not{a}}}^{m \text{ factors}}}{\underbrace{\underbrace{\underset{1}{\not{a}} \cdot \underset{1}{\not{a}} \cdot \underset{1}{\not{a}} \ldots \underset{1}{\not{a}}}_{m \text{ factors}} \cdot \underbrace{a \cdot a \cdot a \ldots a}_{n\text{-}m \text{ factors}}}_{n \text{ factors}}}$$

Note that $n = m + (n - m)$.

$$= \frac{1}{a^{n-m}}$$

Thus if $m < n$, then

$$\frac{a^m}{a^n} = \frac{1}{a^{n-m}}$$

EXAMPLE 2

(a) $\frac{2^6}{2^3} = 2^{6-3} = 2^3$ or $\frac{64}{8} = 8$

(b) $\frac{(-5)^3}{(-5)^3} = 1$ or $\frac{-125}{-125} = 1$

(c) $\frac{3^2}{3^4} = \frac{1}{3^{4-2}} = \frac{1}{3^2}$ or $\frac{9}{81} = \frac{1}{9}$

EXAMPLE 3

(a) $\frac{y^8}{y^5} = y^{8-5} = y^3$ or

$$\frac{\overset{1}{\not{y}} \cdot \overset{1}{\not{y}} \cdot \overset{1}{\not{y}} \cdot \overset{1}{\not{y}} \cdot \overset{1}{\not{y}} \cdot y \cdot y \cdot y}{\underset{1}{\not{y}} \cdot \underset{1}{\not{y}} \cdot \underset{1}{\not{y}} \cdot \underset{1}{\not{y}} \cdot \underset{1}{\not{y}}} = y \cdot y \cdot y = y^3$$

(b) $\frac{y^4}{y^{10}} = \frac{1}{y^{10-4}} = \frac{1}{y^6}$ or

$$\frac{\overset{1}{\not{y}} \cdot \overset{1}{\not{y}} \cdot \overset{1}{\not{y}} \cdot \overset{1}{\not{y}}}{\underset{1}{\not{y}} \cdot \underset{1}{\not{y}} \cdot \underset{1}{\not{y}} \cdot \underset{1}{\not{y}} \cdot y \cdot y \cdot y \cdot y \cdot y \cdot y} = \frac{1}{y \cdot y \cdot y \cdot y \cdot y \cdot y} = \frac{1}{y^6}$$

EXAMPLE 4
Divide:

$$\frac{15m^8n^{12}p^2}{-5m^{10}n^{12}}$$

Solution.

$$\frac{15m^8n^{12}p^2}{-5m^{10}n^{12}} = \frac{-3p^2}{m^{10-8}}$$
$$= \frac{-3p^2}{m^2}$$

Prime Factors

When numbers or numerical coefficients are somewhat complicated, it is best to first write out their prime factorization before dividing.

EXAMPLE 5
Simplify:

$$\frac{288}{320}$$

Solution.

$$288 = 32 \cdot 9$$
$$= 2^5 \cdot 3^2$$
$$320 = 64 \cdot 5$$
$$= 2^6 \cdot 5$$
$$\frac{288}{320} = \frac{2^5 \cdot 3^2}{2^6 \cdot 5}$$
$$= \frac{3^2}{2 \cdot 5}$$
$$= \frac{9}{10}$$

EXAMPLE 6
Simplify:

$$\frac{360x^4yz^2}{-108xy^3z^2}$$

Solution.

$$360 = 2^3 \cdot 3^2 \cdot 5$$
$$108 = 2^2 \cdot 3^3$$

$$\frac{360x^4yz^2}{-108xy^3z^2} = \frac{2^3 \cdot 3^2 \cdot 5x^4yz^2}{-2^2 \cdot 3^3xy^3z^2}$$
$$= \frac{-2 \cdot 5x^3}{3y^2}$$
$$= \frac{-10x^3}{3y^2}$$

Powers of a Polynomial

Powers of a polynomial can appear in both numerator and denominator of a rational expression.

EXAMPLE 7

Simplify: $\frac{(x + a)^5}{(x + a)^3}$

Solution.

$$\frac{(x + a)^5}{(x + a)^3} = (x + a)^{5-3}$$
$$= (x + a)^2$$

The answer can be left in this form.

EXAMPLE 8

Simplify: $\frac{6x(m - n)}{3x(m - n)^4}$

Solution.

$$\frac{6x(m - n)}{3x(m - n)^4} = \frac{2}{(m - n)^{4-1}}$$
$$= \frac{2}{(m - n)^3}$$

EXERCISES

Simplify each expression. (Multiply out the numbers that remain in numerator or denominator.)

SAMPLE. Simplify: $\frac{3^5x^2}{3^2x^4}$	***Solution.*** $\frac{3^5x^2}{3^2x^4} = \frac{3^3}{x^2} = \frac{27}{x^2}$

1. $\frac{5^4}{5^3}$ 2. $\frac{3}{3^2}$ 3. $\frac{7^4 \cdot 11^2}{7^2 \cdot 11^3}$ 4. $\frac{-2^5 \cdot 3^4}{2^3 \cdot 3^3}$ 5. $\frac{a^4}{a^4}$

6. $\frac{b^3}{-b}$ 7. $\frac{-x^{10}}{-x^7}$ 8. $\frac{y^2}{y^5}$ 9. $\frac{z^{14}}{z^8}$ 10. $\frac{m^6}{m^{18}}$

11. $\frac{ab}{a^2b}$ 12. $\frac{x^4y}{xy^4}$ 13. $\frac{m^8n}{mn^2}$ 14. $\frac{-m^4n^3}{-m^4n^3}$ 15. $\frac{a^2bc}{a}$

16. $\frac{r^2s^2t^3}{rs^2t^4}$ 17. $\frac{x^{10}y^{17}z}{x^8y^{16}z^{16}}$ 18. $\frac{-pqrs}{pr^3s^2}$ 19. $\frac{2ab^2}{4ab}$ 20. $\frac{-3xyz}{5xy}$

21. $\frac{-7abc}{14c^2}$ 22. $\frac{20x^3y^3z}{-10x^2z^2}$ 23. $\frac{9a^2bc}{15a^4bc^2}$ 24. $\frac{36x^{10}yz^8}{12x^7y^2z^6}$ 25. $\frac{40m^2n^7}{20m^4n^8}$

26. $\frac{32a^5cd}{-8abc}$ 27. $\frac{-18m^{10}n^7p^8}{-12mn^8p^7}$ 28. $\frac{-54ax^7}{-27a^3x^7}$ 29. $\frac{72}{108}$ 30. $\frac{375}{90}$

31. $\frac{132}{288}$ 32. $\frac{-6250}{200}$ 33. $\frac{64}{-84}$ 34. $\frac{-144}{450}$ 35. $\frac{240}{128}$

36. $\frac{-52}{-156}$ 37. $\frac{121a^4}{44a^3}$ 38. $\frac{224x^6}{98x^3}$ 39. $\frac{256y^3}{96y^9}$ 40. $\frac{560a^8b}{160a^4}$

41. $\frac{250a^4b^2c}{375ab^2c}$ 42. $\frac{150a^4x^2yz^3}{36axy^4z}$ 43. $\frac{(x+1)^2}{x+1}$ 44. $\frac{(x+y)^3}{(x+y)^2}$

45. $\frac{(a+b)^4}{(a+b)^2}$ 46. $\frac{(a-b)^5}{a-b}$ 47. $\frac{(x+y)^2}{(x+y)^2}$ 48. $\frac{m+n}{(m+n)^2}$

49. $\frac{a-b}{(a-b)^4}$ 50. $\frac{2^4(c-d)^5}{2(c-d)^8}$ 51. $\frac{4(x+y)}{2(x+y)}$ 52. $\frac{12(x-y)^2}{4(x-y)}$

53. $\frac{-36(a+b)^7}{6(a+b)^4}$ 54. $\frac{18(m-n)^5}{24(m-n)^3}$ 55. $\frac{(a+b)(x+y)}{a+b}$ 56. $\frac{(x+y)(r+s)^2}{(x+y)(r+s)}$

57. $\frac{(x+y)^4(x-y)}{(x+y)^2(x-y)^3}$ 58. $\frac{4(a+b)^7(a-b)^3}{8(a+b)(a-b)^3}$ 59. $\frac{36(a+b)^{10}(x+y)^7}{144(a+b)^9(x+y)(x-y)}$

60. $\frac{-84(x-y)(a+b)^7}{98(x-y)^6(a+b)}$ 61. $\frac{5^2x^2(a+2)}{5^4x^3(a+2)^2}$ 62. $\frac{108x^2y(x-y)^4}{144xy^2(x-y)}$

5.4 MONOMIAL DIVISORS

Consider a rational expression, such as

$$\frac{5a+5b}{10},$$

in which the denominator is a monomial but the numerator contains more than one term. You can often simplify such an expression by first isolating the common factor in the numerator.

EXAMPLE 1

Simplify: $\frac{5a+5b}{10}$

Solution. Isolate the common factor in the numerator:

$$5a+5b=5(a+b)$$

Therefore

$$\frac{5a+5b}{10}=\frac{5(a+b)}{10}$$ Divide numerator and denominator by 5.

$$=\frac{a+b}{2}$$

EXAMPLE 2

Simplify: $\frac{x^2y^2+2x^2}{x^2}$

Solution. Isolate the common factor in the numerator:

$$x^2y^2+2x^2=x^2(y^2+2)$$

Therefore

$$\frac{x^2y^2+2x^2}{x^2}=\frac{x^2(y^2+2)}{x^2}$$ Divide numerator and denominator by x^2.

$$=y^2+2$$

EXAMPLE 3

Simplify: $\dfrac{4a^2b + 8ab^2}{8a^2b^2}$

Solution. Isolate the common factor in the numerator:

$$4a^2b + 8ab^2 = 4ab(a + 2b)$$

Therefore

$$\frac{4a^2b + 8ab^2}{8a^2b^2} = \frac{\overset{1}{\cancel{4}\cancel{a}\cancel{b}}(a + 2b)}{\underset{2ab}{\cancel{8}\cancel{a^2}\cancel{b^2}}}$$

$$= \frac{a + 2b}{2ab}$$

EXAMPLE 4

Simplify: $\dfrac{6x^2y^3 + 24xyz}{-18xyz}$

Solution.

$$\frac{6x^2y^3 + 24xyz}{-18xyz} = \frac{\overset{1}{\cancel{6}\cancel{x}\cancel{y}}(xy^2 + 4z)}{\underset{3z}{-\cancel{18}\cancel{x}\cancel{y}z}}$$

$$= \frac{-(xy^2 + 4z)}{3z}$$

EXAMPLE 5

Simplify: $\dfrac{5a^2b + 10a^2b^2 + 25a^2b^3}{15ab^2}$

Solution.

$$\frac{5a^2b + 10a^2b^2 + 25a^2b^3}{15ab^2}$$

$$= \frac{\overset{a}{\cancel{5}\cancel{a^2}\cancel{b}}(1 + 2b + 5b^2)}{\underset{3b}{\cancel{15}\cancel{a}\cancel{b^2}}}$$

$$= \frac{a(1 + 2b + 5b^2)}{3b}$$

The same method applies when the numerator is a monomial, but the *denominator* contains more than one term. Now, whenever possible, isolate the common factor in the *denominator*.

EXAMPLE 6

Simplify: $\dfrac{9xy^2z}{6xy^2 - 9xyz}$

Solution.

$$\frac{9xy^2z}{6xy^2 - 9xyz} = \frac{\overset{3y}{\cancel{9}\cancel{x}\cancel{y^2}z}}{\underset{1}{\cancel{3}\cancel{x}\cancel{y}}(2y - 3z)}$$

$$= \frac{3yz}{2y - 3z}$$

EXERCISES

Simplify each rational expression.

1. $\frac{2x + 2y}{2}$
2. $\frac{3a - 3b}{3}$
3. $\frac{5m + 10n}{5}$
4. $\frac{30a + 60b}{30}$
5. $\frac{ax + ay}{a}$
6. $\frac{xy - xz}{x}$
7. $\frac{mx + my}{m}$
8. $\frac{2ax + 2ay}{2a}$
9. $\frac{x^2 + x^2y}{x^2}$
10. $\frac{a^2x - a^2y}{a^2}$
11. $\frac{2a + 2b}{4}$
12. $\frac{9x + 18y}{3}$
13. $\frac{10a - 5b}{20}$
14. $\frac{35x^2 + 28y^2}{14}$
15. $\frac{ax + ay}{a}$
16. $\frac{a^2x - a^2y}{a}$
17. $\frac{ab + ac^2}{a^2}$
18. $\frac{cx^2 + cy^2}{cx}$
19. $\frac{x^2y + x^2y^2}{x^2}$
20. $\frac{4x^2 + 8xy}{4x}$
21. $\frac{6x^2y + 9xy^2}{6xy}$
22. $\frac{25a^2 - 25b^2}{5a}$
23. $\frac{18a^2b + 36ab^2}{6ab}$
24. $\frac{10mnr + 20m^2n}{10mnr}$
25. $\frac{15ab^2 + 10ab^3}{10ab^2}$
26. $\frac{-4x^2 - 4y^2}{-2}$
27. $\frac{30abc - 40a^2c^2}{10a^2c}$
28. $\frac{4x^3y^2z - 8x^2y^2z}{8xy^2z}$
29. $\frac{-24r^2s - 30rst}{18rs}$
30. $\frac{45a^3b^4c - 30a^2bc}{15abc}$
31. $\frac{2x + 2y + 2z}{2}$
32. $\frac{3x + 3y - 3z}{6}$
33. $\frac{5a - 10b + 5c}{25}$
34. $\frac{9ax + 9ay - 9az}{3a}$
35. $\frac{6r^2s + 12r^2t - 6r^2u}{36r^2}$
36. $\frac{au - av + a^2w}{a^2}$
37. $\frac{5a^2x - 10ax^2 + 15a^2x^2}{20a^2x^2}$
38. $\frac{12ax - 9bx + 18cx}{6x^2}$
39. $\frac{20a^2b^3c^2 + 25a^4b^2c - 30a^2b^2c^2}{25a^2b^3c}$

40. $\dfrac{40x^2yz - 30x^2y^3z + 20xy^2z^2}{10x^2y^2z^2}$

41. $\dfrac{12a^2b^7c^6 - 16a^3b^2c^5 + 20a^4b^4c^4}{40a^2b^4c^4}$

42. $\dfrac{6x^2yz^2 - 12x^2y^2z^2 + 18x^2yz^2 - 24xyz^2}{12x^2y^2z^2}$

43. $\dfrac{2x}{4x + 6x^2}$

44. $\dfrac{5xy}{10x^2y + 15xy^2}$

45. $\dfrac{4xyz}{8xy^2 - 12xz^2}$

46. $\dfrac{25ab^2c^3}{50a^3bc^2 - 25ab^2c^6}$

47. $\dfrac{100a^2xy}{10a^2y - 25a^2x + 50a^2xy}$

48. $\dfrac{-3a^2b^3c^{10}}{6a^2b^3c + 9ab^4c^4 - 27a^3b^3c^5}$

5.5 FACTORING AND SIMPLIFYING

Suppose both numerator and denominator of a rational expression contain more than one term. By factoring the numerator or denominator (possibly both), you can often simplify the given expression.

EXAMPLE 1

Simplify: $\dfrac{5x - 5y}{5x + 5y}$

Solution. Isolate the common factor in both numerator and denominator:

$$\frac{5x - 5y}{5x + 5y} = \frac{5(x - y)}{5(x + y)} = \frac{x - y}{x + y}$$

Note that in Example 1, you **cannot** divide the individual terms. **The following is wrong:**

$$\frac{\overset{1}{\cancel{x}} - \overset{1}{\cancel{y}}}{\underset{1}{\cancel{x}} + \underset{1}{\cancel{y}}} = \frac{0}{2} = 0$$

For example, if $x = 2$, $\boldsymbol{y = 1}$, then

$$\frac{x - \boldsymbol{y}}{x + \boldsymbol{y}} = \frac{2 - \mathbf{1}}{2 + \mathbf{1}} = \frac{1}{3},$$

and *not* 0, as above.

EXAMPLE 2

Simplify: $\dfrac{2x + 2y}{x^2 - y^2}$

Solution. Isolate the common factor in the numerator.

$$2x + 2y = 2(x + y)$$

The denominator is the difference of squares.

$$x^2 - y^2 = (x + y)(x - y)$$

Therefore divide numerator and denominator by $x + y$:

$$\frac{2x + 2y}{x^2 - y^2} = \frac{2\overset{1}{\cancel{(x + y)}}}{\underset{1}{\cancel{(x + y)}}(x - y)} = \frac{2}{x - y}$$

EXAMPLE 3

Simplify: $\dfrac{a^2 + 3a + 2}{a^2 + 2a + 1}$

Solution.

$$a^2 + 3a + 2 = (a + 1)(a + 2)$$
$$a^2 + 2a + 1 = (a + 1)^2$$

Therefore

$$\frac{a^2 + 3a + 2}{a^2 + 2a + 1} = \frac{\overset{1}{\cancel{(a + 1)}}(a + 2)}{\underset{a+1}{\cancel{(a + 1)^2}}} = \frac{a + 2}{a + 1}$$

EXAMPLE 4

Simplify: $\dfrac{25ax^2 - 25ay^2}{5x^2 + 10xy + 5y^2}$

Solution.

$$25ax^2 - 25ay^2 = 25a(x^2 - y^2) = 25a(x + y)(x - y)$$

$$5x^2 + 10xy + 5y^2 = 5(x^2 + 2xy + y^2) = 5(x + y)^2$$

Therefore

$$\frac{25ax^2 - 25ay^2}{5x^2 + 10xy + 5y^2} = \frac{\overset{5}{\cancel{25}}a\overset{1}{\cancel{(x + y)}}(x - y)}{\cancel{5}\underset{1(x+y)}{\cancel{(x + y)^2}}} = \frac{5a(x - y)}{x + y}$$

EXERCISES

Simplify each rational expression.

1. $\frac{2a + 2b}{2x + 2y}$
2. $\frac{5a - 5b}{10a + 10b}$
3. $\frac{7x + 7y}{7x - 7y}$
4. $\frac{9x + 9y}{3x - 3y}$
5. $\frac{ax + ay}{ax - ay}$
6. $\frac{2ax + 2ay}{4ax - 4ay}$
7. $\frac{2x + 2y}{2x - 4y}$
8. $\frac{5m + 10n}{10m - 5n}$
9. $\frac{6x + 6y}{5x + 5y}$
10. $\frac{8a - 8b}{5a - 5b}$
11. $\frac{ax + ay}{bx + by}$
12. $\frac{2u + 2v}{4u + 4v}$
13. $\frac{x^2 - y^2}{3x + 3y}$
14. $\frac{5a + 5b}{a^2 - b^2}$
15. $\frac{6x - 6y}{2x^2 - 2y^2}$
16. $\frac{x^2 - 1}{2x + 2}$
17. $\frac{a^2 - 9}{4a + 12}$
18. $\frac{u^2 - 100}{2u + 20}$
19. $\frac{3m + 3n}{5m^2 - 5n^2}$
20. $\frac{5a^2 - 20}{3a + 6}$
21. $\frac{a^2 - 4b^2}{2a - 4b}$
22. $\frac{x^2 - 25y^2}{2x + 10y}$
23. $\frac{4x^2 - 9y^2}{12x + 18y}$
24. $\frac{100a^2 - 64b^2}{5a - 4b}$
25. $\frac{x^2 + 2x + 1}{x^2 + 4x + 3}$
26. $\frac{a^2 + 5a + 6}{a^2 + 4a + 4}$
27. $\frac{y^2 - 9}{y^2 + 4y + 3}$
28. $\frac{z^2 + 7z + 10}{z^2 - 25}$
29. $\frac{u^2 - 36}{u^2 + 7u + 6}$
30. $\frac{r^2 - 1}{2 - 2r^2}$
31. $\frac{m^2 + 7m + 12}{m^2 + 5m + 6}$
32. $\frac{a^2 - 8a + 12}{a^2 - 36}$
33. $\frac{x^2 + 2x - 8}{x^2 - 16}$
34. $\frac{x^2 - y^2}{x^2 + 2xy + y^2}$
35. $\frac{3a^2 - 3b^2}{6a^2 + 18ab + 12b^2}$
36. $\frac{9u^2 - 36}{3u^2 - 9u + 6}$
37. $\frac{5x^2 + 35x + 60}{10x^2 + 10x - 60}$
38. $\frac{a^2u^2 - a^2v^2}{au^2 + 6auv + 5av^2}$
39. $\frac{2x^2 + 11x + 15}{x^2 + 6x + 9}$
40. $\frac{3a^2 - 2ab - b^2}{4a^2 - 4b^2}$

5.6 POLYNOMIAL DIVISION

Degree

Consider a *polynomial in a single variable,* such as

$$x^4 + 5x^3 - x^2 + x + 1.$$

The **degree of a term** is simply the exponent of the variable of that term, if there is a variable; the **degree of a nonzero numerical term** is 0. (The degree of 0 is not defined.) For the above polynomial,

the degree of x^4 is 4,
the degree of $5x^3$ is 3,
the degree of $-x^2$ is 2,
the degree of x is 1,
the degree of 1, the numerical term, is 0.

The **degree of a polynomial** is the highest degree of any of its terms. Thus the degree of $x^4 + 5x^3 - x^2 + 1$ is 4.

A *polynomial in a single variable* is said to be in **standard form** if its terms are arranged in order of decreasing degree. Thus,

$$x^4 + 5x^3 - x^2 + 1$$

is in standard form, but

$$5x^3 + x^4 + 1 - x^2$$

is not.

EXAMPLE 1

(a) Express

$$t^4 - t^3 + 2t^6 + t^8 - 1 + t^2$$

in standard form.

(b) What is the degree of this polynomial?

Solution.

(a) Rearrange the terms in order of decreasing degree:

$$t^8 + 2t^6 + t^4 - t^3 + t^2 - 1$$

is now in standard form.

(b) The degree of this polynomial is 8, the highest degree of any of its terms.

The Division Process

There is a method of dividing polynomials that resembles long division.

Recall that to divide 651 by 21, you proceed as follows:

$$\begin{array}{r} 31 \\ 21\overline{)651} \\ \underline{63} \\ 21 \\ \underline{21} \end{array}$$

For polynomials P, Q, S, where $S \neq 0$,

$$\frac{P}{S} = Q$$

if

$$P = S \cdot Q.$$

You can also express division of polynomials by writing

$$\begin{array}{r} Q \\ S\overline{)P} \end{array}$$

Here S is called the **divisor,** P the **dividend,** and Q the **quotient.**

EXAMPLE 2

Divide:

$$x + 2\overline{)2x^2 + 7x + 6}$$

Solution.

$$\begin{array}{rl}
 & \textit{quotient} \\
 & (\text{A}) \rightarrow 2x \;\; + 3 \leftarrow (\text{C}) \\
\textit{divisor} \rightarrow x + 2 & \overline{)2x^2 + 7x + 6} \leftarrow \textit{dividend} \\
(\text{B}) & \left\{ \begin{array}{r} \underline{2x^2 + 4x} \qquad \\ 3x + 6 \end{array} \right. \\
(\text{D}) & \left\{ \begin{array}{r} \underline{3x + 6} \\ 0 \end{array} \right.
\end{array}$$

(A)

$$\frac{2x^2}{x} = 2x$$

Divide $2x^2$, the first term of the dividend, by x, the first term of the divisor, to obtain $2x$, the first term of the quotient.

(B)

$$\begin{array}{rl}
 & 2x \\
x + 2 & \overline{)\;2x^2 + 7x + 6} \\
 & \underline{\mp 2x^2 \mp 4x} \quad \vdots \\
 & \qquad\quad 3x + 6 \qquad \leftarrow \textit{1st difference polynomial}
\end{array}$$

Now multiply $2x$ by the divisor, $x + 2$, and *subtract* the product, $2x^2 + 4x$, from $2x^2 + 7x$, the first two terms of the divisor. The difference is $3x$. Bring down 6, the next term of the dividend to obtain the **1st difference polynomial,** $3x + 6$.

(C)

$$\frac{3x}{x} = 3$$

Next, divide $3x$, the first term of the 1st difference polynomial, by x, the first term of the divisor. The second term of the quotient is 3.

(D)

$$\begin{array}{lrl}
\textit{quotient} \rightarrow & & 2x \;\; + 3 \\
\textit{divisor} \rightarrow & x + 2 & \overline{)\;2x^2 + 7x + 6} \leftarrow \textit{dividend} \\
 & & \underline{\mp 2x^2 \mp 4x} \\
 & & \qquad\quad 3x + 6 \leftarrow \textit{1st difference polynomial} \\
 & & \qquad\quad \underline{\mp 3x \mp 6} \\
 & & \qquad\qquad\quad 0 \leftarrow \textit{2nd difference polynomial}
\end{array}$$

Multiply 3 by the divisor, $x + 2$, to obtain $3x + 6$. When you subtract this from the 1st difference polynomial, you obtain 0. There are no more terms to bring down. Thus 0 is the 2nd difference polynomial, and $x + 2$ divides $2x^2 + 7x + 6$ (evenly). The quotient is $2x + 3$.

CHECK:

You can *check* a division example *by multiplying the quotient by the divisor. The product should be the dividend, as you see below.*

$$\begin{array}{rr} \textit{quotient}\rightarrow & 2x+3 \\ \textit{divisor}\rightarrow & \times\; x+2 \\ \hline & 2x^2+3x \quad\quad \\ & 4x+6 \\ \hline \textit{dividend}\rightarrow & 2x^2+7x+6 \end{array}$$

EXAMPLE 3

Divide and check:

$$t+5\overline{)3t^2+13t-10}$$

Solution.

$$\begin{array}{rl} \textit{quotient}\rightarrow & \quad 3t - 2 \\ \textit{divisor}\rightarrow \; t+5 & \overline{)\;3t^2+13t-10} \leftarrow \textit{dividend} \\ & \underline{\mp 3t^2 \mp 15t} \\ & \quad\quad - 2t-10 \leftarrow \textit{1st difference polynomial} \\ & \quad\quad \underline{\pm 2t \pm 10} \\ & \quad\quad\quad\quad 0 \leftarrow \textit{2nd difference polynomial} \end{array}$$

The quotient is $3t-2$.

CHECK:

$$\begin{array}{rr} \textit{quotient}\rightarrow & 3t-2 \\ \textit{divisor}\rightarrow & \times\; t+5 \\ \hline & 3t^2-2t \quad\quad \\ & 15t-10 \\ \hline \textit{dividend}\rightarrow & 3t^2+13t-10 \end{array}$$

EXAMPLE 4

Divide and check:

$$3x+1\overline{)6x^2+17x+5}$$

Solution.

$$\begin{array}{rl} \textit{quotient}\rightarrow & \quad 2x + 5 \\ \textit{divisor}\rightarrow \; 3x+1 & \overline{)\;6x^2+17x+5} \leftarrow \textit{dividend} \\ & \underline{\mp 6x^2 \mp 2x} \\ & \quad\quad 15x+5 \leftarrow \textit{1st difference polynomial} \\ & \quad\quad \underline{\mp 15x \mp 5} \\ & \quad\quad\quad\quad 0 \leftarrow \textit{2nd difference polynomial} \end{array}$$

The quotient is $2x+5$.

CHECK:

$$\begin{array}{rr}
\textit{quotient}\rightarrow & 2x + 5 \\
\textit{divisor}\rightarrow & \underline{\times 3x + 1} \\
 & 6x^2 + 15x \\
 & \underline{2x + 5} \\
\textit{dividend}\rightarrow & 6x^2 + 17x + 5
\end{array}$$

When you divide polynomials in a single variable, always express them in standard form. Add 0's for missing terms, as in Example 5, which follows.

Continue the process until you have used up all of the terms of the dividend. There may be more than two difference polynomials.

EXAMPLE 5

Divide

$$6 - 7x + x^3$$

by $x - 2$.

Solution. Express the dividend in standard form. Also, add $0x^2$ for the missing term of degree 2:

$$x^3 + 0x^2 - 7x + 6$$

$$\begin{array}{rrl}
\textit{quotient}\rightarrow & & x^2 + 2x - 3 \\
\textit{divisor}\rightarrow & x - 2 & \overline{)\, x^3 + 0x^2 - 7x + 6} \\
 & & \underline{\mp x^3 \pm 2x^2} \\
 & & \quad 2x^2 - 7x \qquad \leftarrow \textit{1st difference polynomial} \\
 & & \quad \underline{\mp 2x^2 \pm 4x} \\
 & & \qquad -3x + 6 \qquad \leftarrow \textit{2nd difference polynomial} \\
 & & \qquad \underline{\pm 3x \mp 6} \\
 & & \qquad\qquad 0 \qquad \leftarrow \textit{3rd difference polynomial}
\end{array}$$

The quotient is $x^2 + 2x - 3$.

EXAMPLE 6

Divide:

$$x + 1 \overline{)\, x^3 + x^2 + x + 1}$$

Solution.

$$\begin{array}{rl}
 & x^2 \qquad + 1 \\
x + 1 & \overline{)\, x^3 + x^2 + x + 1} \\
 & \underline{\mp x^3 \mp x^2} \\
 & \qquad\quad x + 1 \\
 & \qquad\quad \underline{\mp x \mp 1} \\
 & \qquad\qquad 0
\end{array}$$

The 1st difference polynomial is 0. Bring down 2 terms from the dividend because the divisor has 2 terms.

The quotient is $x^2 + 1$.

EXERCISES

In exercises 1–8: (a) Express each polynomial in standard form. (b) What is the degree of the polynomial?

1. $x^2 + 1 + 2x$
2. $y^3 - 1 + 2y^2 + 4y$
3. $t^3 + t - t^5$
4. $u - u^2 + u^3 - u^4$
5. $1 - t$
6. $x + x^2 + x^3 + x^4$
7. $z^7 - z^4 + z + z^{10}$
8. $x^9 - x^8 + 1 - x^4 - x^{12}$

In exercises 9–20, divide and check:

9. $x + 1\overline{)x^2 + 5x + 4}$
10. $y + 2\overline{)y^2 + 8y + 12}$
11. $x + 3\overline{)x^2 + 8x + 15}$
12. $x - 1\overline{)x^2 + 5x - 6}$
13. $t - 1\overline{)t^2 - 10t + 9}$
14. $u + 4\overline{)u^2 + 12u + 32}$
15. $t + 9\overline{)t^2 + 14t + 45}$
16. $x - 8\overline{)x^2 - 3x - 40}$
17. $x + 4\overline{)x^3 + 5x^2 + 5x + 4}$
18. $x + 2\overline{)2x^3 + 5x^2 + 3x + 2}$
19. $t + 4\overline{)12t^2 + 3t^3 - 4 - t}$
20. $y^2 + 2\overline{)y^4 + 3y^2 + 2}$

In exercises 21–36, divide as indicated:

21. $x + 3\overline{)x^2 + 12x + 27}$
22. $t - 5\overline{)t^2 - 3t - 10}$
23. $z^2 - 1\overline{)z^4 - 1}$
24. $t + 3\overline{)t^3 + 27}$
25. Divide $x^2 - 20 + x$ by $5 + x$.
26. Divide $9 + 3a + 3a^2 + a^3$ by $3 + a^2$.
27. Divide $9x^2 + x^3 + 17x + 6$ by $x^2 + 7x + 3$.
28. Divide $15 + 8a^2 + a^4$ by $a^2 + 3$.
29. $x + 5\overline{)x^3 + 6x^2 + 6x + 5}$
30. $y + 1\overline{)2y^2 + 3y + 1}$
31. $a - 5\overline{)2a^3 - 11a^2 + 7a - 10}$
32. $2b + 1\overline{)4b^3 + 4b^2 + 5b + 2}$
33. $x^2 + x - 2\overline{)x^3 + 3x^2 - 4}$
34. $2y^2 + 3y + 1\overline{)2y^3 + 9y^2 + 10y + 3}$
35. $x^2 + x + 1\overline{)x^4 + 3x^3 + 4x^2 + 3x + 1}$
36. $x^2 + x - 1\overline{)2x^4 + 3x^3 + 4x^2 + 4x - 5}$

5.7 DIVISION WITH A REMAINDER

Observe that 2 does not divide 7 (evenly). When you divide 7 by 2, there is a "remainder."

$$\textit{dividend}\rightarrow \frac{7}{2} = 3\frac{1}{2} \begin{array}{l}\leftarrow \textit{remainder}\\ \leftarrow \textit{divisor}\end{array} \quad (3 = \textit{quotient})$$

divisor→

Thus 3 is the quotient and 1 the remainder. Similarly, division of polynomials often results in a remainder. Here is a simple example.

EXAMPLE 1

Divide: $2x + 1\overline{)4x + 3}$

Solution.

$$\begin{array}{r} 2 \\ 2x + 1\overline{)\ 4x + 3} \\ \underline{\mp 4x \mp 2} \\ 1 \end{array}$$

There are no more terms of the dividend to bring down. Thus 2 is the quotient; 1 is the remainder.

quotient ↓

$$\begin{array}{l}\textit{dividend}\rightarrow\\ \textit{divisor}\rightarrow\end{array} \frac{4x+3}{2x+1} = 2 + \frac{1}{2x+1} \begin{array}{l}\leftarrow \textit{remainder}\\ \leftarrow \textit{divisor}\end{array}$$

Note that the remainder, 1, is a (numerical) polynomial of degree 0, whereas the divisor is a polynomial of degree 1. Thus

$$\underbrace{\text{degree }(1)}_{0} < \underbrace{\text{degree }(2x+1)}_{1}$$

Let P and S be polynomials, $S \neq 0$. Then either S divides P (evenly), as in Section 5.6, or else

$$\frac{P}{S} = Q + \frac{R}{S},$$

where Q and R are polynomials and degree $R <$ degree S. The polynomial Q is called the **quotient** and R the **remainder.** Thus *if S does not divide P, add the rational expression $\frac{R}{S}$ to the quotient Q.*

When you divide polynomials, there are two possibilities for the *final difference polynomial:*

1. This difference polynomial is 0, as in Section 5.6. In this case, there is no remainder.
2. The degree of R, the final difference polynomial, is less than the degree of the divisor, as in Example 1. Then R is the remainder.

Continue the division process until you obtain one of these difference polynomials.

EXAMPLE 2

Find the quotient and remainder:

$$5x + 1\overline{)10x^2 + 12x - 2}$$

Solution.

```
                                remainder
                                    ↓
                                   −4
   quotient→  2x  +  2  +  ────── ←divisor
                           5x + 1
divisor→ 5x + 1⟌ 10x² + 12x − 2       ←dividend
                ∓10x² ∓  2x
                ───────────
                      10x − 2        ←1st difference
                     ∓10x ∓ 2          polynomial
                     ────────
                          −4         ←2nd difference
                                       polynomial
                                       (remainder)
```

The quotient is $2x + 2$ and the remainder is -4. The result is expressed as

$$2x + 2 + \frac{-4}{5x + 1}.$$

To *check* a division example, where there is a remainder:

1. *Multiply the quotient and divisor.*
2. *Add the remainder to this.*

The resulting polynomial should be the *dividend.*

CHECK (for Example 2):

```
 quotient→      2x + 2
  divisor→     ×5x + 1
               ───────
               10x² + 10x
                      2x + 2
               ──────────────
               10x² + 12x + 2
remainder→    +           − 4
               ──────────────
 dividend→     10x² + 12x − 2
```

EXAMPLE 3

Find the quotient and remainder:

$$3x - 2\overline{)12x^3 - 11x^2 + 5x}$$

Also, check the result.

Solution.

$$\begin{array}{rrl}
 & & \textit{quotient}\rightarrow\ 4x^2 - x + 1 + \dfrac{2}{3x-2} \quad \leftarrow\textit{remainder},\ \leftarrow\textit{divisor} \\
\textit{divisor}\rightarrow & 3x-2 \overline{\big)\ 12x^3 - 11x^2 + 5x} & \leftarrow\textit{dividend} \\
 & \underline{\mp 12x^3 \pm 8x^2} & \\
 & -3x^2 + 5x & \leftarrow\textit{1st difference polynomial} \\
 & \underline{\pm 3x^2 \mp 2x} & \\
 & 3x & \leftarrow\textit{2nd difference polynomial} \\
 & \underline{\mp 3x \pm 2} & \\
 & 2 & \leftarrow\textit{3rd difference polynomial}
\end{array}$$

The quotient is $4x^2 - x + 1$ and the remainder is 2. The result is expressed as

$$4x^2 - x + 1 + \frac{2}{3x-2}.$$

CHECK:

$$\begin{array}{rr}
\textit{quotient}\rightarrow & 4x^2 - x + 1 \\
\textit{divisor}\rightarrow & \underline{\times 3x \ \ - 2} \\
 & 12x^3 - 3x^2 + 3x \\
 & \underline{- 8x^2 + 2x - 2} \\
 & 12x^3 - 11x^2 + 5x - 2 \\
\textit{remainder}\rightarrow & \underline{+ \qquad\qquad\qquad 2} \\
\textit{dividend}\rightarrow & 12x^3 - 11x^2 + 5x
\end{array}$$

EXAMPLE 4

Find the quotient and remainder:

$$x^2 \overline{\big)\,x}$$

Solution. 1, the degree of the dividend, x, is less than 2, the degree of the divisor, x^2. Thus, the quotient is 0 and the remainder is x.

$$\frac{x}{x^2} = 0 + \frac{x}{x^2} \quad \leftarrow\textit{remainder}$$

EXERCISES

In exercises 1–10: (a) What is the quotient? (b) What is the remainder? (c) Show that

degree of remainder < degree of divisor.

1. $\dfrac{8x+3}{2x} = 4 + \dfrac{3}{2x}$

2. $\dfrac{10x-1}{5x} = 2 + \dfrac{-1}{5x}$

3. $\dfrac{9x+4}{3x+1} = 3 + \dfrac{1}{3x+1}$

4. $\dfrac{16x+7}{4x+2} = 4 + \dfrac{-1}{4x+2}$

5. $\dfrac{2x^2+1}{x} = 2x + \dfrac{1}{x}$

6. $\dfrac{2x^2+x+1}{x} = 2x + 1 + \dfrac{1}{x}$

7. $\dfrac{x^2 + 4x + 8}{x + 3} = x + 1 + \dfrac{5}{x + 3}$

8. $\dfrac{x^3 + 3x}{x^2 + 1} = x + \dfrac{2x}{x^2 + 1}$

9. $\dfrac{2x^3 - 4x + 5}{x^2 - 2} = 2x + \dfrac{5}{x^2 - 2}$

10. $\dfrac{x^4 + 4x^3 + 6x^2 + 8x + 3}{x^2 + x + 2} = x^2 + 3x + 1 + \dfrac{x + 1}{x^2 + x + 2}$

In exercises 11–22: (a) Find the quotient, (b) find the remainder, and (c) check the result.

11. $x + 1\overline{)3x + 2}$

12. $x + 2\overline{)9x - 4}$

13. $x + 5\overline{)5x + 6}$

14. $x + 5\overline{)20x + 28}$

15. $1 + x\overline{)1 + 10x}$

16. $x + 2\overline{)x^2 + 5x + 8}$

17. $x + 4\overline{)x^2 + 7x + 10}$

18. $x - 1\overline{)x^2 - 2x}$

19. $x + 1\overline{)x^3 + 3x^2 + 3x + 5}$

20. $x^2 + x + 1\overline{)2x^3 + 7x^2 + 6x + 10}$

21. $x^2 - x\overline{)3x^3 + 2x^2 + 5x + 2}$

22. $x^2 + 1\overline{)x^4 - x^2 + 3}$

In exercises 23–38: (a) Find the quotient, (b) find the remainder, and (c) express your result in the form

$$\text{QUOTIENT} + \frac{\text{REMAINDER}}{\text{DIVISOR}}$$

23. $x + 3\overline{)x^2 + 8x + 20}$

24. $x - 5\overline{)x^2}$

25. $2x + 1\overline{)4x^2 + 6x + 5}$

26. $3x + 2\overline{)12x^2 + 11x + 11}$

27. $x + 10\overline{)x^2 + 15x + 9}$

28. $x - 8\overline{)x^2 - 12x + 1}$

29. $x + 1\overline{)x^3 - 3x^2 + 2x + 1}$

30. $x + 2\overline{)x^3 - 2x^2 + x + 2}$

31. $x - 1\overline{)2x^3 - 5x}$

32. $x - 1\overline{)6x^4 - 5}$

33. $x^2 + x + 1\overline{)4x^3 + x^2 - 2x + 10}$

34. $x^2 - 2\overline{)x^4 + x^2}$

35. $x^2 + 1\overline{)x^4 - 1}$

36. $x^2 + x + 1\overline{)x^4 - x^2 + x + 2}$

37. $x^2 + 1\overline{)2x}$

38. $x^3 - x\overline{)x^2 + x}$

What Have You Learned in Chapter 5?

You have learned how to simplify fractions.

You can evaluate rational expressions by substituting numbers for variables.

You can simplify rational expressions by dividing by powers of numbers and variables, as well as by applying factoring techniques.

And you have learned a method of dividing polynomials that resembles long division.

Let's Review Chapter 5.

5.1 Fractions

1. Cross-multiply to show that $\frac{3}{5}$ and $\frac{-9}{-15}$ are equal.

In exercises 2–4, simply each fraction by dividing by factors common to both numerator and denominator. Express the simplified fraction with positive denominator.

2. $\frac{6}{9}$

3. $\frac{15}{-18}$

4. $\frac{-32}{-40}$

5.2 Evaluating Rational Expressions

5. Find (a) the numerator and (b) the denominator of the rational expression $\frac{2y-1}{y^2+3y-5}$.

6. Evaluate $\frac{3x+2}{x-9}$ when $x = 10$.

7. Evaluate $\frac{xy}{x+y+3}$ when $x = 2$ and $y = 3$.

8. Evaluate $\frac{x^2+1}{x^2-2}$ when (a) $x = 1$, (b) $x = -1$, (c) $x = 2$.

In exercises 9–20, simplify each rational expression. (Multiply out the numbers that remain in numerator or denominator.)

5.3 Division of Monomials

9. $\frac{a^4b}{a^2b^5}$

10. $\frac{2 \cdot 5^4z^5x^2}{5^2z^2x^5}$

11. $\frac{-48a^4bc^7}{-36abc^9}$

12. $\frac{3(x+y)^3(x-y)}{3^4(x+y)(x-y)^2}$

5.4 *Monomial Divisors*

13. $\dfrac{4x - 4y}{2}$

14. $\dfrac{ax^2 + bx^2}{x^3}$

15. $\dfrac{3a^2bc + 6abc^2}{9abc}$

16. $\dfrac{20xy^2z}{10xy^2z - 50xy^3z}$

5.5 *Factoring and Simplifying*

17. $\dfrac{5a + 5b}{25a + 50b}$

18. $\dfrac{m^2 - 9}{m^2 + 6m + 9}$

19. $\dfrac{s^3 - s^2}{s^4 - s^3}$

20. $\dfrac{2x^2 - 2y^2}{x^2 - 2xy + y^2}$

5.6 *Polynomial Division*

21. (a) Express $-x^4 + 7x^6 + x^2 + x^2 + 5x^5$ in standard form.
    (b) What is the degree of this polynomial?
22. Divide and check:

$$y + 2\overline{)y^2 + 7y + 10}$$

23. Divide:

$$x + 4\overline{)x^3 + 3x^2 - 3x + 4}$$

24. Divide $a^4 + 4a^2 - 5$ by $a^2 + 5$.

5.7 *Division with a Remainder*

25. It can be shown that

$$\frac{5x + 7}{x + 1} = 5 + \frac{2}{x + 1}$$

(a) What is the quotient? (b) What is the remainder? (c) Show that the degree of the remainder is less than the degree of the divisor.

26. (a) Find the quotient, (b) find the remainder, and (c) check the result:

$$x + 2\overline{)4x^2 - 7x + 9}$$

In exercises 27 and 28: (a) Find the quotient, (b) find the remainder, and (c) express your result in the form

$$\text{QUOTIENT} + \frac{\text{REMAINDER}}{\text{DIVISOR}}.$$

27. $a + 7\overline{)a^2 + 4a - 32}$

28. $x + 4\overline{)x^3 - x^2 + 1}$

And these from Chapters 1–4:

29. (a) Add: $(x + 1) + (x + 3)$ (b) Simplify: $\dfrac{(x + 1) + (x + 3)}{8}$
30. (a) Simplify: $4a^2 + 7 - a(3a + 2) - (8 - 2a)$

(b) Simplify: $\dfrac{4a^2 + 7 - a(3a + 2) - (8 - 2a)}{6a + 6}$

31. (a) Simplify: $(y + 1)(y + 3) + 1$ (b) Simplify: $\dfrac{(y + 1)(y + 3) + 1}{y + 2}$

32. Evaluate $\dfrac{3a - 10}{a - 1}$ (a) when $a = 2$, (b) when $a = 3$. (c) For which of these numbers, 2 or 3, is the evaluation larger?

Try These Exam Questions for Practice.

In questions 1–4, simplify each fraction or rational expression.

1. $\dfrac{36}{42}$

2. $\dfrac{3x^2(x + 1)^4}{9x^2(x + 1)^3}$

3. $\dfrac{5x^4 + 10x^2}{5x^3}$

4. $\dfrac{x^2 - 9}{x^2 + 5x + 6}$

5. Evaluate $\dfrac{5x + y}{x - 2y}$ when $x = 2$ and $y = -2$.

6. Divide and check:

$$x + 3\overline{)x^2 + 9x + 18}$$

7. (a) Find the quotient, (b) find the remainder, and (c) check the result:

$$x + 2\overline{)x^2 + 7x + 2}$$

Rational Expressions and Their Arithmetic

6.1 MULTIPLICATION OF RATIONAL EXPRESSIONS

Products of Fractions

Suppose you cut a pie into quarters, as in Figure 6.1(a). You then decide to cut each piece in half. Observe that each new portion is $\frac{1}{8}$ of the pie. (See Figure 6.1(b).) Thus

$$\frac{1}{2} \text{ of } \frac{1}{4} \text{ equals } \frac{1}{8}.$$

(The word "of" here indicates multiplication.)

$$\frac{1}{2} \cdot \frac{1}{4} = \frac{1}{8}$$

Note that

$$\frac{1}{2} \cdot \frac{1}{4} = \frac{1 \cdot 1}{2 \cdot 4} = \frac{1}{8}.$$

To multiply two fractions or rational expressions:

1. Multiply their numerators.
2. Multiply their denominators.

$$\frac{a}{b} \cdot \frac{c}{d} = \frac{ac}{bd}$$

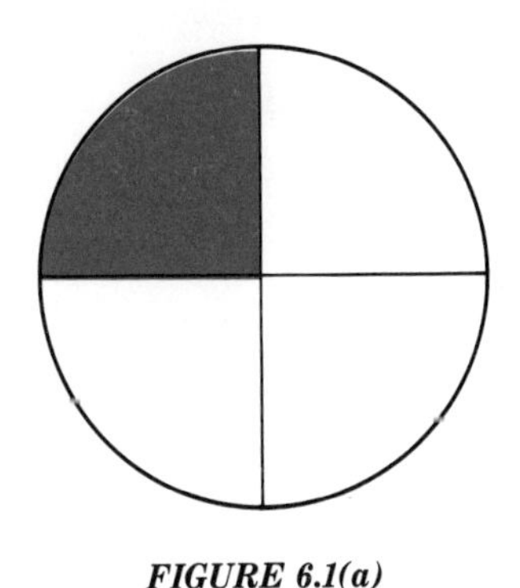

FIGURE 6.1(a)

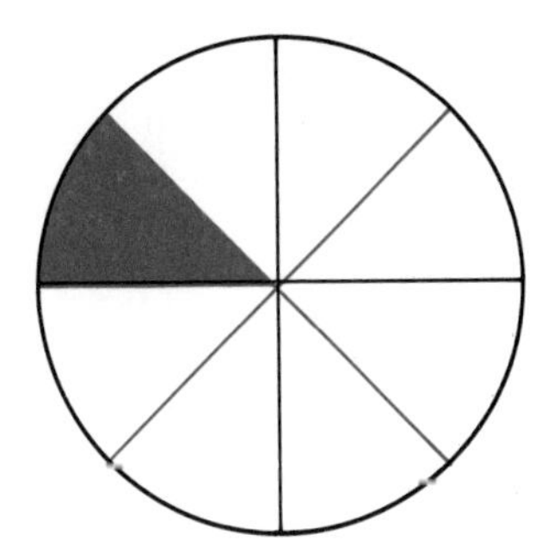

FIGURE 6.1(b). $\frac{1}{2} \cdot \frac{1}{4} = \frac{1}{8}$

First consider products of fractions.

EXAMPLE 1

$$\frac{2}{3} \cdot \frac{1}{5} = \frac{2 \cdot 1}{3 \cdot 5}$$

$$= \frac{2}{15}$$

EXAMPLE 2

$$\frac{-3}{4} \cdot \frac{7}{8} = \frac{(-3)7}{4 \cdot 8}$$

$$= \frac{-21}{32}$$

It is best to divide by factors common to the numerators and denominators before multiplying. You will then be working with smaller numbers.

EXAMPLE 3

Multiply: $\frac{2}{5} \cdot \frac{15}{4}$

Solution.

$$\frac{\overset{1}{\cancel{2}}}{\underset{1}{\cancel{5}}} \cdot \frac{\overset{3}{\cancel{15}}}{\underset{2}{\cancel{4}}} = \frac{3}{2}$$

The same procedure applies to products of three or more factors.

1. Divide by factors common to numerators and denominators.
2. Multiply the resulting numerators.
3. Multiply the resulting denominators.

EXAMPLE 4

Multiply: $\frac{-5}{8} \cdot \frac{12}{25} \cdot \frac{10}{-3}$

Solution.

$$\frac{\overset{\overset{1}{-1}}{\cancel{-5}}}{\underset{\underset{1}{2}}{\cancel{8}}} \cdot \frac{\overset{\overset{1}{3}}{\cancel{12}}}{\underset{\underset{1}{5}}{\cancel{25}}} \cdot \frac{\overset{\overset{1}{2}}{\cancel{10}}}{\underset{\underset{1}{-1}}{\cancel{-3}}} = 1$$

EXAMPLE 5

Find $\frac{3}{4}$ of 12.

Solution. Here the word "of" can be translated as "times." Thus

$$\frac{3}{4} \cdot 12 = \frac{3}{\underset{1}{\cancel{4}}} \cdot \frac{\overset{3}{\cancel{12}}}{1}$$
$$= 9$$

Products of Rational Expressions

EXAMPLE 6

$$\frac{x^3}{3y} \cdot \frac{a+b}{a-b} = \frac{x^3(a+b)}{3y(a-b)}$$

Note that the resulting rational expression is left in factored form.

To multiply two or more rational expressions:

1. First divide by factors common to numerators and denominators.
2. Multiply the resulting numerators.
3. Multiply the resulting denominators.

Leave the resulting rational expression in factored form.

EXAMPLE 7

Multiply: $\frac{4x^3}{x+y} \cdot \frac{(x+y)^2}{8x}$

Solution.

$$\frac{\overset{x^2}{\cancel{4x^3}}}{\underset{1}{\cancel{x+y}}} \cdot \frac{\overset{x+y}{\cancel{(x+y)^2}}}{\underset{2}{\cancel{8x}}} = \frac{x^2(x+y)}{2}$$

You may have to factor a numerator or denominator in order to divide by common factors.

EXAMPLE 8

Multiply:

$$\frac{m+n}{m^2-3m-2} \cdot \frac{m-2}{m^2-n^2}$$

Solution.

$$\frac{m+n}{m^2-3m-2} \cdot \frac{m-2}{m^2-n^2} = \frac{\overset{1}{\cancel{m+n}}}{\underset{1}{(\cancel{m-2})}(m-1)} \cdot \frac{\overset{1}{\cancel{m-2}}}{\underset{1}{(\cancel{m+n})}(m-n)}$$

$$= \frac{1}{(m-1)(m-n)}$$

EXAMPLE 9

Multiply:

$$\frac{x^2-4}{x^2+5x+4} \cdot \frac{2x+8}{(x+2)^2} \cdot \frac{-1}{16x-32}$$

Solution.

$$\frac{x^2-4}{x^2+5x+4} \cdot \frac{2x+8}{(x+2)^2} \cdot \frac{-1}{16x-32} = \frac{\overset{1}{(\cancel{x+2})}\overset{1}{(\cancel{x-2})}}{\underset{1}{(\cancel{x+4})}(x+1)} \cdot \frac{\overset{1}{\cancel{2}}\overset{1}{(\cancel{x+4})}}{\underset{x+2}{\cancel{(x+2)^2}}} \cdot \frac{-1}{\underset{8}{\cancel{16}}\underset{1}{(\cancel{x-2})}}$$

$$= \frac{-1}{8(x+1)(x+2)}$$

EXERCISES

In exercises 1–24, multiply the given fractions.

1. $\frac{1}{2} \cdot \frac{1}{3}$
2. $\frac{1}{4} \cdot \frac{3}{4}$
3. $\frac{2}{3} \cdot \frac{1}{5}$
4. $\frac{-3}{4} \cdot \frac{1}{7}$
5. $\frac{1}{2} \cdot 3$
6. $\frac{1}{2} \cdot 6$
7. $\frac{-1}{3} \cdot 6$
8. $\frac{5}{8} \cdot \frac{2}{15}$
9. $\frac{5}{6} \cdot \frac{3}{10}$
10. $\frac{-2}{7} \cdot \frac{-3}{7}$
11. $\frac{3}{4} \cdot \frac{8}{9}$
12. $\frac{15}{4} \cdot \frac{2}{3}$
13. $\frac{-10}{3} \cdot \frac{9}{20}$
14. $\frac{7}{10} \cdot \frac{25}{49}$

15. $\frac{1}{2} \cdot \frac{1}{3} \cdot \frac{1}{4}$

16. $\frac{1}{3} \cdot \frac{1}{9} \cdot 6$

17. $\frac{2}{3} \cdot 12 \cdot \frac{-1}{5}$

18. $\frac{5}{8} \cdot \frac{1}{10} \cdot \frac{2}{3}$

19. $\frac{-1}{2} \cdot \frac{2}{5} \cdot \frac{1}{-6}$

20. $\frac{-1}{7} \cdot \frac{-2}{5} \cdot \frac{-14}{3}$

21. $\frac{5}{8} \cdot \frac{2}{15} \cdot \frac{16}{3}$

22. $\frac{3}{4} \cdot \frac{49}{16} \cdot \frac{64}{21}$

23. $\frac{10}{7} \cdot \frac{5}{14} \cdot \frac{49}{25} \cdot \frac{2}{3}$

24. $\frac{21}{6} \cdot \frac{5}{8} \cdot \frac{24}{25} \cdot \frac{12}{7}$

In exercises 25–48, multiply the given rational expressions. Leave the resulting rational expression in factored form.

25. $a \cdot \frac{x}{y}$

26. $\frac{-x}{y} \cdot \frac{1}{z}$

27. $\frac{a^2}{c} \cdot \frac{b}{c}$

28 $\frac{2a}{b} \cdot \frac{b^2}{a}$

29. $\frac{4xyz}{3ab} \cdot \frac{9ax}{8yz}$

30. $\frac{a^2bc}{mn^2} \cdot \frac{m^2n^2}{abc^2}$

31. $\frac{16x^2y}{5a^3b} \cdot \frac{25a^2b^2}{4xy^2}$

32. $\frac{18m^2n^3}{7x^2y^4z^5} \cdot \frac{14xy^3z^5}{9m^3n^3}$

33. $\frac{a}{b} \cdot \frac{m}{n} \cdot \frac{x}{y}$

34. $\frac{a^2}{b} \cdot \frac{x}{a} \cdot \frac{b^2}{x}$

35. $\frac{2x}{3a} \cdot \frac{6a^2}{x^2} \cdot \frac{12a}{x}$

36. $\frac{-1}{a^2} \cdot \frac{a}{2} \cdot \frac{4}{a}$

37. $(x + 2) \cdot \frac{1}{x + 4}$

38. $\frac{5x + 5}{x + 2} \cdot \frac{1}{x + 1}$

39. $\frac{3y + 6}{y^2} \cdot \frac{y}{2y + 4}$

40. $\frac{5a + 15b}{x^2y} \cdot \frac{xy^2}{a + 3b}$

41. $\frac{x^2 - 1}{x^2 + 1} \cdot \frac{x^2 + 1}{x + 1}$

42. $\frac{a^2 - b^2}{2cd} \cdot \frac{4c^2}{a + b}$

43. $\frac{a^2 - 4}{3a - 9} \cdot \frac{2a - 6}{a^2 + 4a + 4}$

44. $\frac{(m + 3)^2}{2mn} \cdot \frac{4m^2n}{m^2 - 9}$

45. $\frac{x^2 - 16}{2ax} \cdot \frac{4a^2}{3x + 12} \cdot \frac{3x}{2x - 8}$

46. $\frac{c^2x^2 - c^2}{x^2 + 5x + 6} \cdot \frac{x + 2}{cx + c} \cdot \frac{cx + 3c}{x - 1}$

47. $\frac{a^2 - 9}{4x^3} \cdot \frac{12x}{(a + 3)^2} \cdot \frac{x^2}{a - 3}$

48. $\frac{x^2 - y^2}{4a^3} \cdot \frac{a}{y - x} \cdot \frac{16a^4}{x + y}$

49. Find $\frac{1}{2}$ of 48.

50. Find $\frac{2}{5}$ of 60.

51. Find $\frac{1}{2}$ of $\frac{2}{3}$.

52. Find $\frac{3}{4}$ of $\frac{8}{9}$.

53. Find the product of $\frac{x}{x + 1}$ and $\frac{x^2 - 1}{x^3}$.

54. Find the product of $\frac{a^2}{x + 3}$, $\frac{x^2}{a + 1}$, and $\frac{a + 1}{ax}$.

6.2 DIVISION OF RATIONAL EXPRESSIONS

Motivation

Observe that

divisor ↓   *quotient* ↓

$$8 \div 4 = 2$$
$$8 \div 2 = 4$$
$$8 \div 1 = 8$$

Each time the divisor is *divided by* 2, the quotient is *multiplied by* 2. According to this pattern:

$8 \div \frac{1}{2} = 16$ Note that $8 \cdot \frac{2}{1} = 16$.

$8 \div \frac{1}{4} = 32$ Note that $8 \cdot \frac{4}{1} = 32$.

Thus

to divide 8 by $\frac{1}{2}$, invert $\frac{1}{2}$ to obtain $\frac{2}{1}$, and then multiply;

to divide 8 by $\frac{1}{4}$, invert $\frac{1}{4}$ to obtain $\frac{4}{1}$, and then multiply.

To divide fractions or rational expressions,

$$\frac{a}{b} \div \frac{c}{d},$$

invert the divisor and multiply. Thus

$$\frac{a}{b} \div \frac{c}{d} = \frac{a}{b} \cdot \frac{d}{c} = \frac{ad}{bc}$$

Fractions

EXAMPLE 1

Find: $\frac{1}{2} \div \frac{1}{4}$

Solution. Invert the second fraction, $\frac{1}{4}$, and multiply.

$$\frac{1}{2} \div \frac{1}{4} = \frac{1}{\cancel{2}_1} \cdot \frac{\cancel{4}^2}{1} = 2$$

EXAMPLE 2

Find: $\frac{2}{3} \div \frac{2}{5}$

Solution.

$$\frac{2}{3} \div \frac{2}{5} = \frac{\overset{1}{\cancel{2}}}{3} \cdot \frac{5}{\underset{1}{\cancel{2}}}$$

$$= \frac{5}{3}$$

EXAMPLE 3

Find: $\frac{-5}{6} \div \frac{10}{27}$

Solution.

$$\frac{-5}{6} \div \frac{10}{27} = \frac{\overset{-1}{\cancel{-5}}}{\underset{2}{\cancel{6}}} \cdot \frac{\overset{9}{\cancel{27}}}{\underset{2}{\cancel{10}}}$$

$$= \frac{-9}{4}$$

Rational Expressions

The same procedure applies when you divide rational expressions. Invert the second expression and multiply.

EXAMPLE 4

Find: $\frac{1}{x} \div \frac{1}{x^2}$

Solution. Invert the second expression, $\frac{1}{x^2}$, and multiply:

$$\frac{1}{x} \div \frac{1}{x^2} = \frac{1}{\underset{1}{\cancel{x}}} \cdot \frac{\overset{x}{\cancel{x^2}}}{1} = x$$

EXAMPLE 5

Find: $\frac{ax^2}{x+a} \div \frac{a^2x}{(x+a)^2}$

Solution.

$$\frac{ax^2}{x+a} \div \frac{a^2x}{(x+a)^2} = \frac{\overset{x}{\cancel{a}\cancel{x^2}}}{\underset{1}{\cancel{x+a}}} \cdot \frac{\overset{(x+a)}{\cancel{(x+a)^2}}}{\underset{a}{\cancel{a^2}\cancel{x}}}$$

$$= \frac{x(x+a)}{a}$$

EXAMPLE 6

Find: $\dfrac{x^2 + 2xy + y^2}{x^2y^3} \div \dfrac{x^2 - y^2}{\mathrm{x}^4y^5}$

Solution. First factor both numerators.

$$\frac{x^2 + 2xy + y^2}{x^2y^3} \div \frac{x^2 - y^2}{x^4y^5} = \frac{(x + y)^2}{x^2y^3} \div \frac{(x + y)(x - y)}{x^4y^5}$$

$$= \frac{\overset{(x+y)}{\cancel{(x + y)^2}}}{\underset{1}{\cancel{x^2y^3}}} \cdot \frac{\overset{x^2y^2}{\cancel{x^4y^5}}}{\underset{1}{\cancel{(x + y)}}(x - y)}$$

$$= \frac{x^2y^2(x + y)}{x - y}$$

EXERCISES

In exercises 1–20, divide the given fractions.

1. $\frac{1}{3} \div \frac{1}{6}$
2. $\frac{1}{2} \div \frac{1}{8}$
3. $\frac{-1}{12} \div \frac{1}{3}$
4. $\frac{-1}{2} \div \frac{-1}{3}$
5. $\frac{2}{5} \div \frac{4}{3}$
6. $\frac{4}{5} \div \frac{8}{25}$
7. $\frac{5}{7} \div \frac{15}{14}$
8. $\frac{7}{9} \div \frac{-7}{3}$
9. $\frac{3}{4} \div \frac{3}{16}$
10. $\frac{2}{7} \div 4$
11. $\frac{-3}{5} \div (-9)$
12. $5 \div \frac{1}{5}$
13. $-8 \div \frac{4}{3}$
14. $\frac{15}{16} \div \frac{25}{4}$
15. $\frac{20}{7} \div \frac{10}{21}$
16. $\frac{11}{30} \div \frac{33}{10}$
17. $\frac{5}{144} \div \frac{25}{12}$
18. $\frac{121}{84} \div \frac{132}{49}$
19. $\frac{-55}{64} \div \frac{-125}{128}$
20. $\frac{81}{96} \div \frac{243}{144}$

In exercises 21–44, divide the given rational expressions.

21. $\frac{1}{a} \div \frac{1}{b}$
22. $\frac{a}{b} \div a$
23. $\frac{x}{a^2} \div \frac{x}{a}$
24. $\frac{m^2}{n} \div \frac{m}{n^2}$
25. $\frac{a}{b} \div \frac{b}{a}$
26. $\frac{4x}{y} \div \frac{2x^2}{y}$

27. $\frac{ab}{c} \div \frac{a^2c}{b}$

28. $\frac{x^2y}{2z} \div \frac{xy}{8z^2}$

29. $\frac{abc^3}{d} \div \frac{ab^3}{c^2d}$

30. $\frac{tx^2y}{z} \div \frac{tyz}{x^2}$

31. $\frac{x-a}{2} \div \frac{(x-a)^2}{4}$

32. $\frac{x+1}{x+2} \div \frac{x+3}{x+4}$

33. $\frac{a+1}{a-1} \div \frac{(a+1)^2}{(a-1)^2}$

34. $\frac{b+1}{b-1} \div \frac{b-1}{b+1}$

35. $\frac{x^2-1}{y} \div \frac{x+1}{y^2}$

36. $\frac{3x-3}{2} \div \frac{x-1}{4}$

37. $\frac{4x+4y}{3a-3b} \div \frac{2x+2y}{3}$

38. $\frac{5a+5b}{a-b} \div \frac{a^2-b^2}{2}$

39. $\frac{x^2+3x+2}{ab^2x} \div \frac{ax+a}{bx}$

40. $\frac{(x-1)^2(x+3)}{x+2} \div \frac{x^2+2x-3}{x^2+x-2}$

41. $\frac{x^2+7x+12}{x^2-16} \div \frac{x^2+6x+9}{2x-8}$

42. $\frac{u^2-v^2}{3uv^3} \div \frac{u^2-2uv+v^2}{9u^3v^2}$

43. $\frac{a^2x^2-a^2y^2}{ax+ay} \div \frac{ax-ay}{a^2}$

44. $\frac{x^2-9x+14}{x^2-49} \div \frac{x^2-4}{x+7}$

45. Divide the product of $\frac{ax}{2}$ and $\frac{a^2y}{5}$ by $\frac{axy}{20}$.

46. Divide $\frac{x+y}{3ab}$ by the product of $\frac{x^2-y^2}{9a^2}$ and $\frac{x^2}{6b}$.

47. What fraction must be multiplied by $\frac{1}{3}$ to obtain $\frac{5}{6}$?

48. What fraction must be multiplied by $\frac{2}{3}$ to obtain $\frac{3}{4}$?

49. What rational expression must be multiplied by $\frac{1}{x}$ to obtain $\frac{3}{x^2}$?

50. What rational expression must be multiplied by $\frac{x}{x-a}$ to obtain $\frac{x^2y}{x^2-a^2}$?

6.3 ADDITION OF RATIONAL EXPRESSIONS WITH THE SAME DENOMINATOR

To add or subtract fractions or rational expressions with the *same denominator D:*

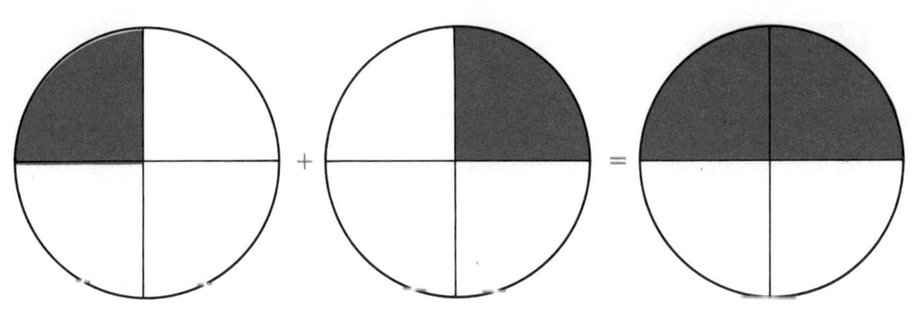

FIGURE 6.2. $\frac{1}{4}+\frac{1}{4}=\frac{1+1}{4}=\frac{2}{4}=\frac{1}{2}$

1. Add or subtract the numerators.
2. The denominator is D.
3. Simplify the resulting expression, if possible.

Fractions

EXAMPLE 1

Add: $\frac{1}{3}+\frac{1}{3}$

Solution.

$$\frac{1}{3}+\frac{1}{3}=\frac{1+1}{3}=\frac{2}{3}$$

EXAMPLE 2

Add: $\frac{1}{6}+\frac{1}{6}$

Solution.

$$\frac{1}{6}+\frac{1}{6}=\frac{1+1}{6}$$
$$=\frac{2}{6}$$

Simplify the sum

$$\frac{2}{6} \text{ to } \frac{1}{3}.$$

EXAMPLE 3

Subtract: $\frac{3}{8}-\frac{1}{8}$

Solution.

$$\frac{3}{8}-\frac{1}{8}=\frac{3-1}{8}$$
$$=\frac{2}{8}$$
$$=\frac{1}{4}$$

EXAMPLE 4

Combine: $\frac{1}{10}+\frac{7}{10}-\frac{3}{10}$

Solution.

$$\frac{1}{10}+\frac{7}{10}-\frac{3}{10}=\frac{1+7-3}{10}$$
$$=\frac{5}{10}$$
$$=\frac{1}{2}$$

Rational Expressions

EXAMPLE 5

Add: $\frac{1}{x}+\frac{2}{x}$

Solution.

$$\frac{1}{x}+\frac{2}{x}=\frac{1+2}{x}$$
$$=\frac{3}{x}$$

EXAMPLE 6

Subtract: $\frac{x}{yz}-\frac{1}{yz}$

Solution.

$$\frac{x}{yz}-\frac{1}{yz}=\frac{x-1}{yz}$$

Factoring the numerator or denominator of the sum or difference may simplify the result.

EXAMPLE 7

Add: $\frac{2x}{x+1}+\frac{2}{x+1}$

Solution.

$$\frac{2x}{x+1}+\frac{2}{x+1}=\frac{2x+2}{x+1}$$
$$=\frac{2\overset{1}{\cancel{(x+1)}}}{\underset{1}{\cancel{x+1}}}$$
$$=2$$

EXAMPLE 8

Subtract: $\frac{x^2}{x+1}-\frac{1}{x+1}$

Solution.

$$\frac{x^2}{x+1} - \frac{1}{x+1} = \frac{x^2-1}{x+1}$$

$$= \frac{\overset{1}{\cancel{(x+1)}}(x-1)}{\underset{1}{\cancel{x+1}}}$$

$$= x - 1$$

EXERCISES

Add or subtract, and simplify:

1. $\frac{1}{5} + \frac{1}{5}$
2. $\frac{2}{7} + \frac{1}{7}$
3. $\frac{2}{3} + \frac{1}{3}$
4. $\frac{2}{3} - \frac{1}{3}$
5. $\frac{3}{4} + \frac{1}{4}$
6. $\frac{3}{5} - \frac{1}{5}$
7. $\frac{5}{8} - \frac{1}{8}$
8. $\frac{9}{10} - \frac{1}{10}$
9. $\frac{\pi}{3} + \frac{\pi}{3}$
10. $\frac{\pi}{2} + \frac{\pi}{2}$
11. $\frac{1}{\pi} + \frac{1}{\pi}$
12. $\frac{3\pi}{2} - \frac{\pi}{2}$
13. $\frac{1}{4} + \frac{1}{4} + \frac{1}{4}$
14. $\frac{1}{6} + \frac{1}{6} + \frac{3}{6}$
15. $\frac{1}{5} + \frac{2}{5} + \frac{2}{5}$
16. $\frac{1}{12} + \frac{5}{12} + \frac{7}{12}$
17. $\frac{7}{9} - \frac{4}{9} + \frac{5}{9}$
18. $\frac{7}{9} - \left(\frac{4}{9} + \frac{5}{9}\right)$
19. $\frac{-3}{4} + \frac{1}{4} + \frac{5}{4}$
20. $\frac{7}{3} - \frac{1}{3} + \frac{2}{3}$
21. $\frac{1}{12} + \frac{1}{12} + \frac{5}{12} + \frac{11}{12}$
22. $\frac{7}{10} - \left(\frac{3}{10} + \frac{1}{10} - \frac{9}{10}\right)$
23. $\left(\frac{7}{6} - \frac{5}{6}\right) - \left(\frac{1}{6} + \frac{1}{6}\right)$
24. $\frac{11}{20} - \left(\frac{3}{20} - \frac{1}{20} + \frac{7}{20}\right)$
25. $\frac{1}{x} + \frac{1}{x}$
26. $\frac{5}{x} + \frac{2}{x}$
27. $\frac{4}{y} - \frac{3}{y}$
28. $\frac{3}{y^2} + \frac{1}{y^2}$
29. $\frac{x}{2} + \frac{1}{2}$
30. $\frac{a}{xy} + \frac{b}{xy}$
31. $\frac{10}{x+1} + \frac{1}{x+1}$
32. $\frac{7}{x+y} - \frac{5}{x+y}$

33. $\frac{6a}{x+2}+\frac{2a}{x+2}$

34. $\frac{5x}{x+3}-\frac{2x}{x+3}$

35. $\frac{6}{a}+\frac{1}{a}+\frac{2}{a}$

36. $\frac{5}{x^2}+\frac{3}{x^2}-\frac{1}{x^2}$

37. $\frac{9}{x}+\frac{1}{x}-\frac{4}{x}$

38. $\frac{3}{b+1}+\frac{5}{b+1}+\frac{2}{b+1}$

39. $\frac{x}{x+2}+\frac{2}{x+2}$

40. $\frac{a}{a-4}-\frac{4}{a-4}$

41. $\frac{3a}{a+1}+\frac{3}{a+1}$

42. $\frac{6x}{x-1}-\frac{6}{x-1}$

43. $\frac{1}{x-1}-\frac{x}{x-1}$

44. $\frac{a^2}{x^2-a^2}-\frac{x^2}{x^2-a^2}$

45. $\frac{x}{x^2-4}-\frac{2}{x^2-4}$

46. $\frac{a}{a^2-9}-\frac{3}{a^2-9}$

47. $\frac{b^2}{b+1}+\frac{1}{b+1}$

48. $\frac{4m^2}{m+2}-\frac{16}{m+2}$

49. $\frac{y^2}{y+1}+\frac{2y}{y+1}+\frac{1}{y+1}$

50. $\frac{a^2}{a+2}+\frac{5a}{a+2}+\frac{6}{a+2}$

6.4 LEAST COMMON MULTIPLES

Recall that ab is a *multiple* of both a and b. In order to add or subtract fractions or rational expressions with *different denominators,* first convert to equal expressions with the *same denominator.* The new denominator will be a multiple of each of the given denominators. Each denominator is an integer or polynomial. In this section you will learn how to find the smallest multiple of several integers or polynomials.

lcm of Several Integers

Definition

> *The* ***least common multiple (lcm) of two or more integers*** *is the smallest positive integer that is a multiple of each of them.*

Write

$$lcm\ (2, 5)$$

for

the least common multiple of 2 and 5.

EXAMPLE 1

Find: *lcm* (2, 5)

Solution.

$$lcm\ (2, 5) = 10$$

In fact,

$$10 = 2 \cdot 5$$

Thus 10 is a multiple of both 2 and 5. None of the integers

$$1, 2, 3, 4, 5, 6, 7, 8, 9$$

is a multiple of *both* 2 and 5. For example, 4 is a multiple of 2, but not of 5.

EXAMPLE 2

Find: *lcm* (2, 6)

Solution. Observe that 6 is a multiple of 2:

$$6 = 3 \cdot 2$$

And 6 is a multiple of 6:

$$6 = 1 \cdot 6$$

By considering the numbers

$$1, 2, 3, 4, 5,$$

check that 6 is the *smallest* positive multiple of *both* 2 and 6. Thus

$$lcm\ (2, 6) = 6$$

Sometimes, the least common multiple can be found almost immediately, as in the above examples. At other times, it is best to consider the prime factors of the integers.

To find the *lcm* of two or more integers:

1. Express them in terms of prime factors.
2. The *lcm* is the product of the *highest* powers of all primes that occur.

EXAMPLE 3

Find: *lcm* (8, 12)

Solution.

$$8 = 2^3$$

$$12 = 2^2 \cdot 3$$

The only primes that occur are 2 and 3. The highest power of 2 that occurs is 2^3. The only power of 3 that occurs is 3 itself. Thus

$$\begin{aligned} lcm\ (8, 12) &= 2^3 \cdot 3 \\ &= 8 \cdot 3 \\ &= 24 \end{aligned}$$

EXAMPLE 4

Find: *lcm* $(-25, -30)$

Solution.

$$-25 = -5^2$$
$$-30 = -(2 \cdot 3 \cdot 5)$$

The highest power of 2 that occurs is 2. The highest power of 3 that occurs is 3. The highest power of 5 that occurs is 5^2.

$$\begin{aligned} lcm\,(-25, -30) &= 2 \cdot 3 \cdot 5^2 \\ &= 6 \cdot 25 \\ &= 150 \end{aligned}$$

Note that the *lcm* is always a *positive* integer.

EXAMPLE 5

Find: *lcm* (40, 50, 64)

Solution.

$$40 = 2^3 \cdot 5$$
$$50 = 2 \cdot 5^2$$
$$64 = 2^6$$

The highest power of 2 that occurs is 2^6. The highest power of 5 that occurs is 5^2. Thus

$$\begin{aligned} lcm\,(40, 50, 64) &= 2^6 \cdot 5^2 \\ &= 64 \cdot 25 \\ &= 1600 \end{aligned}$$

lcm of Several Rational Expressions

Definition

> *The* ***least common multiple of two or more polynomials*** *is the product of the highest powers of all factors that occur when you factor each of these polynomials.*

EXAMPLE 6

Find: *lcm* $(2x^2y, 4xy^3)$

Solution. Write

$$4xy^3 = 2^2xy^3.$$

The highest power of 2 that occurs is 2^2, or 4. The highest power of x that occurs is x^2. The highest power of y that occurs is y^3. Thus

$$lcm\,(2x^2y, 4xy^3) = 4x^2y^3$$

EXAMPLE 7

Find: $lcm\ (x + y, x - y)$

Solution. $x + y$ and $x - y$ are each in factored form. Thus

$$lcm\ (x + y, x - y) = (x + y)(x - y)$$
$$[\text{or } x^2 - y^2]$$

EXAMPLE 8

Find: $lcm\ (x^2 - 1, x^2 + 2x + 1)$

Solution. Factor each polynomial:

$$x^2 - 1 = (x + 1)(x - 1)$$
$$x^2 + 2x + 1 = (x + 1)^2$$
$$lcm\ (x^2 - 1, x^2 + 2x + 1) = (x + 1)^2(x - 1)$$

EXAMPLE 9

Find:

$$lcm\ (9x + 27, x^2 + 2x - 3, 3x^2 - 6x + 3)$$

Solution. Factor each polynomial:

$$\begin{aligned} 9x + 27 &= 9(x + 3) \\ &= 3^2(x + 3) \end{aligned}$$

$$x^2 + 2x - 3 = (x + 3)(x - 1)$$

$$\begin{aligned} 3x^2 - 6x + 3 &= 3(x^2 - 2x + 1) \\ &= 3(x - 1)^2 \end{aligned}$$

$$\begin{aligned} lcm\ (9x + 27, x^2 + 2x - 3, 3x^2 - 6x + 3) &= 3^2(x + 3)(x - 1)^2 \\ &= 9(x + 3)(x - 1)^2 \end{aligned}$$

EXERCISES

In exercises 1–22, find the lcm of the given integers.

1. 2, 3
2. 3, 5
3. 2, 4
4. 2, 8
5. −3, 6
6. −3, −12
7. 5, 25
8. 10, 50
9. 4, 6
10. 6, 9
11. 10, 15
12. 4, 10
13. 12, 30
14. 6, 8
15. 12, 27
16. 50, 125

17. 2, 3, 5
18. 4, 8, 16
19. 2, 3, 4
20. 6, 9, 16
21. 50, 75, 100
22. 48, 64, 72

In exercises 23–50, find the *lcm* of the given polynomials.

23. a, x
24. x, y
25. $5, x$
26. x^2, y^2
27. x, x^2
28. a, a^3
29. $x + 1, (x + 1)^2$
30. $a, a + b$
31. a^2x, ax^2
32. xy^2, xy^3
33. $5xyz, x^2z$
34. $4a^2b, 8abc$
35. $4axy, 6a^2bx$
36. $12mn^2, 8m^2n$
37. $x + a, x - a$
38. $(x + a)^2, (x + a)(x - a)$
39. $x^2 - 4, x + 2$
40. $y^2 - 9, (y - 3)^2$
41. $a^2 - 25, a^2 + 10a + 25$
42. $b^2 + 3b + 2, b^2 + 5b + 6$
43. $4x - 4, 8x^2 - 16x + 8$
44. $z^2 - 100, 7z - 70$
45. $abx - aby, a^2b^2$
46. $t^2 + 6t + 8, (t + 2)^3$
47. $abx^2y^3, a^2b^7xy, ax^4yz$
48. $(x - 1)^7(x + 1), 4x + 4, (x - 1)^9$
49. $a^2 - 4, 5a - 10, 25a + 50$
50. $a^2x^2 - a^2y^2, 4x + 4y, 6x - 6y$

6.5 ADDITION AND SUBTRACTION

Least Common Denominators

In order to add or subtract fractions or rational expressions with *different denominators,* first find equal expressions with the same denominator. This denominator should be the *lcm* of the original denominators.

Definition

> *The* ***least common denominator (lcd) of several fractions or rational expressions*** *is the lcm of the individual denominators.*

Write

$$lcd\left(\frac{1}{2}, \frac{2}{3}\right)$$

for the least common denominator of $\frac{1}{2}$ and $\frac{2}{3}$.

EXAMPLE 1

(a) Find: $lcd\left(\frac{1}{2}, \frac{2}{3}\right)$

(b) Write equal fractions with this *lcd* as denominator.

Solution.
(a) To find the *lcd,* determine the *lcm* of the denominators.

$$lcm\ (2, 3) = 6$$

Therefore

$$lcd\left(\frac{1}{2}, \frac{2}{3}\right) = 6$$

(b) To obtain equal fractions with denominator 6, multiply numerator and denominator of each fraction as indicated.

$$\frac{1}{2} = \frac{1 \cdot 3}{2 \cdot 3} = \frac{3}{6}$$

$$\frac{2}{3} = \frac{2 \cdot 2}{3 \cdot 2} = \frac{4}{6}$$

Recall that when the numerator and denominator of a fraction are each multiplied by the same number, an equal fraction is obtained.

EXAMPLE 2

(a) Find: $lcd\left(\frac{5}{12}, \frac{-1}{18}\right)$

(b) Write equal fractions with this *lcd* as denominator.

Solution.

(a)
$$lcd\left(\frac{5}{12}, \frac{-1}{18}\right) = lcm\ (12, 18)$$

$$12 = 2^2 \cdot 3$$

$$18 = 2 \cdot 3^2$$

$$\begin{aligned} lcm\ (12, 18) &= 2^2 \cdot 3^2 \\ &= 4 \cdot 9 \\ &= 36 \end{aligned}$$

Therefore

$$lcd\left(\frac{5}{12}, \frac{-1}{18}\right) = 36$$

(b)
$$\frac{5}{12} = \frac{5 \cdot 3}{12 \cdot 3} = \frac{15}{36}$$

$$\frac{-1}{18} = \frac{-1 \cdot 2}{18 \cdot 2} = \frac{-2}{36}$$

EXAMPLE 3

(a) Find: $lcd\left(\frac{1}{x}, \frac{5a}{y^2}\right)$

(b) Write equal rational expressions with this *lcd* as denominator.

Solution.

(a) As in the case of fractions, find the *lcm* of the denominators x and y^2.

$$lcm\ (x, y^2) = xy^2$$

Therefore

$$lcd\left(\frac{1}{x}, \frac{5a}{y^2}\right) = xy^2$$

(b) To find equal rational expressions with denominator xy^2, multiply numerator and denominator of each expression as indicated.

$$\frac{1}{x} = \frac{1 \cdot y^2}{x \cdot y^2} = \frac{y^2}{xy^2}$$

$$\frac{5a}{y^2} = \frac{5a \cdot x}{y^2 \cdot x} = \frac{5ax}{xy^2}$$

EXAMPLE 4

(a) Find: $lcd\left(\frac{x}{x^2 - a^2}, \frac{a}{x^2 + 2ax + a^2}\right)$

(b) Write equal expressions with this *lcd* as denominator.

Solution.

(a) First factor the given denominators to find their *lcm*.

$$x^2 - a^2 = (x + a)(x - a)$$

$$x^2 + 2ax + a^2 = (x + a)^2$$

$$lcm\ (x^2 - a^2, x^2 + 2ax + a^2) = (x + a)^2(x - a)$$

Thus

$$lcd\left(\frac{x}{x^2 - a^2}, \frac{a}{x^2 + 2ax + a^2}\right) = (x + a)^2(x - a)$$

(b)

$$\begin{aligned}\frac{x}{x^2 - a^2} &= \frac{x}{(x + a)(x - a)} \\ &= \frac{x \cdot (x + a)}{(x + a)(x - a) \cdot (x + a)} \\ &= \frac{x^2 + ax}{(x + a)^2(x - a)}\end{aligned}$$

and

$$\begin{aligned}\frac{a}{x^2 + 2ax + a^2} &= \frac{a}{(x + a)^2} \\ &= \frac{a \cdot (x - a)}{(x + a)^2 \cdot (x - a)} \\ &= \frac{ax - a^2}{(x + a)^2(x - a)}\end{aligned}$$

As you will see, *when adding rational expressions, you must usually multiply out the numerators before adding.*

Combining Fractions

To add or subtract fractions or rational expressions with *different denominators:*

1. First find their *lcd.*
2. Find equal expressions with this *lcd* as denominator.
3. Add or subtract the numerators.
4. Simplify the resulting expression, if possible.

Steps (1) and (2) are what you found in Examples 1–4. Steps (3) and (4) are the rules for adding expressions with the same denominator (Section 6.3).

EXAMPLE 5

Add: $\frac{3}{4} + \frac{5}{8}$

Solution.

$$lcd\left(\frac{3}{4}, \frac{5}{8}\right) = lcm\ (4, 8) = 8$$

$$\frac{3}{4} = \frac{3 \cdot 2}{4 \cdot 2} = \frac{6}{8}$$

$\left(\frac{5}{8}\text{ already has denominator 8.}\right)$

$$\frac{3}{4} + \frac{5}{8} = \frac{6}{8} + \frac{5}{8}$$
$$= \frac{6 + 5}{8}$$
$$= \frac{11}{8}$$

EXAMPLE 6

Subtract: $\frac{9}{20} - \frac{7}{12}$

Solution.

$$20 = 2^2 \cdot 5$$
$$12 = 2^2 \cdot 3$$

$$lcd\left(\frac{9}{20}, \frac{7}{12}\right) = lcm\ (20, 12)$$
$$= 2^2 \cdot 5 \cdot 3$$
$$= 4 \cdot 15 = 60$$

$$\frac{9}{20} = \frac{9 \cdot 3}{20 \cdot 3} = \frac{27}{60}$$

$$\frac{7}{12} = \frac{7 \cdot 5}{12 \cdot 5} = \frac{35}{60}$$

$$\frac{9}{20} - \frac{7}{12} = \frac{27}{60} - \frac{35}{60} = \frac{27 - 35}{60}$$
$$= \frac{-8}{60}$$
$$= \frac{-2}{15}$$

EXAMPLE 7

Find the value of: $\frac{5}{24} + \frac{1}{18} - \frac{7}{36}$

Solution.

$$\begin{aligned} 24 &= 2^3 \cdot 3 \\ 18 &= 2 \cdot 3^2 \\ 36 &= 2^2 \cdot 3^2 \end{aligned}$$

$$\begin{aligned} lcd\left(\frac{5}{24}, \frac{1}{18}, \frac{7}{36}\right) &= lcm\ (24, 18, 36) \\ &= 2^3 \cdot 3^2 \\ &= 8 \cdot 9 \\ &= 72 \end{aligned}$$

$$\frac{5}{24} = \frac{5 \cdot 3}{24 \cdot 3} = \frac{15}{72}$$

$$\frac{1}{18} = \frac{1 \cdot 4}{18 \cdot 4} = \frac{4}{72}$$

$$\frac{7}{36} = \frac{7 \cdot 2}{36 \cdot 2} = \frac{14}{72}$$

$$\begin{aligned} \frac{5}{24} + \frac{1}{18} - \frac{7}{36} &= \frac{15}{72} + \frac{4}{72} - \frac{14}{72} \\ &= \frac{15 + 4 - 14}{72} \\ &= \frac{5}{72} \end{aligned}$$

Combining Rational Expressions

EXAMPLE 8

Add: $\frac{1}{ax^2} + \frac{2}{a^2x}$

Solution.

$$\begin{aligned} lcd\left(\frac{1}{ax^2}, \frac{2}{a^2x}\right) &= lcm\ (ax^2, a^2x) \\ &= a^2x^2 \end{aligned}$$

$$\frac{1}{ax^2} = \frac{1 \cdot a}{ax^2 \cdot a} = \frac{a}{a^2x^2}$$

$$\frac{2}{a^2x} = \frac{2 \cdot x}{a^2x \cdot x} = \frac{2x}{a^2x^2}$$

$$\begin{aligned} \frac{1}{ax^2} + \frac{2}{a^2x} &= \frac{a}{a^2x^2} + \frac{2x}{a^2x^2} \\ &= \frac{a + 2x}{a^2x^2} \end{aligned}$$

EXAMPLE 9

Subtract: $\frac{2}{x - 4} - \frac{1}{x + 4}$

Solution.

$$lcd\left(\frac{2}{x-4}, \frac{1}{x+4}\right) = lcm\ (x-4, x+4)$$
$$= (x-4)(x+4) \qquad [\text{or } x^2-16]$$

$$\frac{2}{x-4} = \frac{2(x+4)}{(x-4)(x+4)} = \frac{2x+8}{x^2-16}$$

$$\frac{1}{x+4} = \frac{1(x-4)}{(x+4)(x-4)} = \frac{x-4}{x^2-16}$$

$$\frac{2}{x-4} - \frac{1}{x+4} = \frac{2x+8}{x^2-16} - \frac{x-4}{x^2-16}$$
$$= \frac{2x+8-x+4}{x^2-16}$$
$$= \frac{x+12}{x^2-16}$$

Note that $-(x-4) = -x+4$.

EXERCISES

In exercises 1–20: (a) Find the *lcd*, and (b) write equal fractions or rational expressions with this *lcd* as denominator.

1. $\frac{1}{3}$ and 1
2. $\frac{2}{3}$ and $\frac{-1}{5}$
3. $\frac{1}{2}$ and $\frac{5}{6}$
4. $\frac{-3}{4}$ and $\frac{-1}{12}$
5. $\frac{5}{4}$ and $\frac{7}{10}$
6. $\frac{3}{8}$ and $\frac{5}{12}$
7. $\frac{7}{20}$ and $\frac{3}{28}$
8. $\frac{1}{44}$ and $\frac{3}{40}$
9. $\frac{5}{36}$ and $\frac{1}{27}$
10. $\frac{1}{8}, \frac{-1}{12}$, and $\frac{2}{15}$
11. $\frac{1}{x}$ and $\frac{1}{y}$
12. $\frac{1}{a^2}$ and $\frac{a}{b}$
13. $\frac{2}{x+a}$ and -1
14. $\frac{a}{x^2y}$ and $\frac{b}{xy^2z}$
15. $\frac{x-a}{x^2y^3z^2}$ and $\frac{a}{xy^4z}$
16. $\frac{2}{x(x-a)}$ and $\frac{4}{x(x+a)}$
17. $\frac{1}{x^2-1}$ and $\frac{x}{(x-1)^2}$
18. $\frac{3}{x^2+6x+9}$ and $\frac{-1}{x^2+4x+3}$
19. $\frac{1}{ax-2a}, \frac{a}{x^2-4}, \frac{-1}{ax+2a}$
20. $\frac{a}{x^2-25}, \frac{b}{(x+5)^2}, \frac{c}{ax+5a}$

In exercises 21–56, add or subtract, as indicated.

21. $\frac{1}{2}+\frac{1}{3}$
22. $\frac{1}{4}-\frac{1}{5}$

23. $\frac{1}{2}+\frac{1}{4}$

24. $\frac{1}{8}-\frac{1}{4}$

25. $\frac{1}{2}+\frac{5}{6}$

26. $\frac{3}{4}-\frac{1}{12}$

27. $\frac{2}{9}+1$

28. $\frac{1}{10}-1$

29. $\frac{7}{8}+\frac{1}{12}$

30. $\frac{7}{10}+\frac{1}{25}$

31. $\frac{5}{24}+\frac{7}{30}$

32. $\frac{3}{64}-\frac{1}{48}$

33. $\frac{1}{96}+\frac{5}{9}$

34. $\frac{13}{40}-\frac{3}{100}$

35. $\frac{1}{6}+\frac{2}{9}+1$

36. $\frac{5}{8}-\frac{1}{4}+\frac{1}{16}$

37. $\frac{1}{10}+\frac{7}{5}-\frac{2}{25}$

38. $\frac{10}{21}-\left(\frac{3}{7}+\frac{2}{9}\right)$

39. $\frac{1}{x}+\frac{1}{y}$

40. $2-\frac{1}{b}$

41. $\frac{a}{x}+\frac{a}{x^2}$

42. $\frac{x}{2}+\frac{y}{3}$

43. $\frac{2}{xyz}+xyz$

44. $\frac{a}{s^2t}-\frac{b}{st^3}$

45. $\frac{1}{x+a}+\frac{2}{(x+a)^2}$

46. $\frac{4}{y-b}-\frac{1}{y+b}$

47. $\frac{1}{at+a}-\frac{a}{t+1}$

48. $\frac{3}{x+5}+\frac{1}{x^2+10x+25}$

49. $\frac{x}{x^2-4}-\frac{1}{x^2-2x}$

50. $\frac{a}{x^2+5x+6}-\frac{a}{x^2-4}$

51. $\frac{1}{a^2-9a}+\frac{9}{a^2-81}$

52. $\frac{1}{x^2-x}+\frac{2}{x^2-x^3}$

53. $\frac{1}{x-1}+\frac{1}{x-2}+\frac{1}{x-3}$

54. $\frac{5}{a^2-25}-\left(\frac{1}{a^2+10a+25}-\frac{1}{a^2-10a+25}\right)$

55. $\frac{1}{a^2}\left(\frac{x}{a^2-4a}+\frac{1}{a^2-16}\right)$

56. $\frac{1}{x^2+6x+9}-\left(\frac{1}{x+3}+\frac{1}{x^2-9}\right)$

57. $\frac{4}{x-1}+\frac{3}{1-x}$

58. $\frac{a}{a-3}-\frac{1}{3-a}$

6.6 COMPLEX EXPRESSIONS

Complex Fractions

Definition

> *A **complex fraction** is one that contains other fractions in its numerator or denominator. (There may be fractions in both numerator and denominator.)*

$$\frac{\frac{1}{4}}{3},\quad \frac{-2}{\frac{2}{3}+1},\quad \text{and}\quad \frac{\frac{3}{8}}{\frac{7}{10}}$$

are each complex fractions. You can simplify a complex fraction by first expressing it in terms of division.

For example, the numerator of $\frac{\frac{1}{4}}{3}$ is $\frac{1}{4}$; the denominator is 3.

Thus

$$\frac{\frac{1}{4}}{3} = \frac{1}{4} \div 3$$

Similarly, the numerator of $\frac{-2}{\frac{2}{3}+1}$ is -2; the denominator is $\frac{2}{3}+1$.

Thus

$$\frac{-2}{\frac{2}{3}+1} = -2 \div \left(\frac{2}{3}+1\right)$$

And the numerator of $\frac{\frac{3}{8}}{\frac{7}{10}}$ is $\frac{3}{8}$; the denominator is $\frac{7}{10}$.

Therefore

$$\frac{\frac{3}{8}}{\frac{7}{10}} = \frac{3}{8} \div \frac{7}{10}$$

EXAMPLE 1

Simplify:

$$\frac{\frac{1}{2}}{5}$$

Solution.

$$\begin{aligned}\frac{\frac{1}{2}}{5} &= \frac{1}{2} \div 5\\ &= \frac{1}{2} \cdot \frac{1}{5}\\ &= \frac{1}{10}\end{aligned}$$

EXAMPLE 2

Simplify:

$$\frac{\frac{2}{3}}{\frac{4}{9}}$$

Solution.

$$\frac{\frac{2}{3}}{\frac{4}{9}} = \frac{2}{3} \div \frac{4}{9}$$

$$= \frac{\overset{1}{\cancel{2}}}{\underset{1}{\cancel{3}}} \cdot \frac{\overset{3}{\cancel{9}}}{\underset{2}{\cancel{4}}}$$

$$= \frac{3}{2}$$

EXAMPLE 3

Simplify:

$$\frac{1 + \frac{1}{2}}{\frac{1}{4}}$$

Solution.

$$\frac{1 + \frac{1}{2}}{\frac{1}{4}} = \left(1 + \frac{1}{2}\right) \div \frac{1}{4}$$

$$= \frac{2+1}{\cancel{2}} \cdot \frac{\overset{2}{\cancel{4}}}{1}$$

$$= 3 \cdot 2$$

$$= 6$$

Complex Rational Expressions

Definition

> *A **complex rational expression** is one that contains other rational expressions in numerator or denominator (possibly in both).*

As with fractions, you simplify a complex rational expression by first rewriting it in terms of division.

EXAMPLE 4

Simplify:

$$\frac{\frac{1}{x}}{\frac{1}{y}}$$

Solution.

$$\frac{\frac{1}{x}}{\frac{1}{y}} = \frac{1}{x} \div \frac{1}{y} = \frac{1}{x} \cdot \frac{y}{1} = \frac{y}{x}$$

EXAMPLE 5

Simplify:

$$\frac{\frac{1}{x-4}}{\frac{x}{x^2-16}}$$

Solution.

$$\frac{\frac{1}{x-4}}{\frac{x}{x^2-16}} = \frac{1}{x-4} \div \frac{x}{x^2-16}$$

$$= \frac{1}{\cancel{x-4}} \cdot \frac{\overset{(x+4)\cancel{(x-4)}}{\cancel{x^2-16}}}{x}$$

$$= \frac{x+4}{x}$$

EXAMPLE 6

Simplify:

$$\frac{\frac{2}{a-2} - \frac{1}{a+2}}{\frac{a}{a^2-4}}$$

Solution.

$$\frac{\frac{2}{a-2} - \frac{1}{a+2}}{\frac{a}{a^2-4}} = \left(\frac{2}{a-2} - \frac{1}{a+2}\right) \div \frac{a}{a^2-4}$$

$$= \frac{2(a+2) - 1(a-2)}{\cancel{(a-2)(a+2)}} \cdot \frac{\overset{\cancel{(a-2)(a+2)}}{\cancel{a^2-4}}}{a}$$

$$= \frac{2a+4-a+2}{a}$$

$$= \frac{a+6}{a}$$

EXERCISES

Simplify:

1. $\frac{\frac{1}{2}}{2}$

2. $\frac{5}{\frac{1}{3}}$

3. $\frac{\frac{3}{4}}{-3}$

4. $\frac{-7}{\frac{1}{3}}$

5. $\frac{\frac{1}{2}}{\frac{1}{6}}$

6. $\frac{\frac{2}{5}}{\frac{1}{10}}$

7. $\dfrac{\frac{-1}{8}}{\frac{3}{4}}$

8. $\dfrac{\frac{-2}{9}}{\frac{-1}{3}}$

9. $\dfrac{\frac{4}{3}}{\frac{16}{9}}$

10. $\dfrac{\frac{2}{7}}{\frac{1}{14}}$

11. $\dfrac{1+\frac{1}{2}}{\frac{1}{2}}$

12. $\dfrac{2+\frac{1}{3}}{\frac{5}{3}}$

13. $\dfrac{1-\frac{1}{4}}{\frac{1}{8}}$

14. $\dfrac{6-\frac{3}{4}}{\frac{-1}{2}}$

15. $\dfrac{\frac{1}{2}+\frac{1}{3}}{\frac{1}{6}}$

16. $\dfrac{1-\frac{1}{2}}{1+\frac{1}{4}}$

17. $\dfrac{\frac{3}{4}+\frac{1}{12}}{\frac{5}{6}+\frac{1}{8}}$

18. $\dfrac{\frac{1}{9}-\frac{1}{6}}{\frac{2}{3}+\frac{1}{9}}$

19. $\dfrac{\frac{3}{5}-\frac{1}{10}}{\frac{2}{25}-\frac{1}{50}}$

20. $\dfrac{\frac{1}{12}-\frac{1}{144}}{\frac{1}{3}-\frac{2}{9}}$

21. $\dfrac{a}{\frac{x}{y}}$

22. $\dfrac{\frac{a}{x}}{y}$

23. $\dfrac{\frac{-1}{x}}{\frac{y}{2}}$

24. $\dfrac{\frac{2}{x}}{\frac{4}{y}}$

25. $\dfrac{\frac{a}{b^2}}{\frac{b}{a^2}}$

26. $\dfrac{\frac{xy}{z}}{\frac{y}{x^2}}$

27. $\dfrac{\frac{-ab}{c^2}}{\frac{a}{c^2}}$

28. $\dfrac{\frac{r^2s}{t^2}}{\frac{t}{r^2s^3}}$

29. $\dfrac{\frac{6x}{y^2}}{\frac{12y}{x}}$

30. $\dfrac{\frac{8a^2}{b}}{\frac{12a}{b^2}}$

31. $\dfrac{\frac{20a^2b}{cd}}{\frac{15ad}{bc}}$

32. $\dfrac{\frac{1}{x+1}}{\frac{x}{x-1}}$

33. $\dfrac{\frac{x^2-1}{x^2-4}}{\frac{x+1}{x+2}}$

34. $\dfrac{\frac{a^2-3a}{a+3}}{\frac{a^2}{a^2-9}}$

35. $\dfrac{\frac{1}{a}+1}{a^2}$

36. $\dfrac{1-\frac{1}{x}}{\frac{1}{x^2}}$

37. $\dfrac{\frac{1}{x+1}-\frac{1}{x-1}}{\frac{2}{x^2-1}}$

38. $\dfrac{\frac{x}{x-a}+\frac{a}{x+a}}{\frac{ax}{x^2-a^2}}$

39. $\dfrac{\frac{1}{a^2-b^2}+\frac{1}{a+b}}{\frac{a}{b^2-a^2}}$

40. $\dfrac{\frac{a^3-a^2}{a^2-1}}{\frac{a^3}{a^2+2a+1}}$

6.7 DECIMALS

Fractions to Decimals

Recall that

$$10^1 = 10, \quad 10^2 = 100, \quad 10^3 = 1000.$$

A rational number of the form

$$\frac{N}{10^m} \qquad (m = 1, 2, 3, \ldots)$$

can easily be expressed in decimal notation. For example,

$$\frac{7}{10} = .7, \quad \frac{-91}{100} = -.91, \quad \frac{13}{10} = 1.3$$

Here the dot is called the **decimal point.** The *first* **digit,**

0, 1, 2, 3, 4, 5, 6, 7, 8, or 9,

to the *right* of the decimal point stands for *tenths;* the *second* digit to the *right* of the decimal point stands for *hundredths;* the *third* digit to the *right* of the decimal point stands for *thousandths,* etc. Thus

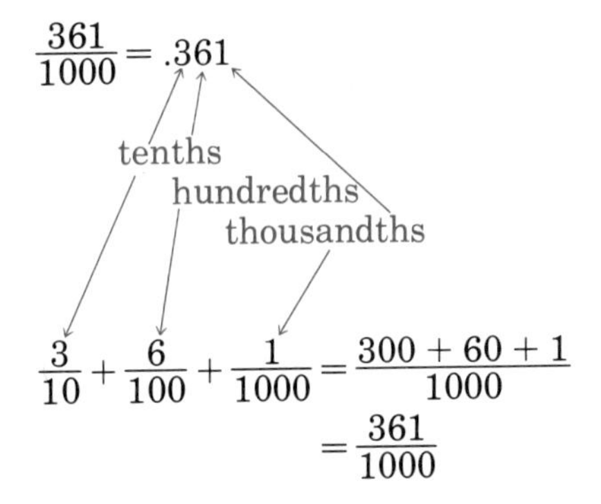

To express a fraction such as $\frac{3}{100}$, write 3 in the hundredths place and place a 0 in the tenths place. Thus

$$\frac{3}{100} = .03$$

A mixed number, such as $41\frac{7}{10}$ can be expressed by first writing *the integral part to the left of the decimal point.* Thus

$$41\frac{7}{10} = 41.7$$

EXAMPLE 1

Express each fraction as a decimal:

(a) $\frac{9}{10}$ (b) $\frac{9}{100}$ (c) $\frac{91}{100}$ (d) $\frac{91}{1000}$

Solution.

(a) $\frac{9}{10} = .9$ (b) $\frac{9}{100} = .09$ (c) $\frac{91}{100} = .91$ (d) $\frac{91}{1000} = .091$

A rational number such as $\frac{3}{4}$ can be represented as a decimal by *dividing numerator by denominator:*

$$4\overline{)3.00}\quad .75$$

←Line up the decimal points.
←Add as many 0's as necessary to the right of the decimal point.

Thus $$\frac{3}{4} = .75$$

EXAMPLE 2

Express

(a) $\frac{1}{2}$, (b) $\frac{7}{8}$ as decimals.

Solution.

(a) Divide numerator, 1, by denominator, 2:

$$2\overline{)1.0}\quad .5$$

Thus $$\frac{1}{2} = .5$$

(b) Divide numerator, 7, by denominator, 8:

$$8\overline{)7.000}\quad .875$$

Thus $$\frac{7}{8} = .875$$

The rational number $\frac{1}{3}$ can be expressed as an **infinite repeating decimal.** Divide numerator by denominator:

$$3\overline{)1.000000}\quad .333333$$

Thus $$\frac{1}{3} = .333333\ldots$$ Here the digit 3 repeats, as indicated by the 3 dots.

The rational number $\frac{2}{11}$ can be expressed as an infinite repeating decimal in which *two digits,* 1 and 8, *repeat:*

$$11\overline{)2.000000}\quad .181818$$

Thus $\frac{2}{11} = .181818\ldots$ Again, 3 dots indicate this repetition.

EXAMPLE 3

Express as infinite repeating decimals:

(a) $\frac{2}{9}$ (b) $\frac{1}{6}$

Solution.

(a) $$9\overline{)2.000000}\quad .222222$$

Thus $\frac{2}{9} = .222222\ldots$

(b) $$6\overline{)1.0000000}\quad .1666666$$

Here just the 6 repeats. Thus

$$\frac{1}{6} = .1666666\ldots$$

Every rational number can be written as either a (regular) decimal or as an infinite repeating decimal.

Decimals to Fractions

Decimals can also be converted to fractions.

EXAMPLE 4

Express each decimal as a fraction or mixed number:
(a) .7 (b) .25 (c) .009 (d) 1.17

Solution.

(a) $$.7 = \frac{7}{10}$$

(b) $$.25 = \frac{25}{100}$$

This can be simplified by dividing numerator and denominator by 25. Thus

$$.25 = \frac{25}{100} = \frac{1}{4}$$

(c) $$.009 = \frac{9}{1000}$$

(d) $$1.17 = 1 + \frac{17}{100}$$

$$= 1\tfrac{17}{100} \qquad \text{(as a mixed number)}$$

$$= \frac{117}{100} \qquad \text{(as an "improper" fraction)}$$

Addition and Subtraction

To add or subtract decimals, line up the decimal points and add or subtract in columns. Thus

$$\begin{array}{r} .5 \\ .6 \\ \hline 1.1 \end{array} \quad \text{because} \quad \frac{5}{10} + \frac{6}{10} = \frac{11}{10} = 1.1$$

EXAMPLE 5

Add: $\begin{array}{r} 4.723 \\ .845 \\ \underline{12.535} \end{array}$

Solution.

$$\begin{array}{r} 4.723 \\ .845 \\ 12.535 \\ \hline 18.103 \end{array}$$

Observe that

$$\frac{7}{10} = \frac{70}{100}$$

In decimal notation,

$$.7 = .70$$

In general, you can add any number of 0's to the right of the last digit after the decimal point. Thus

$$.63 = .630 = .6300$$

EXAMPLE 6

Subtract: $.7 - .43$

Solution.

$$\begin{array}{r} .70 \\ -.43 \\ \hline .27 \end{array}$$

Multiplication

$$.3 \times 10 = \frac{3}{10} \times 10 = 3 \qquad [3.]$$

To multiply a decimal by 10, *move the decimal point* 1 *place to the right. To multiply by* 10^n, *move the decimal point n places to the right.*

EXAMPLE 7

Multiply:
(a) $.174 \times 10$
(b) $.174 \times 100$
(c) $.174 \times 1000$

Solution.
(a)
$$.174 \times 10 = 1.74 \qquad [1.74]$$
(b) $100 = 10^2$. Thus
$$.174 \times 100 = 17.4 \qquad [17.4]$$
(c) $1000 = 10^3$. Thus
$$.174 \times 1000 = 174 \qquad [174.]$$

When one or both factors of a product is a decimal, count the total number of decimal places in the factors to place the decimal point in the product. Thus

$$.3 \times .11 = \frac{3}{10} \times \frac{11}{100} = \frac{33}{1000} = .033$$

1 decimal place (.3) ***2 decimal places*** (.11) 10^1 10^2 10^3 ***3 decimal places*** (.033)

Here you *add a 0 immediately to the right of the decimal point* in order to obtain 3 decimal places.

EXAMPLE 8

Multiply:
(a) $4 \times .25$
(b) $.04 \times .025$

Solution.
(a)

$$\begin{array}{rl} .25 & \leftarrow 2 \text{ decimal places} \\ 4 & \leftarrow 0 \text{ decimal places} \\ \hline 1.00 & \leftarrow 2 \text{ decimal places} \end{array}$$

Thus
$$.25 \times 4 = 1$$

(b)

$$\begin{array}{rl} .025 & \leftarrow 3 \text{ decimal places} \\ .04 & \leftarrow 2 \text{ decimal places} \\ \hline .001\,00 & \leftarrow 5 \text{ decimal places} \end{array}$$

Add two 0's immediately to the right of the decimal point to make 5 decimal places in the product. Note that the 0's to the right of the 1 can be dropped because:

$$.00100 = .001$$

Thus

$$.025 \times .04 = .001$$

Division

$$.3 \div 10 = \frac{3}{10} \times \frac{1}{10} = \frac{3}{100} = .03 \qquad [.0\,3]$$

To divide a decimal by 10, *move the decimal point* 1 *place to the left. To divide by* 10^n, *move the decimal point n places to the left.*

EXAMPLE 9

Divide:
(a) $3.92 \div 10$
(b) $3.92 \div 100$
(c) $3.92 \div 1000$

Solution.
(a) $3.92 \div 10 = .392$ [.3 92]
(b) $3.92 \div 100 = .0392$ [.03 92]
(c) $3.92 \div 1000 = .00392$ [.003 92]

To divide by an n-placed decimal, multiply both dividend and divisor by 10^n, so that the divisor will then be an integer.

EXAMPLE 10

Divide: $.027\overline{)16.2}$

Solution.
Note that the divisor, .027, is a 3-placed decimal.

$$\frac{16.2}{.027} = \frac{16.2 \times 1000}{.027 \times 1000} = \frac{16\,200.}{027.}$$

$$\begin{array}{r} 600.\\ 027.\overline{)16\,200.}\\ \underline{16\,2}\\ 00 \end{array}$$

←Line up the decimal
←points in the
quotient and dividend.

The quotient is 600.

EXERCISES

In exercises 1–20, express each fraction as a decimal.

1. $\frac{3}{10}$
2. $\frac{33}{100}$
3. $\frac{-3}{10}$
4. $\frac{33}{10}$
5. $\frac{3}{1000}$
6. $\frac{33}{1000}$
7. $\frac{21}{100}$
8. $\frac{21}{10000}$
9. $\frac{713}{10}$
10. $\frac{713}{100}$
11. $\frac{713}{1000}$
12. $\frac{713}{10000}$
13. $\frac{1}{4}$
14. $\frac{1}{5}$
15. $\frac{-3}{5}$
16. $\frac{5}{8}$
17. $\frac{1}{20}$
18. $\frac{9}{50}$
19. $\frac{-3}{40}$
20. $\frac{3}{200}$

In exercises 21–28, express each fraction as an infinite repeating decimal.

21. $\frac{2}{3}$
22. $\frac{1}{9}$
23. $\frac{1}{11}$
24. $\frac{5}{9}$
25. $\frac{5}{6}$
26. $\frac{4}{3}$
27. $\frac{1}{12}$
28. $\frac{-7}{12}$

In exercises 29–38, express each decimal as a fraction in lowest terms.

29. .9
30. .19
31. .119
32. −.2
33. .02
34. .20
35. .002
36. .75
37. 1.25
38. 10.0003

In exercises 39–44, add:

39. $\begin{array}{r} .3 \\ \underline{.8} \end{array}$

40. $\begin{array}{r} .7 \\ \underline{.9} \end{array}$

41. $\begin{array}{r} .3 \\ .6 \\ \underline{.2} \end{array}$

42. $\begin{array}{r} 1.5 \\ 2.7 \\ \underline{.9} \end{array}$

43. $\begin{array}{r} .9 \\ -.5 \\ -.7 \\ \underline{.8} \end{array}$

44. $\begin{array}{r} .3 \\ -.9 \\ -.7 \\ -.2 \\ \underline{.6} \end{array}$

In exercises 45–48, subtract the bottom decimal from the top one.

45. $\begin{array}{r} .9 \\ \underline{.3} \end{array}$ 46. $\begin{array}{r} .81 \\ \underline{.29} \end{array}$ 47. $\begin{array}{r} .803 \\ \underline{-.085} \end{array}$ 48. $\begin{array}{r} -.93 \\ \underline{-.382} \end{array}$

In exercises 49–58, multiply, as indicated.

49. $.635 \times 10$
50. $.635 \times 100$
51. $.635 \times 1000$
52. $.65 \times 10$
53. $.00065 \times 100$
54. $.4 \times 3$
55. $(-.04) \times .3$
56. $(-.0004) \times (-.3)$
57. $.75 \times .004$
58. $.0075 \times 4000$

In exercises 59–68, divide, as indicated.

59. $\frac{.04}{.02}$
60. $\frac{.0004}{.2}$
61. $\frac{.4}{.0002}$
62. $\frac{-.004}{.02}$
63. $1.5\overline{)1.35}$
64. $.21\overline{)147}$
65. $93\overline{).0186}$
66. $.737\overline{).0008107}$
67. $.038\overline{)19}$
68. $.0072\overline{)14400}$

In exercises 69–80, combine terms, as indicated.

SAMPLE. $.2x + .1x + .4x$	***Solution.*** $\begin{array}{r} .2x \\ .1x \\ \underline{.4x} \\ .7x \end{array}$

69. $.6x + .3x$
70. $.9y - .3y$
71. $.12ab + .34ab + .45ab$
72. $.36x^2 + .05x^2 - .9x^2$

SAMPLE. $(.3x)(.2x^2y)$	***Solution.*** $(.3x)(.2x^2y) = [(.3)(.2)][x(x^2y)]$ $= .06x^3y$

73. $(.3x)(.5y)$
74. $(.2x^2)(.3x)(.9x)$
75. $(.5a)(.1a^2b)(.01b^3)$
76. $(-.2x^2)(-.04x)(-.01x^2)$
77. $\frac{.4x^2}{.2x}$
78. $\frac{.09a^2b}{.3ab}$
79. $\frac{1.6ab^2}{.4ab^2}$
80. $\frac{2.8(x^2 - y^2)}{.14(x - y)}$

What Have You Learned in Chapter 6?

You can add, subtract, multiply and divide fractions and rational expressions.
You can simplify complex expressions.
You can convert fractions to decimals, and decimals to fractions.
And you can combine decimals.

Let's Review Chapter 6.

6.1 Multiplication of Rational Expressions

In exercises 1–4, multiply the given fractions or rational expressions.

1. $\frac{3}{5} \cdot \frac{5}{8}$
2. $\frac{-2}{3} \cdot \frac{-1}{9} \cdot \frac{6}{5}$
3. $\frac{a^2}{bc} \cdot \frac{b^3c}{a}$
4. $\frac{(x+3)^2}{4yz} \cdot \frac{8y^2}{x^2-9}$

6.2 Division of Rational Expressions

In exercises 5–8, divide the given fractions or rational expressions.

5. $\frac{1}{3} \div \frac{-1}{9}$
6. $\frac{12}{25} \div \frac{3}{50}$
7. $\frac{4x^2y^2}{3ab} \div \frac{12xy}{9a^2b}$
8. $\frac{x^2-y^2}{x+2} \div \frac{x+y}{x^2+3x+2}$

6.3 Addition of Rational Expressions with the Same Denominator

In exercises 9–12, add or subtract, and simplify:

9. $\frac{1}{6} + \frac{1}{6}$
10. $\frac{5}{8} - \frac{3}{8}$
11. $\frac{2x}{x+3} + \frac{6}{x+3}$
12. $\frac{x}{x^2-9} - \frac{3}{x^2-9}$

6.4 Least Common Multiples

In exercises 13 and 14, find the *lcm* of the given integers.

13. 10 and 15
14. 2, 4, and 6

In exercises 15 and 16, find the *lcm* of the given polynomials.

15. x^2 and xy
16. $x^2 - 4$ and $x^2 - 3x + 2$

6.5 *Addition and Subtraction*

17. (a) Find *lcd* $\left(\frac{1}{3}, \frac{2}{5}\right)$. (b) Write equal fractions with this *lcd* as denominator.

In exercises 18–20, add or subtract, as indicated.

18. $\frac{3}{4} + \frac{1}{2}$

19. $\frac{3}{x} - \frac{1}{x^2}$

20. $\frac{x}{x+a} + \frac{a}{x^2 - a^2}$

6.6 *Complex Expressions*

In exercises 21–24, simplify each expression:

21. $\dfrac{\frac{3}{8}}{\frac{1}{4}}$

22. $\dfrac{1 - \frac{1}{3}}{\frac{5}{6}}$

23. $\dfrac{\frac{a}{b}}{\frac{c}{b^2}}$

24. $\dfrac{\frac{x-a}{x^2-1}}{\frac{x^2-a^2}{x+1}}$

6.7 *Decimals*

25. Express $\frac{4}{5}$ as a decimal.
26. Add: $.37 + 52 + .08$
27. Multiply: $.38 \times 1.1$
28. Divide: $.63\overline{).00189}$

And these from Chapters 1–5:

29. Simplify: $\frac{-64}{80}$
30. Simplify: $\frac{x^3 - 4x^2}{x^2 - 5x + 4}$
31. Evaluate $6x + .2y$ when $x = .1$ and $y = .2$.
32. Rearrange the following fractions so that you can write "<" between any two of them.

$$\frac{1}{4}, \frac{1}{2}, \frac{3}{8}, \frac{5}{16}, \frac{5}{8}, \frac{7}{16}$$

33. Rearrange the following decimals so that you can write "<" between any two of them.

$$.7, .07, .77, .71, .69, .9$$

34. Rearrange the following numbers so that you can write "<" between any two of them. $\frac{3}{4}, .8, .7, \frac{3}{5}, \frac{2}{3}, .65$

Try These Exam Questions for Practice.

In questions 1–5, combine and simplify.

1. $\frac{3}{4} \cdot \frac{8}{9}$

2. $\frac{x^2}{x+1} \div \frac{x}{x^2-1}$

3. $\frac{1}{2} + \frac{5}{4} - \frac{3}{8}$

4. $\frac{2x}{x-4} - \frac{8}{x-4}$

5. $\frac{1}{10ax^2} + \frac{3}{20a^2x^2} - \frac{1}{a^2x^2}$

6. Simplify: $\dfrac{\frac{a^2}{b}}{\frac{a^3}{b^4}}$

7. Express $\frac{3}{20}$ as a decimal.

8. Multiply: (2.5)(.008)

Equations

7.1 ROOTS

Definition

> *An **equation** is a statement of equality. Thus*
>
> $$x + 2 = 5$$
>
> *is an equation Here $x + 2$ is the **left side** and 5 is the **right side.***

$$3 + 2 = 5$$

is also an equation because it is a statement of equality.

Definition

> *ROOT. A number is a **root** (or **solution**) of an equation (with one variable) if a true statement results when the number replaces the variable.*

EXAMPLE 1

(a) 3 is a root of the equation

$$x + 2 = 5$$

because when you replace x by 3, you obtain the *true* statement

$$3 + 2 = 5.$$

(b) 4 is *not* a root of the equation because

$$4 + 2 = 5$$

is *false*.

In this section, you will *check* whether or not a given number is a root of an equation. *To check this, replace the variable of the equation by the number to see whether a true statement results. If the variable occurs more than once in the equation, each time it occurs, replace it by the same number.*

When you check a root of an equation, write $\stackrel{?}{=}$ instead of =. At the end of the check, write $\stackrel{\checkmark}{=}$ if the number is a root; if not, write $\stackrel{X}{=}$

EXAMPLE 2

Is 10 a root of the equation

$$2x - 7 = x + 3?$$

Solution. Replace each occurrence of x by 10:

$$\begin{aligned} 2(10) - 7 &\stackrel{?}{=} 10 + 3 \\ 20 - 7 &\stackrel{?}{=} 13 \\ 13 &\stackrel{\checkmark}{=} 13 \end{aligned}$$

Thus 10 is a root of the given equation.

EXAMPLE 3

Is -1 a root of the equation

$$1 - 3y = 5 - y?$$

Solution. Replace each occurrence of the variable y by -1:

$$\begin{aligned} 1 - 3(-1) &\stackrel{?}{=} 5 - (-1) \\ 1 + 3 &\stackrel{?}{=} 5 - 1 \\ 4 &\stackrel{X}{=} 6 \end{aligned}$$

Thus -1 is *not* a root of the given equation.

EXAMPLE 4

Is $\frac{1}{4}$ a root of the equation

$$8x + 3 = 6 - 4x?$$

Solution. Replace each occurrence of x by $\frac{1}{4}$.

$$\begin{aligned} 8\left(\frac{1}{4}\right) + 3 &\stackrel{?}{=} 6 - 4\left(\frac{1}{4}\right) \\ 2 + 3 &\stackrel{?}{=} 6 - 1 \\ 5 &\stackrel{\checkmark}{=} 5 \end{aligned}$$

Thus $\frac{1}{4}$ is a root of the given equation.

EXERCISES

Check whether the given number is a root of the equation.

SAMPLE. $2x + 1 = 11;\ \boxed{5}$	***Solution.*** $2(5) + 1 \overset{?}{=} 11$ $11 \overset{\checkmark}{=} 11$ 5 is a root.

1. $x + 4 = 10;\ \boxed{6}$
2. $3x = 6;\ \boxed{2}$
3. $x - 2 = 5;\ \boxed{7}$
4. $4 - x = 2;\ \boxed{6}$
5. $3x + 1 = 4;\ \boxed{1}$
6. $5x - 2 = 12;\ \boxed{2}$
7. $1 - x = -1;\ \boxed{2}$
8. $\frac{x}{4} = 3;\ \boxed{12}$
9. $\frac{x+1}{5} = 2;\ \boxed{10}$
10. $-3x = 9;\ \boxed{3}$
11. $2x + 7 = 1;\ \boxed{-3}$
12. $1 - 5x = 1;\ \boxed{\frac{1}{5}}$
13. $\frac{1}{x} = 4;\ \boxed{4}$
14. $\frac{2}{x} = -2;\ \boxed{-1}$
15. $x + 1 = 2x;\ \boxed{1}$
16. $4x - 2 = 3x;\ \boxed{4}$
17. $1 + x = \frac{x}{2};\ \boxed{3}$
18. $\frac{x-1}{2} = x;\ \boxed{2}$
19. $\frac{1}{x-1} = \frac{x}{5};\ \boxed{\frac{1}{4}}$
20. $x = 5x;\ \boxed{0}$
21. $\frac{1}{x} = \frac{1}{4};\ \boxed{\frac{1}{4}}$
22. $\frac{2}{x} = \frac{1}{2};\ \boxed{4}$
23. $\frac{-3}{x} = \frac{-1}{9};\ \boxed{-30}$
24. $\frac{x}{2} = \frac{1}{4};\ \boxed{\frac{1}{2}}$
25. $x = -x;\ \boxed{-1}$
26. $\frac{x}{3} = x;\ \boxed{0}$
27. $\frac{1}{3x-2} = -1;\ \boxed{-1}$
28. $\frac{1}{\frac{1}{x}} = \frac{1}{4};\ \boxed{\frac{1}{4}}$
29. $2x - 1 = 0;\ \boxed{.5}$
30. $10x + 2 = 9;\ \boxed{.7}$
31. $20x + 1 = 2;\ \boxed{.5}$
32. $25x - 2 = -1;\ \boxed{.04}$

7.2 SOLVING SIMPLE EQUATIONS

Equivalent Equations

Up to now, you have *checked* whether or not a given number is the root of an equation. If you are given an equation, you can find its root(s) by simplifying the equation.

Definition

> *EQUIVALENT EQUATIONS. Equations are said to be* ***equivalent*** *if they have the same root (or roots).*

EXAMPLE 1

$$2x = 8,$$

$$x + 1 = 5,$$

and

$$x = 4$$

are all equivalent. Each has the single root 4. Clearly the last of these is the simplest equation because the variable is on one side all by itself.

Definition

> *To* ***solve an equation*** *means to find its roots.*

Each equation you will be solving in this chapter will have a single root.

You solve an equation by transforming it into simpler and simpler equivalent equations. Your goal is to obtain an equation in which one side is a variable and the other a number. For example, if after simplifying you obtain

$$x = 2,$$

then 2 is the root of the given equation.

Addition Property

When you add the same expression to both sides of an equation, you obtain an equivalent equation.

EXAMPLE 2

Solve:

$$x - 4 = 6$$

Check your answer.

Solution.

$$\begin{aligned} x - 4 &= 6 && \text{Add 4 to both sides.} \\ x - 4 + 4 &= 6 + 4 \\ x &= 10 \end{aligned}$$

The root is 10.

CHECK. Replace x by 10 in the given equation.

$$\begin{aligned} 10 - 4 &\stackrel{?}{=} 6 \\ 6 &\stackrel{\checkmark}{=} 6 \end{aligned}$$

Subtracting a number a is the same as adding a. Thus *when you subtract the same expression from both sides of an equation, you obtain an equivalent equation.*

EXAMPLE 3

Solve and check: $x + 8 = 13$

Solution.

$$x + 8 = 13 \qquad \text{Subtract 8 from both sides.}$$
$$x + 8 - 8 = 13 - 8$$
$$x = 5$$

CHECK:

$$5 + 8 \stackrel{?}{=} 13$$
$$13 \stackrel{\checkmark}{=} 13$$

Multiplication Property

When you multiply or divide both sides of an equation by the same nonzero number, you obtain an equivalent equation.

EXAMPLE 4

Solve: $\frac{x}{3} = 6$

Solution.

$$\frac{x}{3} = 6 \qquad \text{Multiply both sides by 3.}$$
$$\frac{x}{3} \cdot 3 = 6 \cdot 3$$
$$x = 18$$

EXAMPLE 5

Solve: $5y = 20$

Solution.

$$5y = 20 \qquad \text{Divide both sides by 5.}$$
$$\frac{5y}{5} = \frac{20}{5}$$
$$y = 4$$

The Multiplication Property enables you to multiply or divide both sides of an equation by a *nonzero* number. You cannot multiply or divide both sides by 0. Also, complications may arise if you multiply or divide by a *variable.*

EXAMPLE 6

The equation

$$1 = 2$$

is obviously a *false* statement.

(a) If you multiply both sides of the equation by 0, you obtain the *true* statement:

$$1 \cdot 0 = 2 \cdot 0$$

or

$$0 = 0$$

Thus if you multiply both sides of an equation by 0, you always obtain a true statement, even if the original statement is false.

(b) If you multiply both sides of the equation

$$1 = 2$$

by x, you obtain

$$1x = 2x.$$

Note that 0 is a root of this equation because

$$1 \cdot 0 = 2 \cdot 0.$$

But 0 was not a root of the original equation

$$1 = 2.$$

Thus multiplying both sides of an equation by a variable may add extra roots.

Using Both Properties

Both the Addition and Multiplication Properties frequently apply in the same example. *In simplifying an equation, bring variables to one side, numerical terms to the other.*

EXAMPLE 7

Solve: $2x - 1 = 14 - 3x$

Solution. Here, bring variables to the left, numerical terms to the right.

$$2x - 1 = 14 - 3x \qquad \text{Add } 3x + 1 \text{ to both sides.}$$
$$2x - 1 + 3x + 1 = 14 - 3x + 3x + 1$$
$$5x = 15 \qquad \text{Divide both sides by 5.}$$
$$\frac{5x}{5} = \frac{15}{5}$$
$$x = 3$$

EXAMPLE 8

Solve and check: $1 - 4x = 2x - 2$

Solution. Here, bring variables to the right, numerical terms to the left.

$$\begin{aligned} 1 - 4x &= 2x - 2 && \text{Add } 4x + 2 \text{ to both sides.} \\ 1 - 4x + 4x + 2 &= 2x - 2 + 4x + 2 \\ 3 &= 6x && \text{Divide both sides by 6.} \\ \frac{3}{6} &= \frac{6x}{6} \\ \frac{1}{2} &= x \end{aligned}$$

CHECK.

$$\begin{aligned} 1 - 4\left(\frac{1}{2}\right) &\stackrel{?}{=} 2\left(\frac{1}{2}\right) - 2 \\ 1 - 2 &\stackrel{\checkmark}{=} 1 - 2 \\ [\text{or} \quad -1 &\stackrel{\checkmark}{=} -1] \end{aligned}$$

EXERCISES

In exercises 1–26, solve the equation.

1. $x - 1 = 6$
2. $y - 3 = 4$
3. $x + 5 = 8$
4. $x + 10 = 2$
5. $\frac{x}{2} = 7$
6. $\frac{t}{5} = 6$
7. $2x = 4$
8. $7x = 21$
9. $-3y = -9$
10. $-2z = 6$
11. $\frac{x}{4} = -16$
12. $\frac{x}{6} = \frac{1}{2}$
13. $\frac{x}{5} = \frac{-1}{5}$
14. $t + 3 = -2$
15. $-t = 7$
16. $6 - t = -9$
17. $2t + 3 = 11$
18. $\frac{t}{2} + 1 = 7$
19. $\frac{t + 1}{2} = 5$
20. $\frac{x - 1}{3} = 4$
21. $5x - 3 = 7$
22. $\frac{u - 4}{3} = 2u$
23. $4x - 1 = 3 - 2x$
24. $\frac{x}{2} - 4 = 1 - \frac{x}{2}$
25. $3x - 1 + \frac{x}{2} = 2 - x$
26. $4x + 2x + 1 = 1 - 2x + 4$

In exercises 27–38, (a) solve the equation and (b) check the root.

27. $3x = 12$

28. $x - 5 = 9$

29. $\frac{x}{6} = -2$

30. $5 - x = 1$

31. $2x + 1 = 3$

32. $7 - 3x = 1$

33. $1 + 4x = 2$

34. $2x - 5 = x + 5$

35. $6x - 3 = 5x + 2$

36. $\frac{x+1}{4} = x - 5$

37. $9 - 3x = -2x + 3$

38. $1 + 7x = 4x + 4$

7.3 EQUATIONS WITH PARENTHESES

Parentheses are necessary to clarify the meaning of arithmetic and algebraic expressions. Often one or both sides of an equation contains parentheses.

EXAMPLE 1

Solve: $4 - (2x - 8) = 2x$

Solution.

$$4 - (2x - 8) = 2x$$

Use the Distributive Law to remove parentheses on the left.

$$4 - 2x + 8 = 2x$$

Next, simplify the left side.

$$12 - 2x = 2x$$
$$12 - 2x + 2x = 2x + 2x$$
$$12 = 4x$$
$$\frac{12}{4} = \frac{4x}{4}$$
$$3 = x$$

In the next example, divide both sides by a factor, rather than simplify the left side (by means of the Distributive Law).

EXAMPLE 2

Solve: $4(5x + 2) = -8$

Solution.

$$4(5x + 2) = -8$$

Divide both sides by 4.

$$\frac{4(5x + 2)}{4} = \frac{-8}{4}$$
$$5x + 2 = -2$$
$$5x + 2 - 2 = -2 - 2$$
$$5x = -4$$
$$\frac{5x}{5} = \frac{-4}{5}$$
$$x = \frac{-4}{5}$$

EXAMPLE 3

Solve and check: $\frac{1}{3}[x - 2(x + 3)] = -1$

Solution.

$$\frac{1}{3}[x - 2(x + 3)] = -1 \quad \text{Multiply both sides by 3.}$$
$$3 \cdot \frac{1}{3}[x - 2(x + 3)] = 3(-1)$$
$$x - 2(x + 3) = -3 \quad \text{Remove parentheses on the left. (Use the Distributive Law.)}$$
$$x - 2x - 6 = -3 \quad \text{Simplify the left side.}$$
$$-x - 6 = -3$$
$$-x - 6 + 6 = -3 + 6$$
$$-x = 3 \quad \text{Multiply both sides by } -1.$$
$$x = -3$$

CHECK.

$$\frac{1}{3}[-3 - \underbrace{2(\underbrace{(-3) + 3}_{0})}_{0}] \stackrel{?}{=} -1$$
$$\frac{1}{3}(-3) \stackrel{?}{=} -1$$
$$-1 \stackrel{\checkmark}{=} -1$$

EXAMPLE 4

Solve:

$$(x + 1)(x + 2) = (1 - x)^2 + 11$$

Solution.

$$(x + 1)(x + 2) = (1 - x)^2 + 11 \quad \text{Multiply out.}$$
$$x^2 + 3x + 2 = 1 - 2x + x^2 + 11$$
$$x^2 + 3x + 2 - x^2 + 2x - 2 = 1 - 2x + x^2 + 11 - x^2 + 2x - 2$$
$$5x = 10$$
$$x = 2$$

EXERCISES

In exercises 1–30, solve each equation.

1. $7 - (x + 2) = 1$
2. $4 - (y - 2) = 8$
3. $5 - (9 - x) = \frac{x}{3}$
4. $1 - 3(x - 1) = 2$
5. $6(t + 1) = 12$
6. $8 - 2(z - 2) = 4$
7. $5(t + 2) = t + 2$
8. $y + 6 = 10(y + 2) - 5$
9. $2(x + 3) = 3x + 5$
10. $4\left(u + \frac{1}{2}\right) = 6u + 1$

11. $7\left(w - \frac{2}{3}\right) = 6w - 3$

12. $2\left(x - \frac{3}{4}\right) = -4x$

13. $3y - 5 = 1 - (y - 12)$

14. $-7(t + 5) = t - 3$

15. $(6 + 5u) - (3u - 2) = -1$

16. $5 - (z - 3) + 6(z - 1) = 0$

17. $1 - [x - (1 - x)] = 0$

18. $2[4 - (3z - 5)] = 3z$

19. $\frac{1}{2}[y - (5 - 2y)] = 1$

20. $-2[x - (4x - 5)] = 5x$

21. $2\left(7 - \frac{2x - 1}{3}\right) = \frac{x}{2}$

22. $x - [3 - (2x - 5)] = 0$

23. $2y - (1 - [y - (3 - 2y)]) = 1$

24. $5(x - 2) - 3(10 - x) = 0$

25. $x^2 - (x + 1)^2 = 3$

26. $(x + 2)^2 - (x + 1)^2 = 9$

27. $(x + 7)^2 = (x + 5)^2 + 28$

28. $(y + 1)^2 - 3 = (y - 2)^2$

29. $(y + 2)(y + 3) = (y + 1)(y - 2)$

30. $t^2 = (t - 5)^2 - 25$

In exercises 31–40, solve and check.

31. $x - (12 - 2x) = 0$

32. $10 - (y - 6) = y + 2$

33. $-3(y + 2) = y - 10$

34. $4 - (z + 1) = 2 - 3z$

35. $1 - [x - (2 - 3x)] = 4$

36. $2x + 3(x - 2) = 19$

37. $(x + 4)^2 = x^2$

38. $(y + 7)^2 = (y + 4)^2 + 12y + 3$

39. $(2 - x)^2 + 5 = x^2 + 1$

40. $[4 - (6 - x)]^2 = x^2 - 8$

7.4 EQUATIONS WITH RATIONAL EXPRESSIONS

Proportions

Definition

*A **proportion** is a statement that two fractions are equal.*

Thus the statements

$$\frac{2}{4} = \frac{1}{2}$$

and

$$\frac{9}{12} = \frac{3}{4}$$

are each proportions.

Proportions are generally written in the form

$$\frac{a}{b} = \frac{c}{d},$$

and are often read,

"a is to b as c is to d."

EXAMPLE 1

Find the value of a in the proportion:

$$\frac{a}{10} = \frac{6}{15}$$

Solution.

$$\frac{a}{10} = \frac{6}{15} \qquad \text{Multiply both sides by 10.}$$

$$a = \frac{\overset{2}{\cancel{6}} \cdot \overset{2}{\cancel{10}}}{\underset{\underset{1}{\cancel{3}}}{\cancel{15}}}$$

$$a = 4$$

EXAMPLE 2

Find the value of b in the proportion:

$$\frac{8}{b} = \frac{2}{3}$$

Solution.

$$\frac{8}{b} = \frac{2}{3} \qquad \text{Cross-multiply. } \frac{8}{b} \times \frac{2}{3}$$

$$8 \cdot 3 = 2b \qquad \text{Divide by 2.}$$

$$4 \cdot 3 = b$$

$$12 = b$$

Proportions occur naturally in everyday situations.

EXAMPLE 3

The width and length of an 8×10 photograph are enlarged proportionally. If the length of the enlargement is 15 inches, what is its width?

Solution. The width of the original photograph is 8 inches; the length is 10 inches. Set up the proportion:

$$\frac{8}{10} = \frac{c}{15} \qquad \text{Cross-multiply.}$$

$$8 \cdot 15 = 10c$$

$$120 = 10c$$

$$12 = c$$

The width of the enlargement is 12 inches. (See Fig. 7.1 on page 187.)

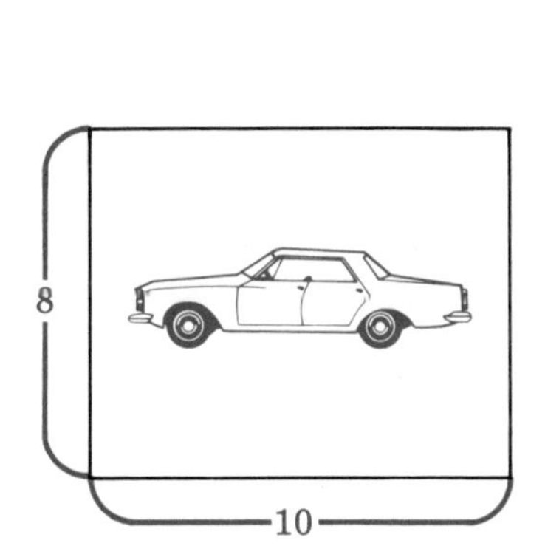

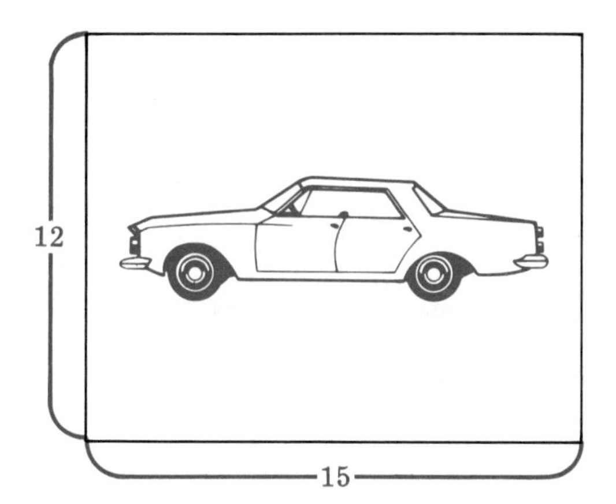

FIGURE 7.1

Numerical Denominators

Suppose

$$\frac{a}{b} = \frac{c}{d}.$$

Then you can *cross-multiply* to obtain

$$ad = bc.$$

EXAMPLE 4

Solve: $\frac{x}{2} = \frac{x+6}{5}$

Solution.

$$\frac{x}{2} = \frac{x+6}{5} \qquad \text{Cross-multiply.}$$
$$5x = 2(x+6)$$
$$5x = 2x + 12$$
$$3x = 12$$
$$x = 4$$

Note that when you cross-multiplied, you actually multiplied both sides by $2 \cdot 5$. In fact, $2 \cdot 5$ is the *lcd* of $\frac{x}{2}$ and $\frac{x+6}{5}$. Equations that involve fractions or rational expressions can often be simplified by multiplying both sides by the *lcd* of these expressions.

EXAMPLE 5

Solve: $\frac{x}{2} = \frac{x}{3} + 1$

Solution.

$$lcd\left(\frac{x}{2}, \frac{x}{3}, 1\right) = 6$$

$$\begin{aligned} \frac{x}{2} &= \frac{x}{3} + 1 \\ 6 \cdot \frac{x}{2} &= 6\left(\frac{x}{3} + 1\right) \\ 3x &= 2x + 6 \\ x &= 6 \end{aligned}$$

EXAMPLE 6

Solve and check: $\frac{2y-4}{8} = \frac{3y+2}{4}$

Solution.

$$lcd\left(\frac{2y-4}{8}, \frac{3y+2}{4}\right) = 8$$

$$\begin{aligned} \frac{2y-4}{8} &= \frac{3y+2}{4} \\ 8\left(\frac{2y-4}{8}\right) &= 8\left(\frac{3y+2}{4}\right) \\ 2y - 4 &= 2(3y+2) && \text{Factor the left side.} \\ 2(y-2) &= 2(3y+2) \\ y - 2 &= 3y + 2 \\ y - 2 - y - 2 &= 3y + 2 - y - 2 \\ -4 &= 2y \\ -2 &= y \end{aligned}$$

CHECK.

$$\begin{aligned} \frac{2(-2)-4}{8} &\stackrel{?}{=} \frac{3(-2)+2}{4} \\ \frac{-8}{8} &\stackrel{?}{=} \frac{-4}{4} \\ -1 &\stackrel{\checkmark}{=} -1 \end{aligned}$$

Polynomial Denominators

EXAMPLE 7

Solve: $\frac{t}{t+3} = 2$

Solution.

$$\begin{aligned} \frac{t}{t+3} &= 2 && \text{Multiply both sides by } t+3, \text{ the } lcd \text{ [or cross-multiply].} \\ t &= 2(t+3) && \text{Use the Distributive Law.} \\ t &= 2t + 6 \\ -6 &= t \end{aligned}$$

EXAMPLE 8

Solve and check: $\frac{4}{y+3} = \frac{1}{y-3}$

Solution.

$$\frac{4}{y+3} = \frac{1}{y-3}$$

Multiply both sides by $(y+3)(y-3)$, the *lcd* [or cross-multiply].

$$\begin{aligned} 4(y-3) &= 1(y+3) \\ 4y - 12 &= y + 3 \\ 3y &= 15 \\ y &= 5 \end{aligned}$$

CHECK.

$$\frac{4}{5+3} \stackrel{?}{=} \frac{1}{5-3}$$

$$\frac{4}{8} \stackrel{?}{=} \frac{1}{2}$$

$$8 \stackrel{\checkmark}{=} 8$$

Cross-multiply.

EXAMPLE 9

Solve: $\frac{6}{x} + \frac{4}{x+1} = \frac{9}{x}$

Solution.

$$lcd\left(\frac{6}{x}, \frac{4}{x+1}, \frac{9}{x}\right) = x(x+1)$$

$$\begin{aligned} \frac{6}{x} + \frac{4}{x+1} &= \frac{9}{x} \\ \left(\frac{6}{x} + \frac{4}{x+1}\right)x(x+1) &= \frac{9}{x}\,x(x+1) \\ 6(x+1) + 4x &= 9(x+1) \\ 6x + 6 + 4x &= 9x + 9 \\ 10x + 6 &= 9x + 9 \\ x &= 3 \end{aligned}$$

Simplify.

Equations With Decimals

When decimals occur in an equation, it is best to multiply both sides of the equation by a power of 10, and thus eliminate decimals from the equation.

EXAMPLE 10

Solve: $.06x + .055x = 230$

Solution. Write

$$.06 = .060$$

Thus

$$.060x + .055x = 230$$

Multiply both sides by 1000 (or 10^3).

$$1000(.060x + .055x) = 1000 \times 230$$
$$60x + 55x = 230000$$
$$115x = 230000$$
$$x = 2000$$

EXERCISES

In exercises 1–8, find the value of a, b, c, or d in each proportion.

1. $\frac{a}{9} = \frac{2}{3}$
2. $\frac{1}{b} = \frac{4}{20}$
3. $\frac{12}{30} = \frac{c}{5}$
4. $\frac{7}{11} = \frac{-21}{d}$
5. $\frac{9}{b} = \frac{6}{10}$
6. $\frac{a}{12} = \frac{-25}{30}$
7. $\frac{-15}{9} = \frac{10}{d}$
8. $\frac{108}{144} = \frac{c}{48}$
9. A salesman receives a commission of $3.60 on a $200 sale. At this rate how much does he receive on a $450 sale?
10. A hostess makes 40 cups of coffee for 28 guests. How many cups should she make for 42 guests?
11. A typist makes 3 errors on 5 pages. At this rate how many errors will she make on 30 pages?
12. Grapefruits sell at three for 80¢. How much does a dozen grapefruits cost?
13. Which is a better buy—a 6-ounce box of soap flakes at 75¢ or an 8-ounce box at 96¢?
14. Which is a better buy—a 12-ounce can of tomatoes at 54¢ or a pound-can at 75¢?

In exercises 15–50, solve each equation.

15. $\frac{x}{4} = \frac{x-2}{2}$
16. $\frac{x+2}{5} = \frac{x}{4}$
17. $\frac{y}{10} = \frac{y-10}{5}$
18. $\frac{x-2}{5} = x + 2$
19. $\frac{2x}{7} = \frac{x+3}{5}$
20. $\frac{t}{2} = \frac{t-2}{3}$
21. $\frac{y+5}{5} = \frac{y-5}{15}$
22. $\frac{x+5}{2} = \frac{1-x}{4}$

23. $\frac{x}{2} - 3 = \frac{x}{5}$

24. $\frac{x}{3} - 1 = \frac{x}{4}$

25. $\frac{x}{5} + \frac{x}{2} = 7$

26. $\frac{x}{3} - \frac{x}{9} = 2$

27. $\frac{x+1}{2} + \frac{x-1}{4} = 4$

28. $\frac{x}{3} + \frac{x+2}{4} = 4$

29. $\frac{2x}{3} - 1 = \frac{x+1}{2}$

30. $\frac{x+5}{2} - \frac{x}{7} = 0$

31. $\frac{y+3}{5} = \frac{2y+3}{7}$

32. $\frac{z+2}{4} = \frac{z-1}{3}$

33. $\frac{z-10}{10} = \frac{z-8}{12}$

34. $\frac{z-2}{3} = \frac{4-z}{4}$

35. $\frac{u}{u-1} = 2$

36. $\frac{v}{v+6} = -1$

37. $\frac{x}{x+10} = 0$

38. $\frac{y+1}{y-2} = 2$

39. $\frac{z-1}{z+1} = 2$

40. $\frac{z+4}{z-4} = 3$

41. $\frac{4}{y+1} = \frac{1}{y-5}$

42. $\frac{2}{u-3} = \frac{5}{u}$

43. $\frac{-1}{u+2} = \frac{2}{u+5}$

44. $\frac{3}{u-4} = \frac{2}{u-1}$

45. $\frac{1}{z} - \frac{1}{2} = \frac{1}{3}$

46. $\frac{7}{t} - \frac{3}{2} = \frac{5}{2t}$

47. $\frac{4}{x-2} - \frac{1}{x} = \frac{5}{x-2}$

48. $\frac{18}{v+2} - \frac{5}{v+1} = \frac{10}{v+1}$

49. $\frac{9}{y^2-1} - \frac{3}{y+1} = \frac{-4}{y-1}$

50. $\frac{2}{t+2} - \frac{10}{t^2-4} = \frac{1}{2-t}$

In exercises 51–56, solve and check:

51. $\frac{2x}{5} = \frac{x+1}{3}$

52. $\frac{x}{4} + \frac{x}{2} = 9$

53. $\frac{t}{t-9} = 2$

54. $\frac{y+3}{y-2} = 2$

55. $\frac{2}{u+2} = \frac{5}{1-u}$

56. $\frac{2}{u} + \frac{2}{u+3} = \frac{6}{u+3}$

In exercises 57–64, solve each equation.

57. $.2x + .5x = 14$

58. $x + .4x = 42$

59. $.23x + .15x = 7.6$

60. $.001x + .012x = 390$

61. $.06x + .075x = 27$

62. $.3y + 9 = 5.1$

63. $.005t - .01t = 1$

64. $5u = 1.1 - .5u$

7.5 LITERAL EQUATIONS

Definition

> *A **literal equation** is one that contains at least two variables and that must be solved for one variable in terms of the others.*

EXAMPLE 1

$$x + y = 5$$

is a literal equation. You can solve for x in terms of y by subtracting y from both sides. Thus

$$x = 5 - y$$

You can also solve for y in terms of x by subtracting x from both sides of the given equation:

$$y = 5 - x$$

To solve a literal equation, bring terms containing *the variable for which you are solving* to one side. Bring *all other terms* to the other side.

EXAMPLE 2

Solve

$$5x - 3a + 1 = 0$$

for x. Check your result.

Solution.

$$5x - 3a + 1 = 0$$

Leave $5x$ on the left side. Bring all other terms to the right side. Thus add $3a - 1$ to both sides.

$$\begin{aligned} 5x - 3a + 1 + 3a - 1 &= 0 + 3a - 1 \\ 5x &= 3a - 1 \\ \frac{5x}{5} &= \frac{3a - 1}{5} \\ x &= \frac{3a - 1}{5} \end{aligned}$$

CHECK. Substitute $\frac{3a - 1}{5}$ for x in the given equation:

$$\begin{aligned} 5\left(\frac{3a - 1}{5}\right) - 3a + 1 &\stackrel{?}{=} 0 \\ 3a - 1 - 3a + 1 &\stackrel{?}{=} 0 \\ 0 &\stackrel{\checkmark}{=} 0 \end{aligned}$$

EXAMPLE 3

Solve

$$4x - 3y = 12$$

(a) for x,
(b) for y.

Solution.

(a)

$$4x - 3y = 12 \qquad \text{Add } 3y \text{ to both sides.}$$

$$4x = 12 + 3y$$

$$\frac{4x}{4} = \frac{12 + 3y}{4}$$

$$x = \frac{12 + 3y}{4}$$

(b)

$$4x - 3y = 12 \qquad \text{Subtract } 4x \text{ from both sides.}$$

$$-3y = 12 - 4x$$

$$\frac{-3y}{-3} = \frac{12 - 4x}{-3}$$

$$y = \frac{4x - 12}{3}$$

Note that
$12 - 4x = -(4x - 12)$

EXAMPLE 4

Solve

$$2x - 5y + 3z = 20$$

for z.

Solution.

$$2x - 5y + 3z = 20$$

Bring the terms containing x and y to the right side. Thus add $-2x + 5y$ to both sides.

$$3z = 20 - 2x + 5y$$

$$z = \frac{20 - 2x + 5y}{3}$$

EXAMPLE 5

Solve

$$xy + x = 1$$

for x.

Solution. Here both terms containing x are on the left. In order to bring y to the right, first factor the polynomial on the left.

$$xy + x = x(y + 1)$$

Thus the given equation becomes

$$x(y + 1) = 1 \qquad \text{Divide both sides by } y + 1.$$

$$x = \frac{1}{y + 1}$$

In Example 5, when you divide both sides by $y + 1$, you must check that $y + 1 \neq 0$ (because division by 0 is undefined), and thus that $y \neq -1$. But if $y = -1$, then from the given equation,

$$\underbrace{x(-1) + x}_{0} = 0$$
$$= 1$$

This cannot be. Thus $y + 1 \neq 0$.

EXAMPLE 6

The formula
$$C = \frac{5}{9}(F - 32)$$

relates the Centigrade and Fahrenheit temperature scales. Here C stands for degrees Centigrade and F for degrees Fahrenheit.

(a) Find C when F is 212, the boiling point of water.
(b) Solve for F in terms of C.
(c) Find F when $C = 37$.

Solution.

(a) Substitute 212 for F in the given formula.

$$C = \frac{5}{9}(212 - 32)$$
$$C = \frac{5}{\cancel{9}} \cdot \overset{20}{\cancel{180}}$$
$$C = 100$$

(b)
$$C = \frac{5}{9}(F - 32). \quad \text{Divide both sides by } \frac{5}{9}$$
$$\frac{9}{5}C = \frac{9}{5} \cdot \frac{5}{9}(F - 32) \quad \left(\text{or multiply by } \frac{9}{5}\right).$$
$$\frac{9}{5}C = F - 32$$
$$\frac{9}{5}C + 32 = F$$

(c) Substitute 37 for C in the formula obtained in part (b).

$$\frac{9}{5} \cdot 37 + 32 = F$$
$$\frac{333}{5} + 32 = F$$
$$66.6 + 32 = F$$
$$98.6 = F$$

***FIGURE 7.2(a). Degrees Farenheit* (F)**

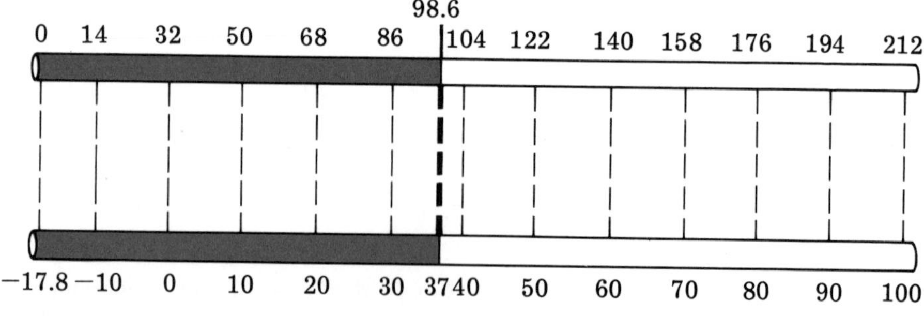

***FIGURE 7.2(b). Degrees Centigrade* (C), $C = \frac{5}{9}(F - 32)$**

Thus 37° Centigrade corresponds to normal body temperature, 98.6° Fahrenheit.

EXERCISES

In exercises 1–12, solve for x in terms of the other variable(s).

1. $5x = y$
2. $6x = 3y$
3. $x + 2 = y$
4. $x - 3 = 2y$
5. $x + y + 4 = 0$
6. $3x - 2y + 1 = 0$
7. $2x - 5t = 7x + 2t$
8. $x + 2t - 1 = 5x + 2$
9. $x + y + z = 1$
10. $3x - y + 2z = 0$
11. $5x + y - z = 2x + y - 3z$
12. $x - 1 + y = z + 3(x - 1)$

In exercises 13–20, solve for the indicated variable.

13. $2x + 3y = 5$ (for y)
14. $5a - 4b = 10$ (for a)
15. $2u - 3v = 6$ (for v)
16. $6m - \frac{n}{2} = 5$ (for n)
17. $2a + b - 3c = 12$ (for a)
18. $4 - 3s + 2t = 5$ (for s)
19. $xy - y = 3$ (for y)
20. $ab + 2a = 1$ (for b)

In exercises 21–32, solve for the indicated variables.

21. $5x - y = 20$ (a) for x, (b) for y
22. $3a - 4b = 7$ (a) for a, (b) for b
23. $5(c - 2b) = 1 + c$ (a) for b, (b) for c
24. $s - 3t = 2s + 3t$ (a) for s, (b) for t
25. $2y - z + 1 = 10z - y$ (a) for y, (b) for z
26. $u - 3v + 2 = 2u - v$ (a) for u, (b) for v
27. $xy + 4y = 1$ (a) for x, (b) for y
28. $3m - mn = 2n$ (a) for m, (b) for n
29. $5x - y + 2z = 10$ (a) for x, (b) for y

30. $u - 2v + 6w = 7$ (a) for v, (b) for w

31. $\frac{a}{b} = 1$ (a) for a, (b) for b

32. $\frac{2x}{y} = -3$ (a) for x, (b) for y

In exercises 33–38: (a) solve for the indicated variable, and (b) check your answer.

33. $2x - y = 10$ (for x)

34. $5a - 2b = 4$ (for b)

35. $\frac{3}{u} = \frac{v}{2}$ (for v)

36. $\frac{12}{x} = \frac{y}{5}$ (for y)

37. $x - y + 2z = 8$ (for y)

38. $\frac{a}{2} + \frac{b}{3} - \frac{c}{4} = 12$ (for c)

39. The formula

$$A = lw$$

expresses the area, A, of a rectangle in terms of the length, l, and width, w. Solve for l.

40. The formula

$$C = 2\pi r$$

expresses the circumference, C, of a circle in terms or r, the length of the radius. Solve for r.

41. (a) The formula

$$A = \frac{bh}{2}$$

expresses the area, A, of a triangle in terms of b, the length of the base, and h, the length of the altitude. Solve for h.

(b) Find h when $A = 100$ square inches and $b = 4$ inches.

42. (a) The formula

$$P = 2l + 2w$$

expresses the perimeter, P, of a rectangle in terms of the length, l, and width, w. Solve for l.

(b) Find l when $P = 60$ feet and $w = 10$ feet. (See Figure 7.3.)

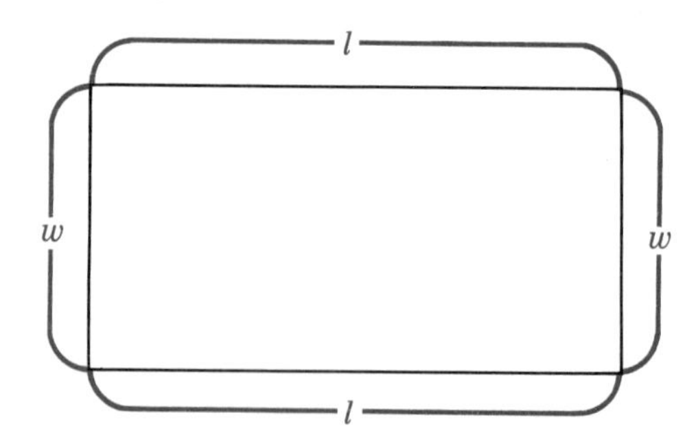

FIGURE 7.3. $p = 2l + 2w$

43. (a) The formula $$d = rt$$

expresses the distance, d, that can be traveled at a constant rate, r, over a period of time, t. Solve for r.

(b) Find r when $d = 300$ miles and $t = 5$ hours.

44. (a) The formula $$V = lwh$$

expresses the volume, V, of a rectangular box in terms of the length, l, the width, w, and the height, h. Solve for h. (See Figure 7.4.)

(b) Find h when $V = 2000$ cubic inches, $l = 20$ inches, and $w = 5$ inches.

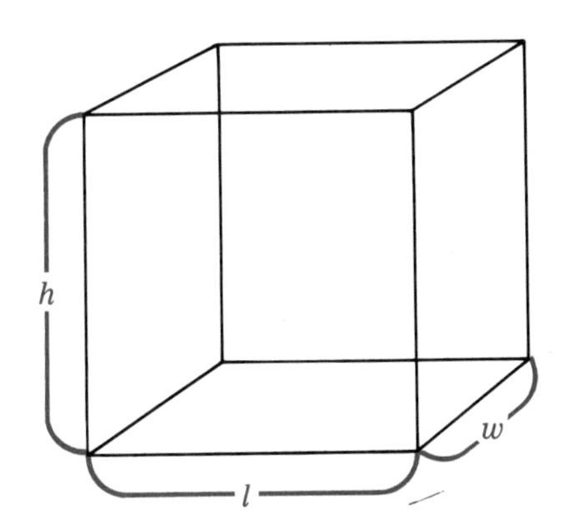

FIGURE 7.4. ***$V = lwh$***

What Have You Learned in Chapter 7?

You have learned that an equation is a statement of equality, and that a number is called a root of an equation if a true statement results when the number replaces the variable.

You can solve equations such as

$$2x - 5 = 3x + 4,$$

$$4x - 2(x + 1) = 7,$$

and

$$\frac{x+4}{6} = \frac{2x+1}{5}.$$

And if you are given a literal equation, such as

$$2x - 3y = 6,$$

you can solve for either one of the variables in terms of the other.

Let's Review Chapter 7.

7.1 Roots

In exercises 1–4, check whether the given number is a root of the equation.

1. $3x + 2 = 11$; $\boxed{3}$
2. $5x - 2 = 2x + 3$; $\boxed{2}$
3. $\frac{x+2}{3} = \frac{x-5}{5}$; $\boxed{10}$
4. $\frac{4}{x} = 16$; $\boxed{\frac{1}{4}}$

7.2 Solving Simple Equations

In exercises 5–7, solve each equation.

5. $3y = -9$
6. $\frac{t-2}{5} = 2$
7. $4y + 1 = 8y - 1$
8. Solve and check: $10 - 4x = 3x - 4$

7.3 Equations with Parentheses

In exercises 9–11, solve each equation.

9. $x - (8 - x) = 16$
10. $4(x - 3) - 3(x - 4) = 6$
11. $(t + 1)^2 - (t + 2)^2 = 3$
12. Solve and check:

$$2x - [7 - (x - 2)] = 6$$

7.4 *Equations with Rational Expressions*

13. Find the value of b in the proportion

$$\frac{3}{b} = \frac{2}{6}.$$

In exercises 14–16, solve each equation.

14. $\frac{x}{7} + \frac{x}{2} = 9$ 15. $\frac{1}{y} + \frac{1}{2} = \frac{3}{4}$ 16. $.3x + .02x = 96$

7.5 *Literal Equations*

17. Solve for x: $2x - y = 3$
18. Solve for y: $3 - 2x + 7y - 9 = 0$
19. Solve $\frac{3a}{b} = -2$ (a) for a, (b) for b.
20. (a) Solve $7u - 3v = 0$ for v. (b) Check your answer.

And these from Chapters 1–6:

21. Simplify: $8 - [(3 - 5) - (7 - 2)] + 5 - (8 - 10)$
22. Simplify: $x^2 - [(1 - x^2) + 4 - 3(x - 2)^2]$
23. Evaluate $3(x + 4) + 2x - 5$, when $x = -3$.
24. Simplify: $$\frac{\frac{x-2}{4}}{\frac{x^2-4}{2}}$$
25. Factor: $t^2 - 7t + 10$
26. Rearrange the following decimals so that you can write "<" between any two of them.

.9, .89, .98, .899, .909, .91

Try These Exam Questions for Practice.

1. Check whether 8 is a root of the equation

$$\frac{x-2}{2} = 3.$$

2. Solve for y: $y - 3 = -2$
3. Solve for x and check: $2x + 5 = 5x - 1$
4. Solve for t: $4t - (2t - 1) = 2$
5. Solve for x: $\frac{4}{x} + \frac{1}{2} = \frac{5}{6}$
6. Solve for u: $.4u + .5u = .36$
7. Solve for y: $x + 2y - 32 = 1$

Word Problems

8.1 INTEGER PROBLEMS

Why Study Integers?

Many practical problems are solved by algebraic methods. Indeed, one of your main goals in this course is to learn how to apply mathematical methods to situations that arise in everyday life. When a problem is stated in words, you must translate the problem into mathematical symbols.

The first type of problem concerns the integers:

$$\ldots -3, -2, -1, 0, 1, 2, 3, \ldots$$

Of course, you do not *directly* worry about integers in everyday conversation. But you constantly *use* integers in describing other concepts, such as money, time, distance, and temperature. For example, you use

100	to describe earning \$100,
−100	to describe losing \$100,
32°F	to describe the freezing point of water,
−10°F	to describe the temperature in Minneapolis on a January morning.

Furthermore, integer problems serve to develop your ability to apply mathematics to other situations.

Translating Words to Symbols

EXAMPLE 1

Let x represent an integer. Express each of the following in mathematical symbols:

(a) one more than the integer
(b) two less than the integer
(c) twice the integer
(d) three more than half the integer
(e) half of three more than the integer

Solution.

(a)

$$\underbrace{\text{one more than}}_{1\,+}\ \underbrace{\text{the integer}}_{x}$$

or

$$x + 1$$

(b)

$$\underbrace{\text{two less than}}_{-2\,+}\ \underbrace{\text{the integer}}_{x}$$

or

$$x - 2$$

(c)

$$\underbrace{\text{twice the integer}}_{2x}$$

(d)

$$\underbrace{\text{three more than}}_{3\,+}\ \underbrace{\text{half the integer}}_{\frac{1}{2}x}$$

or

$$\frac{x}{2} + 3$$

Here you first halve the integer; then you add three.

(e)

$$\underbrace{\text{half of}}_{\frac{1}{2}}\ \underbrace{\text{three more than}}_{(3\,+}\ \underbrace{\text{an integer}}_{x)}$$

or

$$\frac{x+3}{2}$$

Here you first add three to the integer; then you halve it. Do you see that

$$\frac{x}{2} + 3 \neq \frac{x+3}{2} \text{ ?}$$

Recall that $\neq$ is read: *is not equal to*

EXAMPLE 2

Express each in mathematical symbols:
(a) A certain integer is increased by 10.
(b) Twice an integer is decreased by 3.
(c) The sum of two consecutive integers

Solution.
(a) Let x represent the integer.

$$\underbrace{\text{A certain integer}}_{x}\ \underbrace{\text{is increased by 10}}_{+\,10}$$

(b) Let y represent the integer.

$$\underbrace{\text{Twice an integer}}_{2y}\ \underbrace{\text{is decreased by 3}}_{-3}$$

(c) *Consecutive integers differ by* 1. For example, 5 and 6 are consecutive integers. If z represents an integer, the next consecutive integer is $z + 1$.

$$\underbrace{\text{The sum of two consecutive integers}}_{z + (z + 1)}$$

or

$$2z + 1$$

EXAMPLE 3

(a) The sum of two integers is 20. If x represents one of these integers, express the other in terms of x.
(b) The larger of two integers is 4 more than the smaller. If y represents the smaller integer, express the larger one in terms of y.

Solution.
(a) Let x represent one integer. Let $\square$ represent the other integer.

$$\underbrace{\text{The sum of two integers}}\ \text{is}\ 20$$

$$x + \square = 20$$
$$x - x + \square = 20 - x \qquad \text{Subtract } x \text{ from each side.}$$
$$\square = 20 - x$$

Thus $20 - x$ represents the second integer.

(b) Let y represent the smaller integer. You are told that the larger integer is

$$\underbrace{\text{4 more than } y}_{4 + y}$$

Thus

$$y + 4$$

represents the larger integer.

Solving Integer Problems

In each of the following problems, translate the English expressions into mathematical symbols. You can then formulate the problems in terms of an equation. Solve the equation in order to solve the original problem.

EXAMPLE 4

Five times an integer is 40. Find this integer.

Solution. Let x represent this integer. Translate the problem:

$$\underbrace{\text{Five times an integer}}_{5x} \; \underset{=}{\text{is}} \; \underset{40}{40.}$$

Solve the equation by dividing both sides by 5:

$$x = 8$$

Thus 8 is the integer you seek.

EXAMPLE 5

Find two consecutive integers whose sum is 25.

Solution. You want two consecutive integers. Let x represent the smaller integer. The next consecutive integer is then $x + 1$.

Translate the problem:

$$\underbrace{\text{The sum of two consecutive integers}}_{x + (x + 1)} \; \underset{=}{\text{is}} \; \underset{25}{25.}$$

Solve the equation:

$$\begin{aligned} 2x + 1 &= 25 \\ 2x &= 24 \\ x &= 12 \\ x + 1 &= 13 \end{aligned}$$

The two integers are 12 and 13.

EXAMPLE 6

The sum of three consecutive even integers is 54. Find these integers. Check your result.

Solution. The *even* integers are

$$\ldots -8, -6, -4, -2, 0, 2, 4, 6, 8, \ldots$$

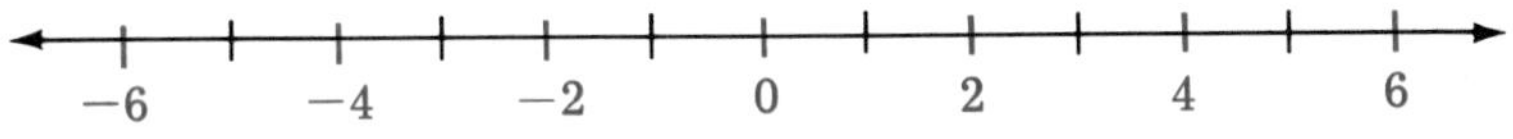

FIGURE 8.1

Consecutive even integers are *two* units apart. Let x represent the smallest of these even integers. Then $x + 2$ is the next consecutive even integer, and $x + 4$ is the largest of the three.

Translate the problem:

$$\underbrace{\text{The sum of three consecutive even integers}}_{x + (x + 2) + (x + 4)} \underset{=}{\text{is}} \underset{54}{54.}$$

Solve the equation:

$$\begin{aligned} 3x + 6 &= 54 \\ 3x &= 48 \\ x &= 16 \\ x + 2 &= 18 \\ x + 4 &= 20 \end{aligned}$$

The three integers are 16, 18, 20.
CHECK.

$$\begin{aligned} 16 + 18 + 20 &\stackrel{?}{=} 54 \\ 54 &\stackrel{\checkmark}{=} 54 \end{aligned}$$

EXAMPLE 7

Six more than an integer is four times this integer. Find the integer.

Solution. Let x represent this integer.

Translate the problem:

$$\underbrace{\text{Six more than an integer}}_{x + 6} \underset{=}{\text{is}} \underbrace{\text{four times this integer.}}_{4x}$$

Solve the equation:

$$\begin{aligned} 6 &= 3x \\ 2 &= x \end{aligned}$$

Thus 2 is the integer you seek.

EXERCISES

1. Let x represent an integer. Express each of the following in terms of x.
   (a) two more than the integer
   (b) six less than the integer
   (c) triple the integer
   (d) one-third of the integer

2. Let y represent an integer. Express each of the following in terms of y.
   (a) twice the integer
   (b) one less than twice the integer
   (c) four more than twice the integer
   (d) one-fourth of four more than twice the integer

3. Let x represent an integer. Express each in terms of x.
   (a) The integer is increased by 9.
   (b) The integer is decreased by 2.
   (c) half of the integer
   (d) the next consecutive integer

4. Suppose the sum of two integers is 50. If x represents one of these integers, express the other in terms of x.

5. The difference between two integers is 19. If x represents the larger integer, express the smaller one in terms of x.

6. The larger of two integers is 8 more than the smaller. If x represents the smaller integer, express the larger one in terms of x.

7. The larger of two integers is 5 more than the smaller. If x represents the larger integer, express the sum of these integers in terms of x.

8. If x represents an integer, express the sum of this integer and three times the next consecutive integer.

9. Ten more than an integer is 23. Find this integer.

10. Three times an integer is -27. Find this integer.

11. A certain integer is decreased by 7, and the result is 25. Find this integer.

12. Half of an integer is 22. What is this integer?

13. Find two consecutive integers whose sum is 31.

14. Find three consecutive integers whose sum is 27. Check your result.

15. Find two consecutive even integers whose sum is 42.

16. Find three consecutive odd integers whose sum is 33.

17. One more than twice an integer is 15. Find this integer. Check your result.

18. One less than three times an integer is 23. Find this integer.

19. Seven more than four times an integer is 51. Find this integer. Check your result.

20. Two less than five times an integer is 28. Find this integer.

21. The sum of two integers is 50. The larger is six more than the smaller. Find these integers. Check your result.

22. Half of an integer plus one-fifth of the next consecutive integer equals 10. Find these integers.

23. Eric and Doug together have \$60. Eric has \$10 more than Doug. How much does Doug have?

24. Mozart wrote five more than four times the number of symphonies that Beethoven wrote. Altogether they wrote 50 symphonies. How many did each write?

8.2 AGE PROBLEMS

In these problems, you must determine a person's age, which is assumed to be an integer. The age may be given at various times—for example, now, three years ago, two years from now. Again, first translate the problem into mathematical symbols.

Translating Words to Symbols

EXAMPLE 1

A mother is 20 years older than her son. If x represents the son's age, express the mother's age in terms of x.

Solution. Let x be the son's age.

$$\underbrace{\text{A mother}}_{\square}\ \underbrace{\text{is}}_{=}\ \underbrace{\text{20 years older than}}_{20\,+}\ \underbrace{\text{her son.}}_{x}$$

The mother's age is $x + 20$.

EXAMPLE 2

Let x represent a man's age *now*.
(a) Express his age three years ago.
(b) Express his age in two years.

Solution.
(a) Three years ago the man was three years *younger*. His age was

$$x - 3.$$

(b) In two years the man will be two years *older* than he is now. His age will be

$$x + 2.$$

EXAMPLE 3

Five years ago a man was four times as old as his daughter. If x represents the daughter's *present* age, express the man's *present* age.

Solution. Let x represent the daughter's age *now*.
The daughter's age 5 *years ago* was

$$x - 5.$$

The man's age 5 years ago was 4 times that of his daughter *(then)*. His age *then* was

$$4(x - 5).$$

Now the man is 5 years older. His *present* age is given by:

$$\begin{aligned} 4(x - 5) + 5 &= 4x - 20 + 5 \\ &= 4x - 15 \end{aligned}$$

Solving Age Problems

In the following "age problems," after you have translated the problem into mathematical symbols, solve the resulting equation, as in the preceding section.

EXAMPLE 4

A father is 25 years older than his 5-year-old son. How old is the father?

Solution. Let x represent the father's age.

Translate the problem:

A father is 25 years older than his 5-year-old son.

$$\underbrace{x}_{\text{A father}} \underbrace{=}_{\text{is}} \underbrace{25 +}_{\text{25 years older than}} \underbrace{5}_{\text{his 5-year-old son}}$$

$$x = 30$$

The father is 30 years old.

EXAMPLE 5

Two years ago a mother was twice as old as her daughter. The daughter is now 22 years old. How old is her mother?

Solution. Let x represent the mother's age *now*.

Two years ago the mother was two years younger. The mother's age two years ago was $x - 2$.

Two years ago the daughter was also two years younger. She is now 22. Two years ago she was 20.

Translate the problem:

Two years ago a mother was twice as old as her daughter.

$$\underbrace{x - 2}_{\text{Two years ago a mother}} \underbrace{=}_{\text{was}} \underbrace{2}_{\text{twice as old as}} \underbrace{(20)}_{\text{her daughter}}$$

Simplify and solve this equation.

$$x - 2 = 40$$
$$x = 42$$

The mother is now 42 years old.

EXAMPLE 6

A woman is now five times as old as her son. Three years ago she was eight times as old as her son. How old is the son now?

Solution. Let x represent the son's age *now*. Then $5x$ represents the woman's age now.

Three years ago, the son's age was $x - 3$. *Three years ago,* the woman's age was $5x - 3$.

Translate the problem. In order to do this, reword the second sentence of the problem as follows:

$$\underbrace{\text{The woman's age three years ago}}_{5x-3} \quad \overset{\text{was}}{=} \quad \underbrace{\text{eight times her son's age three years ago.}}_{8(x-3)}$$

Solve the equation.

$$\begin{aligned} 5x - 3 &= 8x - 24 \\ 5x - 3 - 5x + 24 &= 8x - 24 - 5x + 24 \\ 21 &= 3x \\ 7 &= x \end{aligned}$$

The son is now 7 years old [and his mother 35. Three years ago he was 4 and she was 32, or $8 \cdot 4$.]

EXAMPLE 7

Helen is 9 years older than her sister Sondra. In three years Helen's age will be double Sondra's age. How old is each sister now?

Check your result.

Solution. Let x represent Sondra's age *now*. Then Helen's age (now) is $x + 9$.

In three years, Helen's age will be

$$(x + 9) + 3 \quad \text{or} \quad x + 12,$$

and Sondra's will be

$$x + 3.$$

Translate the problem:

$$\underbrace{\text{In three years Helen's age}}_{x+12} \quad \overset{\text{will be}}{=} \quad \underbrace{\text{double Sondra's age (in three years).}}_{2(x+3)}$$

Solve this equation:

$$\begin{aligned} x + 12 &= 2x + 6 \\ 6 &= x \\ 15 &= x + 9 \end{aligned}$$

Thus Sondra is 6 and Helen is 15.

CHECK.

$$\begin{aligned} 15 + 3 &\stackrel{?}{=} 2(6 + 3) \\ 18 &\stackrel{\checkmark}{=} 18 \end{aligned}$$

EXERCISES

1. Fred is 6 years older than Terri. If x represents Terri's age, express Fred's age.

2. A man is two years older than his wife and twenty five years older than his

son. If x represents the man's age, express (a) his wife's age, (b) his son's age.

3. A woman is 5 years younger than her husband. She is twice as old as her daughter. Let x represent the woman's age. Express (a) her husband's age, (b) her daughter's age.
4. If I am x years old now, how old will I be in 20 years?
5. Three years ago, Maria was y years old. (a) How old is she now? (b) How old will she be in 7 years?
6. Two years ago, Dave was twice as old as his sister. If x represents his sister's present age, express Dave's present age.
7. Gene is 4 years older than Alex. Gene is 17. How old is Alex?
8. Joe is 5 years younger than his brother Aaron, who is 18. How old is Joe?
9. Jerry is 7 years older than Henry, who is 3 years older than Bill. If Bill is 15, how old is Jerry?
10. Two brothers are 3 years apart in age. In 6 years the older brother will be 23. How old is the younger brother now?
11. Three years ago a mother was 5 times as old as her daughter. The daughter is now 8 years old. How old is the mother now? Check your result.
12. In three years Marco will be half as old as his brother. The brother will then be 24. How old is Marco now?
13. A man is five years older than his wife. Fifteen years ago he was twice her age. How old is the man? Check your result.
14. Barbara's grandfather is five times as old as Barbara. In four years the grandfather will be four times as old as Barbara. How old is Barbara?
15. A 31-year-old woman has a 7-year-old son. In how many years will she be double his age? Check your result.
16. A man has a son and daughter. The man is six times as old as his son, who is two years older than his sister. In six years the man will be four times as old as his daughter. How old is the man?

8.3 DISTANCE PROBLEMS

The distance that can be traveled at a *constant rate* equals the rate multiplied by the time spent traveling. For example, if a car travels at the constant rate of 50 miles per hour, in 4 hours it will travel

$$50 \cdot 4 \text{ or } 200 \text{ miles.}$$

Let

$$r = \text{rate},$$
$$t = \text{time},$$
$$d = \text{distance}.$$

Then

$$r \cdot t = d$$

This is known as the **distance formula.** It is convenient to express this formula by means of a table. Thus, for the above example, write

r	$\cdot\ t\ =$	d
50	4	200

Next, suppose an airplane flies at the constant rate of 400 miles per hour. In 6 hours it travels 2400 miles.

r	$\cdot\ t\ =$	d
400	6	2400

The units of measurement must match. Thus, when distance is measured in miles and time in hours, then rate is given in *miles per hour,* which can be expressed as the fraction

$$\frac{\text{miles}}{\text{hour}}$$

Corresponding to the formula

$$r \cdot t = d,$$

you have

$$\frac{\text{miles}}{\text{hour}} \cdot \text{hours} = \text{miles}$$

Throughout this section, *all rates are assumed to be constant.*

Other forms of the distance formula apply when you are given some other information. For example, suppose you are told that a man walks 12 miles at the constant rate of 3 miles per hour. To find the time he walks, divide both sides of the distance formula

$$r \cdot t = d$$

by r to obtain

$$t = \frac{d}{r}$$

Thus

$$t = \frac{12}{3}$$

$$t = 4$$

He walks for 4 hours.

Similarly, suppose you are told that a bus covers 180 miles of a highway in 3 hours. To find its rate, divide both sides of the distance formula by t to obtain

$$r = \frac{d}{t}$$

Then the bus travels at the rate of

$$\frac{180}{3} \text{ or 60 miles per hour.}$$

EXAMPLE 1

A train traveling at the rate of 90 miles per hour goes 495 miles. How long does this take?

Solution. Let t be the number of hours.

r	$t = \frac{d}{r}$	d
90	t	495

$$t = \frac{d}{r}$$

$$t = \frac{495}{90}$$

$$t = \frac{\cancel{3^2} \cdot \cancel{5} \cdot 11}{2 \cdot \cancel{3^2} \cdot \cancel{5}}$$

$$t = \frac{11}{2}$$

The trip takes $5\frac{1}{2}$ hours.

Suppose two cars leave from the same place at the same time along a straight road, one traveling at 60 miles per hour, the other at 50 miles per hour.

(a) If they travel in *the same direction,* then *each hour* the faster car goes 10 (or $60 - 50$) miles further than the slower car. [See Figure 8.2(a).]

(b) If they travel in *opposite directions,* then *each hour* the cars separate 110 (or $60 + 50$) miles. [See Figure 8.2(b).]

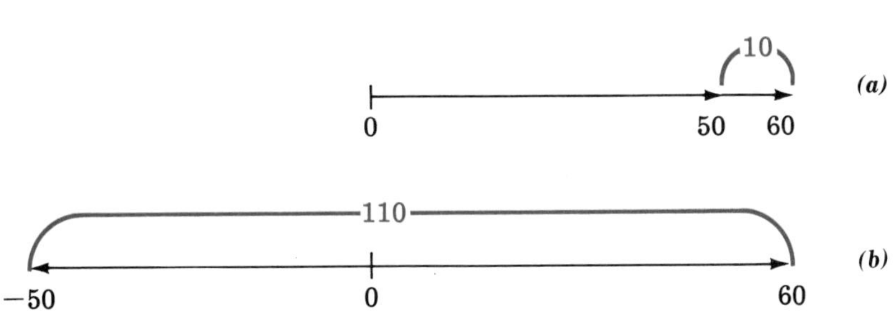

FIGURE 8.2(a). If the cars travel in the same direction, the faster car goes 10 miles further than the slower car in one hour.

FIGURE 8.2(b). If the cars travel in opposite directions, the cars separate 110 miles in one hour.

EXAMPLE 2

Two cars leave a restaurant at the same time headed in the same direction along a straight road. One car travels at 60 miles per hour, the other at 40 miles per hour. How far apart are they after 4 hours?

Solution.

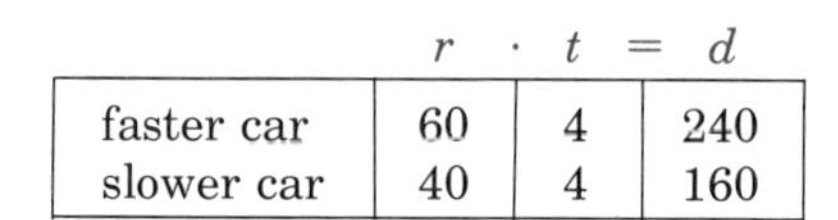

	r ·	t =	d
faster car	60	4	240
slower car	40	4	160

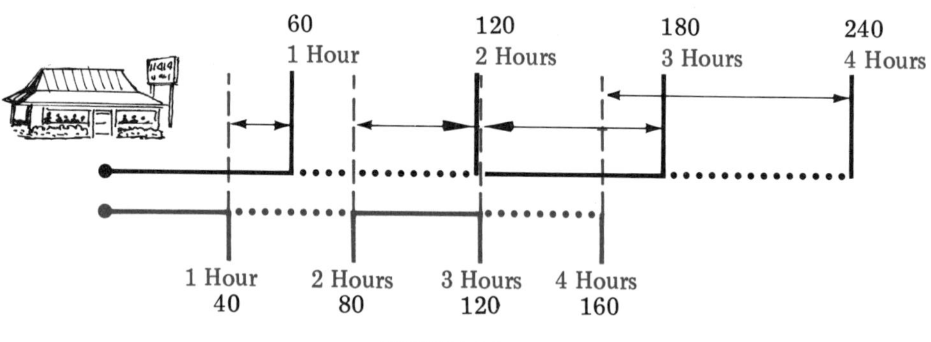

FIGURE 8.3

The cars travel in the same direction. In 1 hour the faster car is 20 (= 60 − 40) miles ahead of the slower car. In 4 hours it is 80 (= 4 · 20) [or 240 − 160] miles ahead of the slower vehicle.

EXAMPLE 3

Two cars leave a ball park at the same time. One car travels eastward at 55 miles per hour; the other car travels westward at 70 miles per hour. How far apart are they after 2 hours?

Solution.

	r ·	t =	d
eastward	55	2	110
westward	70	2	140

The cars travel in opposite directions. Their distance apart, d, is the sum of the distances each has traveled in 2 hours:

$$d = 110 + 140$$
$$d = 250$$

The distance apart is 250 miles. Each hour the cars move apart 55 + 70 (= 125) miles. Thus 250 = 125 · 2. (See Figure 8.4.)

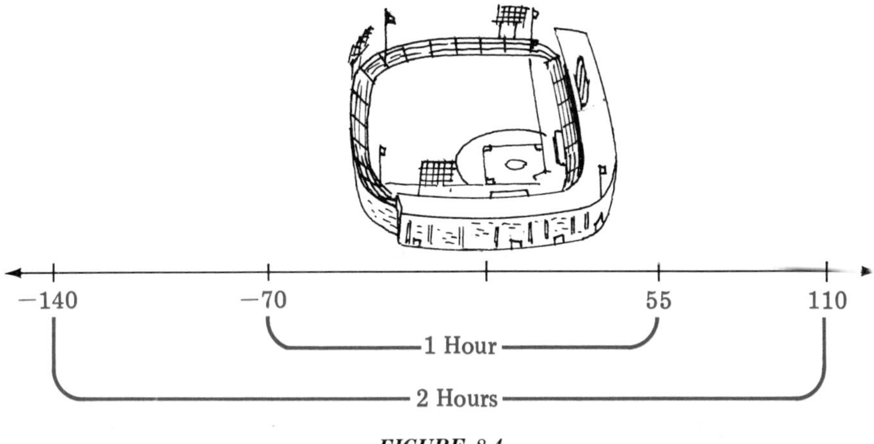

FIGURE 8.4

Next, suppose a man runs to a mailbox and then walks back home. The round trip takes 10 minutes.

If he runs for t minutes, then he walks for $10 - t$ minutes. For example, if he runs for 4 minutes, then he walks for $10 - 4$, or 6 minutes. Assume he takes the same route going and returning. Then *the distance each way is the same.* (See Figure 8.5.)

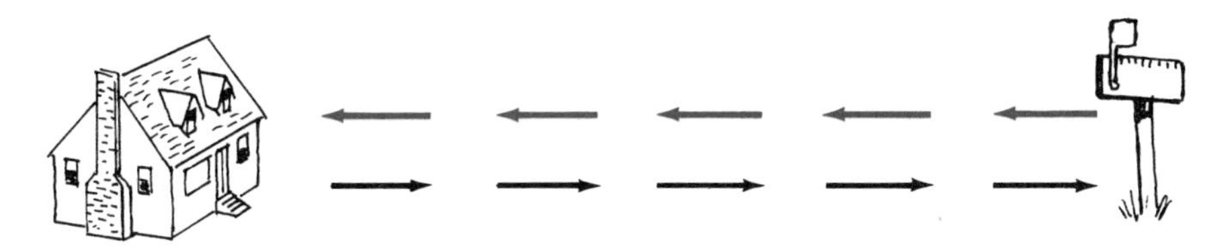

FIGURE 8.5

EXAMPLE 4

A canoe goes upstream at the rate of 5 miles per hour. It returns downstream at the rate of 10 miles per hour. The round trip takes 3 hours.

(a) How long does the canoe travel upstream?

(b) How far upstream does it go?

(c) Check your results.

Solution.

(a) The canoe travels the same distance upstream as downstream. (See Figure 8.6.) Let t be the number of hours it travels upstream. Because the round trip takes 3 hours, it travels for $(3 - t)$ hours downstream.

	r ·	t =	d
upstream	5	t	$5t$
downstream	10	$3-t$	$10(3-t)$

FIGURE 8.6

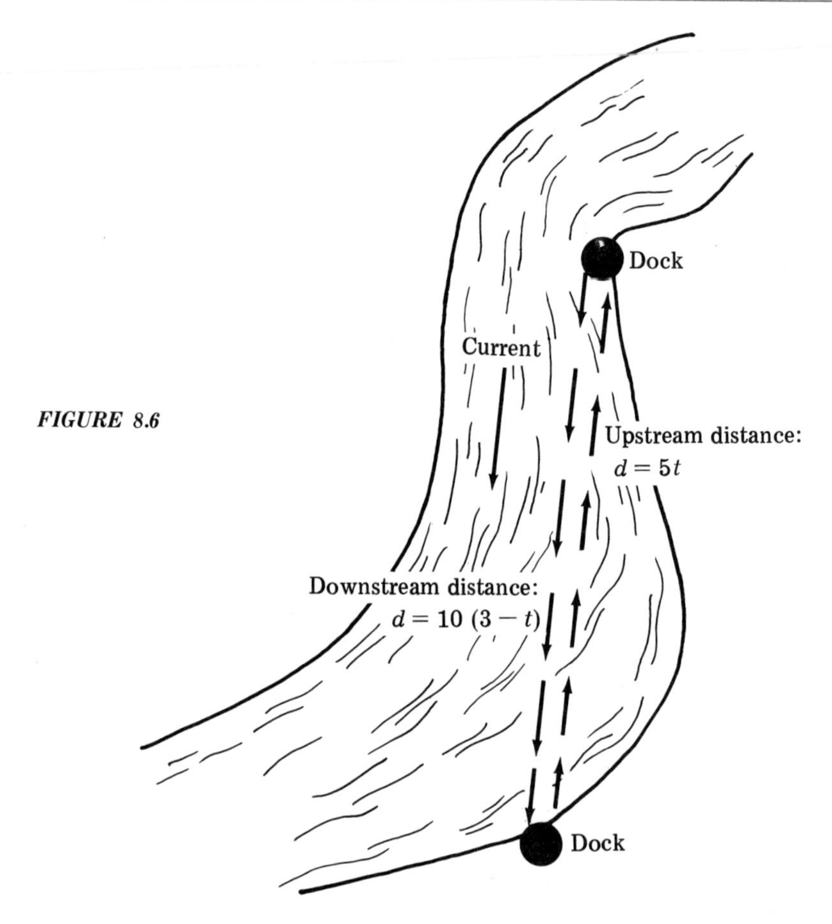

The distance upstream equals the distance downstream.

$$5t = 10(3 - t)$$

Solve this equation:

$$5t = 30 - 10t$$
$$15t = 30$$
$$t = 2$$

The canoe travels for 2 hours upstream (and for 1 hour downstream).

(b) To find the *distance* upstream, use the distance formula with this new piece of information from part (a).

	r ·	t =	d
upstream	5	2	10

The canoe travels 10 miles upstream.

(c) You can check both parts by showing that the distance downstream is also 10 (miles). Recall that the time traveling downstream is 1, or $3 - 2$ (hours).

	r ·	t =	d
downstream	10	1	10

EXERCISES*

1. A car travels at 30 miles per hour for 3 hours. How far does it travel?

2. How long does it take Bob to walk home from his office if he walks 4 miles per hour and lives 6 miles from his office?

3. An automobile travels for two hours at 60 miles per hour. It then slows down and travels at 40 miles per hour in the same direction for the next hour and a half. How far does it travel?

4. A train travels 280 miles in 4 hours. At what rate does it travel?

5. An airplane travels 1050 miles in $3\frac{1}{2}$ hours. At what rate does it travel?

6. A car travels at 45 miles per hour for 4 hours. On the return trip it travels at 60 miles per hour. How long does the return trip take? Check your result.

7. A car traveling at 70 miles per hour along a straight road passes another car cruising at 45 miles per hour. The two cars are headed in the same direction.
   (a) How far apart are they after 1 hour?
   (b) How far apart are they after 3 hours?

8. Two cars leave a toll booth at the same time. They travel in opposite directions along a straight road, one at 65 miles per hour, the other at 60 m.p.h.
   (a) How far apart are they after 1 hour?
   (b) How far apart are they after 4 hours?

9. Two cars leave a town traveling in opposite directions along a straight road. One car goes 50 miles per hour; the other goes 65 miles per hour. How far apart are they after 6 hours? Check your result.

10. Two trains approach one another along (straight) parallel tracks, starting from stations 120 miles apart. One train travels at 100 miles per hour, the second train at 80 m.p.h. How long after they start do they pass each other?

11. Two joggers leave from the same place, headed in the same direction along a straight path. One goes at 8 miles per hour, the other at 6 miles per hour. How far apart are they after 45 minutes? Check your result.

12. A car and a motor scooter travel along a straight highway in the same direction. The scooter travels at three-fourths the rate of the car. At the end of 5 hours they are 75 miles apart. How fast is the car traveling?

13. A canoe goes upstream at the rate of 6 miles per hour. It returns downstream at the rate of 9 miles per hour. If the round trip takes 5 hours, how far upstream does it go? Check your result.

14. A jogger can run at the rate of 8 miles per hour over level ground and at 4 miles per hour over hilly ground. Altogether, it takes him an hour and a half to cover 10 miles. How many of these miles are level?

---

*All rates are constant.

8.4 INTEREST PROBLEMS

Before solving financial problems concerning interest, commissions, mark-ups, and discounts, you must be familiar with some basic facts about percentage.

Percent

Percent means hundredths. For example,

$$17\% \text{ means } .17 \text{ or } \frac{17}{100},$$

$$9\% \text{ means } .09 \text{ or } \frac{9}{100},$$

$$50\% \text{ means } .50 \text{ or } \frac{1}{2},$$

$$100\% \text{ means } 1.00 \text{ or } 1,$$

and

$$150\% \text{ means } 1.50 \text{ or } 1\frac{1}{2}.$$

Note that

$$\frac{1}{2}\% \text{ means } \frac{1}{2} \textit{ of } 1\%.$$

Here, the word "of" indicates multiplication.
Thus

$$\begin{aligned}\frac{1}{2}\% &= \frac{1}{2} \times .01\\ &= \frac{1}{2} \times .010\\ &= .005\end{aligned}$$

Also,

$$\begin{aligned}8\frac{1}{2}\% &= 8\% + \frac{1}{2}\%\\ &= .080\\ &\ \underline{+.005}\\ &= .085\end{aligned}$$

EXAMPLE 1

Find $7\frac{1}{2}\%$ of 2000.

Solution.

$$7\tfrac{1}{2}\% = .075$$

Multiply:

$$\begin{array}{r}.075\\ \underline{2000}\\ 150.000\end{array}$$

$7\frac{1}{2}\%$ of 2000 is 150.

Interest

When money is invested, it earns **interest** over a period of time. To simplify matters, *the basic period of time will always be one year.* The amount of money invested, called the **principal,** times the **yearly interest rate** equals the interest earned in one year. Let

R be the interest rate,
P be the principal,
I be the interest earned.

Then
$$R \cdot P = I$$

Thus \$100 (principal) invested for one year at a rate of 8% earns

$$(.08)(100) \text{ dollars(interest)}$$

or \$8 in one year. Also, \$300 invested for one year at 7% earns

$$(.07)(300) \text{ dollars}$$

or \$21 in one year.

EXAMPLE 2

Jerry deposits \$500 in a bank that pays an annual (once a year) interest rate of 6%. How much interest does he earn if he leaves all his money, including his interest, in for two years?

Solution. There are two different amounts involved—the original principal (for the *first* year) as well as the principal at the end of one year (for the *second* year). Use *subscripts* on P to distinguish between these principals. Thus let P_1 be the principal of \$500 for the *first* year.

There are also different amounts of interest earned for each year. Let I_1 be the interest earned for the *first* year. The interest *rate* for each year is 6%, or .06.

R	$\cdot\ P_1$	$= I_1$
.06	500	I_1

$$I_1 = (.06)(500)$$
$$I_1 = 30$$

Jerry earns \$30 interest the first year. He leaves this in the bank and therefore has \$530 at the end of one year. Thus his principal, P_2, for the *second* year is \$530. Let I_2 be the interest earned during the *second* year. Again the interest rate is .06.

R	$\cdot\ P_2$	$= I_2$
.06	530	I_2

$$I_2 = (.06)(530)$$
$$I_2 = 31.80$$

He earns \$31.80 interest for the second year. For two years he earns

$$\begin{array}{r} \$30.00 \\ +\ 31.80 \\ \hline \$61.80 \end{array}$$

interest.

The formula for interest

$$R \cdot P = I$$

is similar to the distance formula (of the preceding section)

$$r \cdot t = d$$

Just as you could rewrite the distance formula to find time or rate, so too, can you use other forms of the interest formula. To find the rate, use

$$R = \frac{I}{P}$$

For example, if you are told that a man earns \$7 for one year on a principle of \$100, the rate is given by:

$$\begin{aligned} R &= \frac{I}{P} \\ &= \frac{7}{100} \\ &= .07 \end{aligned}$$

The rate is 7%.

To find the principle when you know the interest earned and the interest rate, use

$$P = \frac{I}{R}$$

EXAMPLE 3

How much must Alice invest at $8\frac{1}{2}\%$ in order to earn \$340 interest in one year?

Check your result.

Solution.

$$I = 340$$
$$R = 8\tfrac{1}{2}\% = .085$$

R	$P = \frac{I}{R}$	I
.085	P	340

$$P = \frac{I}{R}$$

$$P = \frac{340}{.085} = \frac{340 \times 1000}{.085 \times 1000} = \frac{340\,000}{85}$$

(Note that numerator and denominator were each multiplied by 1000 to eliminate decimals.)

$$P = 4000$$

Alice must invest $4000 at $8\frac{1}{2}$% to earn $340 interest in one year.
CHECK.

$$(.085)(4000) \stackrel{?}{=} 340$$
$$340 \stackrel{\checkmark}{=} 340$$

EXAMPLE 4

Part of Jose's $5000 is deposited in a bank at 6%. The remainder is invested in higher yielding, though less secure, bonds that pay 10% annually. Altogether, Jose receives $440 interest a year from these two investments. How much has he deposited in the bank?

Solution. Let x be the amount of money in the bank. Then

$$5000 - x$$

is the amount in bonds.

	R	$\cdot$ P	$=$ I
in bank	.06	x	$.06x$
in bonds	.10	$5000 - x$	$.10(5000 - x)$

Altogether he receives $440 interest from these two investments. Thus

$$\underbrace{\text{interest from the bank}}_{.06x} + \underbrace{\text{interest from bonds}}_{.10(5000 - x)} = \$440$$

$$.06x + .10(5000 - x) = 440$$

Multiply both sides by 100.

$$6x + 10(5000 - x) = 44\,000$$
$$\underbrace{6x + 50000 - 10x}_{-4x} = 44\,000$$
$$6000 = 4x$$
$$1500 = x$$

Jose deposits $1500 in the bank (and invests $3500 in bonds).

EXERCISES

In exercises 1–8, express each percent as a decimal.

1. 17%
2. 8%
3. 42%
4. $9\frac{1}{2}$%

5. $10\frac{1}{2}\%$
6. 200%
7. 250%
8. $7\frac{1}{4}\%$
9. Find 10% of 6000.
10. Find 5% of 7500.
11. Find 8% of 12 000.
12. Find $5\frac{1}{2}\%$ of 200.
13. Find $6\frac{1}{2}\%$ of 5000.
14. Find $8\frac{1}{2}\%$ of 6600.
15. How much interest is earned in one year on a principal of $600 if the annual interest rate is 7%?
16. How much interest is earned in one year on a principal of $400 if the annual interest rate is $7\frac{1}{2}\%$?
17. What is the annual interest rate if $45 interest is paid in a year on a principal of $750? Check your result.
18. What is the annual interest rate if $144 is paid in a year on a principal of $1800?
19. How much money must be invested for a year at 9% in order to earn $108 interest? Check your result.
20. How much money must be invested for a year at $8\frac{1}{2}\%$ in order to earn $102 interest?
21. Which earns more interest? $900 invested at 8% or $1000 at $7\frac{1}{2}\%$?
22. How much money must be invested at 8% in order to earn the same as $800 at 7%? Check your result.
23. A sum of $250 is left for 2 years as security on an apartment. The landlord pays an interest rate of 5% annually. How much interest is earned?
24. Maria invests part of her $2000 at 6% and the remainder at 8%. Altogether, she receives $150 interest a year from these two investments. How much does she invest at 6%?
25. Part of a sum of $4500 is invested in 10% bonds and the remainder in tax-free 7% bonds. Altogether, the annual interest from these two investments is $420. How much is invested in 10% bonds? Check your result.
26. If you have three times as much invested at 8% as at $7\frac{1}{2}\%$ and if your annual interest income from these two investments is $315, how much is invested at 8%?
27. Jill invests $100 more at 7% than at 9%. Her annual interest from these two investments is $39. How much does she invest at each rate?
28. Roberto invests the same amount at $7\frac{1}{2}\%$ as at 9%. His annual interest from these two investments is $330. How much does he invest at $7\frac{1}{2}\%$?

8.5 COMMISSIONS, MARK-UPS, AND DISCOUNTS

Commissions

Many salesmen earn commissions based on what they sell. A **commission** is a payment determined by the **sale price.** The **rate of commission** is a *percentage* of the sale price. Let

$$S = \text{sale price},$$
$$R = \text{rate of commission},$$
$$C = \text{commission}.$$

Then

$$R \cdot S = C$$

For example, if the sale price of a suit is $100 and the rate of commission is 6%, the salesman's commission is given by:

R ·	S =	C
.06	100	6

Thus he earns $6 commission for his sale.

EXAMPLE 1

A car salesman earns a $128 commission upon selling a car for $3200. What is his rate of commission?

Solution.

R ·	S =	C
R	3200	128

$$3200R = 128$$
$$R = \frac{128}{3200}$$

To divide, place a decimal point after the 8:

$$3200 \overline{\smash{)}128.00}^{\;.04}$$

The rate of commission, R, is 4%.

EXAMPLE 2

An encyclopedia salesman works on an 8% rate of commission. If a set of encyclopedias sells for $250, how many sets must he sell to earn $300 in commissions?

Check your result.

Solution. Let x be the number of sets he must sell. For each set the sale price is $250. For x sets the total sale is $\$250x$.

R	$\cdot\ S$	$= C$
.08	$250x$	300

Solve:

$$\begin{aligned}(.08)(250x) &= 300\\ 8(250x) &= 30\,000\\ 2000x &= 30\,000\\ x &= 15\end{aligned}$$

He must sell 15 sets to earn $300 in commissions.
CHECK.

R	$\cdot\ S$	$= C$
.08	(250)(15) —3750—	300

$$\begin{aligned}(.08)(3750) &\overset{?}{=} 300\\ 300 &\overset{\checkmark}{=} 300\end{aligned}$$

Mark-Ups

It costs a craftsman $10 to make a pair of sandals. He sells the sandals for $15. The **cost** of the item is $10 and its **sale price** is $15. The **mark-up** is

$15 − $10, or $5.

$15 ---------------------- sale price

$5 ---------------- mark-up

$10 ---------------------- cost

In general,

$$\text{mark-up} = \text{sale price} - \text{cost}$$

The **rate of mark-up** is defined by

$$\text{rate of mark-up} = \frac{\text{mark-up}}{\text{cost}}.$$

In the above illustration, the rate of mark-up is

$$\frac{5}{10}, \text{ or } 50\%.$$

Let

$$\begin{aligned}R &= \text{rate of mark-up},\\ M &= \text{mark-up},\\ C &= \text{cost}.\end{aligned}$$

Then
$$R = \frac{M}{C}$$

or
$$R \cdot C = M$$

EXAMPLE 3

A wholesaler buys baby carriages for $60, marks them up 25%, and sells them to a retail store. The store then marks the carriages up an additional 20%.

(a) At what price does the wholesaler sell the carriages?

(b) At what price does the retail store sell them?

Solution.

(a) Here the cost is $60 and the rate of mark-up is 25%.

R	$\cdot\ C$	$= M$
.25	60	M

$$(.25)(60) = M$$
$$15 = M$$

The mark-up is $15.

$$\text{cost} + \text{mark-up} = \text{sale price}$$

Fill in the left side of the equation, and add to obtain the right side.

$$60 + 15 = 75$$

Thus the wholesaler sells the carriages for $75.

(b) Now $75 is the *cost* to the retailer. The new mark-up is 20%.

R	$\cdot\ C$	$= M$
.20	75	M

$$(.20)75 = M$$
$$15 = M$$

The mark-up (by the retailer) is also $15.

$$\text{cost} + \text{mark-up} = \text{sale price}$$
$$75 + 15 = 90$$

The retail store sells the carriages for $90.

Discounts

During a sale, a store will sell merchandise at a discount. For example, a radio, which normally sells for $100 is marked down to $60. Here, $100 is the **marked price,** $60 is the **sale price,** and $40 is the **discount** (or **reduction**).

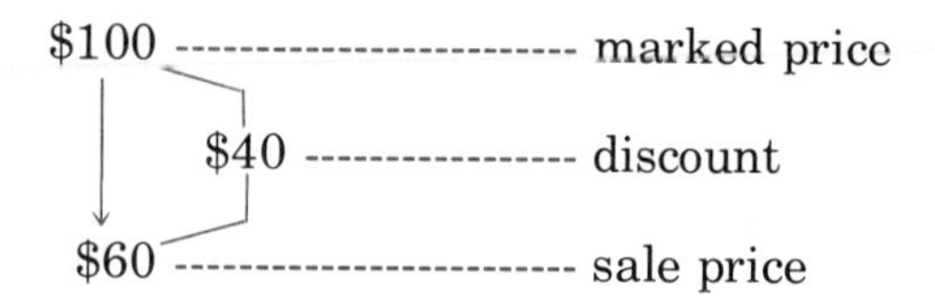

Thus

$$\text{discount} = \text{marked price} - \text{sales price}$$

The **rate of discount** is defined by the formula:

$$\text{rate of discount} = \frac{\text{discount}}{\text{marked price}}$$

The rate of discount for the radio is therefore $\frac{40}{100}$, or 40%. In general,

$$\text{rate of discount} \cdot \text{marked price} = \text{discount}$$

In symbols, $R \cdot M = D$

EXAMPLE 4

All dresses in a boutique are reduced (discounted) by 30%.
(a) What is the discount for a dress marked $60?
(b) What is the sale price for this dress?

Solution.
(a)

R	$\cdot\ M$	$= D$
.30	60	D

$$(.30)(60) = D$$
$$18.00 = D$$

The discount is $18.
(b) Let S be the sale price.

$$\text{discount} = \text{marked price} - \text{sales price}$$
$$D = M - S$$
$$S = M - D$$
$$S = 60 - 18$$
$$S = 42$$

The sales price is $42.

EXERCISES

1. How much commission is earned on a total sales of $2200 at a 9% rate of commission?

2. How much commission is earned on a total sales of $1750 at a 7.5% rate of commission?

3. An automobile salesman earns a commission of $400 by selling a $5000 automobile. What is his rate of commission? Check your result.

4. A cosmetics saleswoman earns a commission of $45 by selling $900 worth of perfume. What is her rate of commission?

5. Suppose a salesman works on a 12% rate of commission. What must his sales be to earn a commission of $300?

6. A book salesman works on a 14% rate of commission. If each book sells for $10.50, how many books must he sell to earn $294 in commissions? Check your result.

7. A real estate agent sells a house for $42 000.
   (a) If the rate of commission is 4%, how much commission does she receive?
   (b) How much money does the seller receive?

8. A shoe salesman receives a weekly salary of $150 and a 6% rate of commission on his sales. One week he sells $950 worth of shoes. How much does the salesman earn that week?

9. A saleswoman receives a 4% commission on the first $1000 of sales each week, 5% on the next $2000, and 6% on all sales thereafter. What is her commission on $8000 of sales for a week?

10. A vendor buys ice cream pops for 20¢ apiece and marks them up at the rate of 25%. For how much does the vendor sell an ice cream pop? Check your result.

11. A wholesaler buys gloves for $8 a pair and marks them up 50%. He sells them to a retailer who then marks them up 20%.
    (a) At what price does the wholesaler sell the gloves?
    (b) At what price does the retailer sell them?

12. Recently, New York City allowed landlords of rent-controlled apartments to increase the rent by $7\frac{1}{2}$% each year. If an apartment was originally rented for $400 per month, what was the monthly rent two years later?

13. A mark-up of $25 on a typewriter is at the rate of 25%. What is the sale price?

14. A tie sells for $7. Its rate of mark-up is 75%. What is the mark-up?

15. A television set is marked at $400 and is on sale with 20% off.
    (a) What is the discount?
    (b) What is the sale price?

16. A coat marked at $120 is on sale with a $30 discount.
    (a) What is the rate of discount?
    (b) What is the sale price?

17. Two sweaters are on sale, each with the same rate of discount. The grey turtle neck is marked at $20 and is on sale for $14. The black cardigan is marked at $22. What is its sale price? Check your result.

18. The price of a car is reduced by 15%. The discount amounts to $450. What was its marked price?

19. A book is on sale for $10. The rate of discount is 20%.
    (a) What is the discount?
    (b) What was the marked price?

20. A kitchen table is on sale for $150. The rate of discount is 25%.
    (a) What is the discount?
    (b) What was the marked price?

What Have You Learned in Chapter 8?

You have learned how to translate a problem stated in words into a problem involving mathematical symbols.

And you have applied your knowledge of solving equations to solving problems concerned with such matters as age, distance, interest, and commissions.

Let's Review Chapter 8.

8.1 Integer Problems

1. Six more than twice an integer is 20. Find the integer.
2. Find three consecutive *even* integers whose sum is 90. Check your answer.

8.2 Age Problems

3. Two sisters are five years apart in age. In two years the younger sister will be 19. How old is the older sister now?
4. A man is twice as old as his daughter. Twenty years ago he was six times her age. How old is the man?

8.3 Distance Problems

5. How long does it take to drive 175 miles at 50 miles per hour?
6. Two cars on a straight highway pass each other, traveling in opposite directions. Each car is traveling at 65 miles per hour. In how many hours will they be 260 miles apart?

8.4 Interest Problems

7. How much interest is earned in one year on a principle of $900 if the annual interest rate is $6\frac{1}{2}\%$?
8. Frank invests $500 more at 8% than at 9%. His total yearly interest from these two investments is $210. How much does he invest at each rate?

8.5 Commission, Mark-Ups and Discounts

9. A salesman works on a $12\frac{1}{2}\%$ rate of commission. What must his sales be in order to earn a commission of $250?
10. An encyclopedia is on sale for $120. The rate of discount is 25%. (a) What is the discount? (b) What is the marked price?

Now try these unclassified problems:

11. A twelve-foot rope is cut into two pieces. The longer piece is twice as long as the shorter one. Find the length of each piece.
12. Carol has three dollars in nickels and dimes. She has the same number of each coin. How many nickels does she have?
13. A television channel has three minutes of programming for every minute of commercials. How long are the commercials on a one and one-half hour program?
14. In a school election 2420 votes are cast for two candidates. One candidate wins by 10 votes. How many votes does she receive?
15. The perimeter of a rectangle is 50 inches. The length is three inches longer than the width. Find the dimensions.
16. How much do you pay for an $8 item if the sales tax is 7%?

And these from Chapters 1–7:

17. How much change do you receive from a $50 bill when you purchase two $12 items and one $24 item?
18. The sum of $2x + 3$ and $x - 7$ is multiplied by $x^2 - 2$. Find the resulting product.
19. Multiply: (.085)(.06)
20. Divide: $.27\overline{).00162}$
21. Solve for x:

$$3x - 2(x + 2) = 2x - 4$$

22. Solve for y in terms of the other variables:

$$2x - 5y + 3z = 1 + 3y$$

Try These Exam Questions for Practice.

1. Ten more than an integer is three times this integer. Find the integer.
2. A man is six years older than his wife. When he married her eighteen years ago, she was three-fourths his age. How old is he now?
3. A Buick traveling at 65 miles per hour passes a Volkswagen traveling at 55 miles per hour along a straight highway. The cars are headed in the same direction. How far apart are they one and one-half hours later?
4. Eight hundred dollars is deposited in a bank that pays an annual interest rate of 5%. How much interest is earned if all the money, including the interest, is left in for two years?

9

Functions

9.1 RECTANGULAR COORDINATES

Coordinate Axes

Real numbers correspond to points on a horizontal line, ***L***. Now you will see how pairs of numbers, in a definite order, correspond to points on a plane.

Draw a vertical line through the origin, 0, on ***L***. (See Figure 9.1.) Because ***L*** is horizontal, the two lines are perpendicular. These lines are called the **coordinate axes.** The horizontal line ***L*** will now be called the x-**axis,** and the vertical line will be called the y-**axis.** The two lines intersect at the origin, which will represent 0 on the y-axis (as well as on the x-axis).

For convenience, choose the same distance unit on the y-axis as on the x-axis. *On the y-axis, positive numbers are located upward from 0, and negative numbers downward* (see Figure 9.2 on the next page).

Every point on the y-axis corresponds to a real number. You will identify points on the y-axis with the real numbers they represent, just as you did with points on the x-axis.

EXAMPLE 1

Locate the following points on the y-axis:

(a) 3 (b) $\frac{1}{2}$ (c) -2 (d) π (e) $\frac{-5}{2}$

Solution. See Figure 9.3 on page 231.

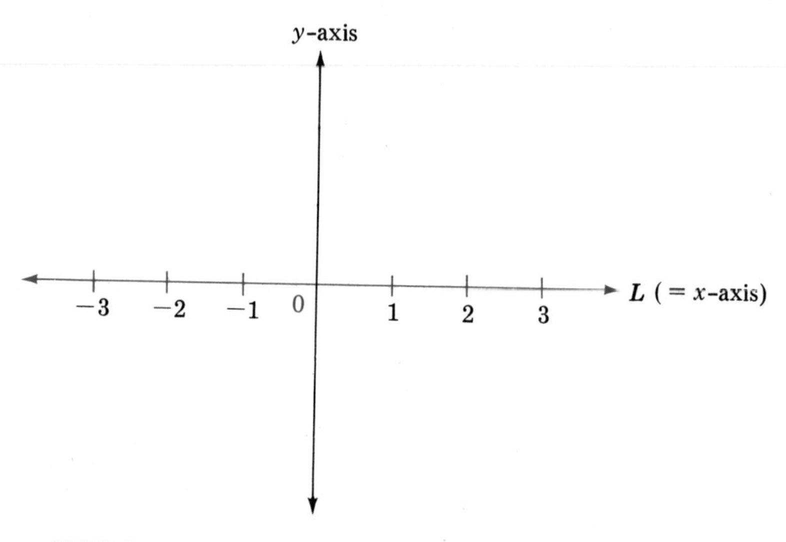

FIGURE 9.1 The coordinate axes

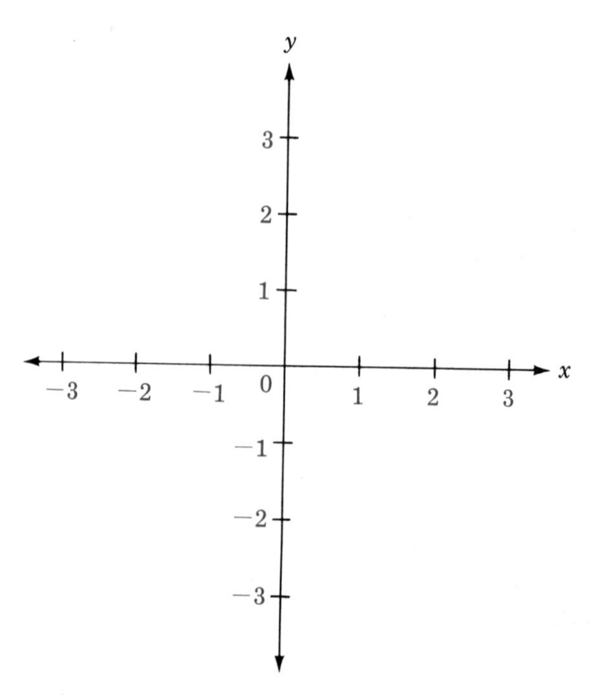

FIGURE 9.2. On the y-axis, positive numbers are plotted upward from 0, and negative numbers downward.

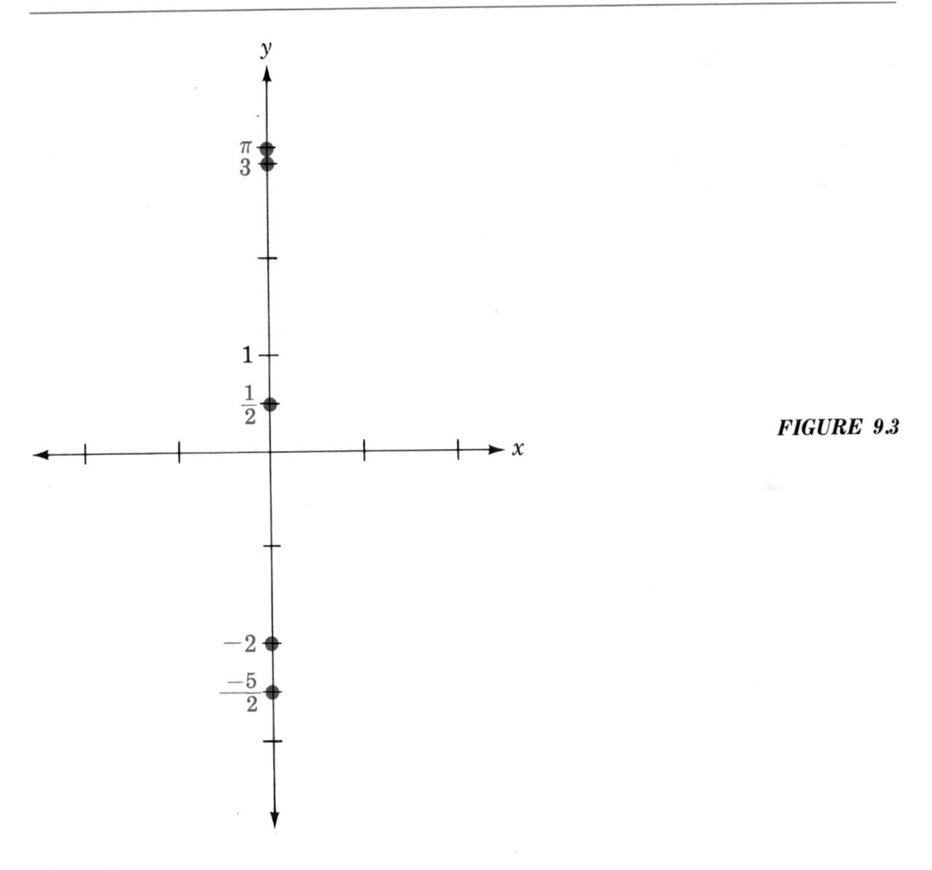

FIGURE 9.3

Coordinates

Every real number corresponds to a point on *each* of the coordinate axes. If you consider two numbers, a and b, then a corresponds to a point on the x-axis and b to a point on the y-axis. Locate these points and draw perpendiculars to the axes through each of them, as in Figure 9.4(a). The intersection of these perpendiculars is the point P of the plane associated with a and b *in this order.*

To indicate the order of a and b, write

$$(a, b).$$

Call (a, b) an **ordered pair.** *The ordered pair (a, b) indicates two numbers a and b in the order written.* The number a is called the **x-coordinate of** (a, b) and b the **y-coordinate of** (a, b).

Thus 2 is the x-coordinate of (2, 5), and 5 is the y-coordinate. On the other hand, if you change the order of the coordinates, then 5 is the x-coordinate and 2 the y-coordinate of (5, 2). Observe that

$$(2, 5) \neq (5, 2).$$

[See Figure 9.4(b).]

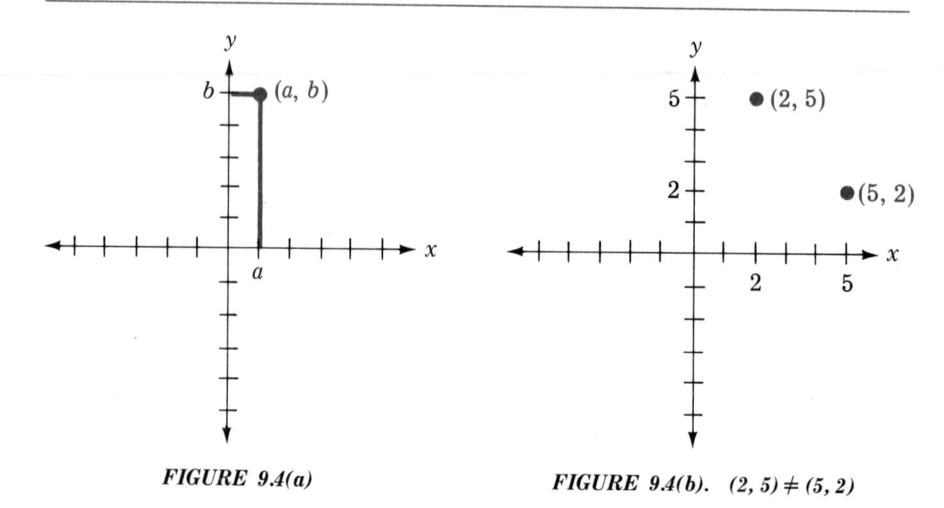

FIGURE 9.4(a)

FIGURE 9.4(b). $(2, 5) \neq (5, 2)$

Every ordered pair of real numbers corresponds to a point on the plane, as indicated. Moreover, every point P on the plane also corresponds to an ordered pair of numbers, (a, b). Together a and b are called the **rectangular coordinates of** P, or simply the **coordinates of** P. You will identify a point P of the plane with its ordered pair of coordinates (a, b), and write

$$P = (a, b).$$

EXAMPLE 2

For each point in Figure 9.5, find
(a) its x-coordinate,
(b) its y-coordinate.
(c) Identify the point with its ordered pair of coordinates.

Solution.

P:	(a) 1,	(b) 2,	(c) $P = (1, 2)$
Q:	(a) 2,	(b) 1,	(c) $Q = (2, 1)$
R:	(a) 4,	(b) 0,	(c) $R = (4, 0)$
S:	(a) 0,	(b) -3,	(c) $S = (0, -3)$
T:	(a) $\frac{1}{2}$,	(b) $\frac{3}{2}$,	(c) $T = \left(\frac{1}{2}, \frac{3}{2}\right)$

Observe that points on the *x-axis* have *y-coordinate* 0 and points on the *y-axis* have *x-coordinate* 0. The origin, O, which is on both coordinate axes, has both of its coordinates equal to 0: (See Fig. 9.6 on page 233.)

$$O = (0, 0)$$

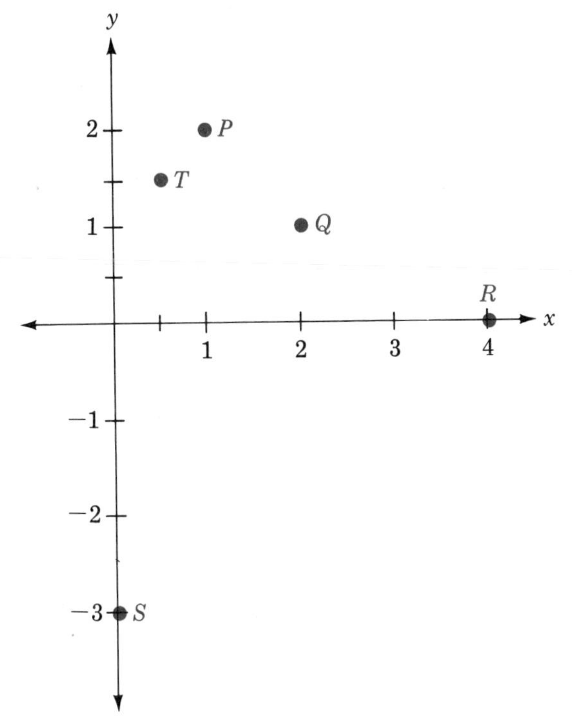

FIGURE 9.5

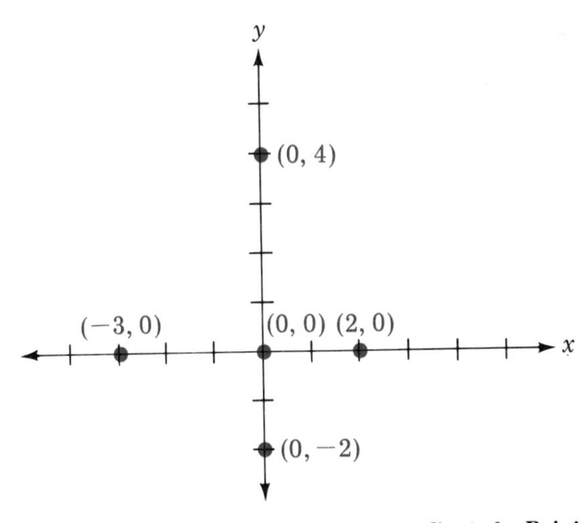

FIGURE 9.6. ***Points on the x-axis have y-coordinate 0. Points on the y-axis have x-coordinate 0.***

EXAMPLE 3

Locate the following ordered pairs on a rectangular coordinate system:

$$(4, 1), (1, 4), (-2, -2), (0, -1), (.5, -.5)$$

Solution. See Figure 9.7.

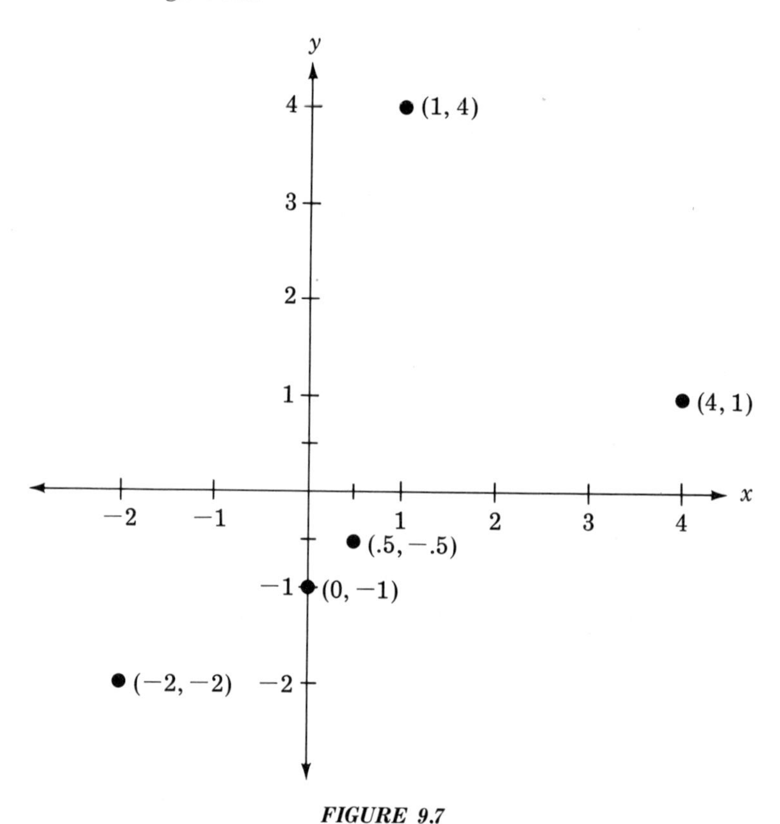

FIGURE 9.7

EXERCISES

1. Draw a rectangular coordinate system. Locate the following points on the y-axis:

   (a) 4, (b) −4, (c) $\frac{-1}{2}$, (d) $\frac{3}{2}$, (e) $-\pi$

2. In Figure 9.8, indicate which numbers are represented by the following points on the y-axis:

   (a) P, (b) Q, (c) R, (d) S, (e) T

3. For each point in Figure 9.9, find
   (a) its x-coordinate,
   (b) its y-coordinate.
   (c) Identify the point with its pair of coordinates.

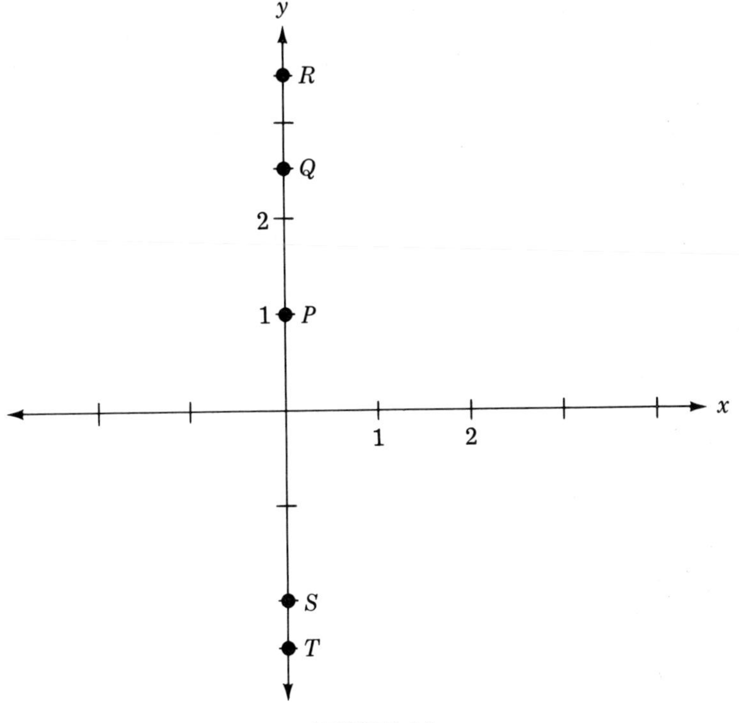

FIGURE 9.8

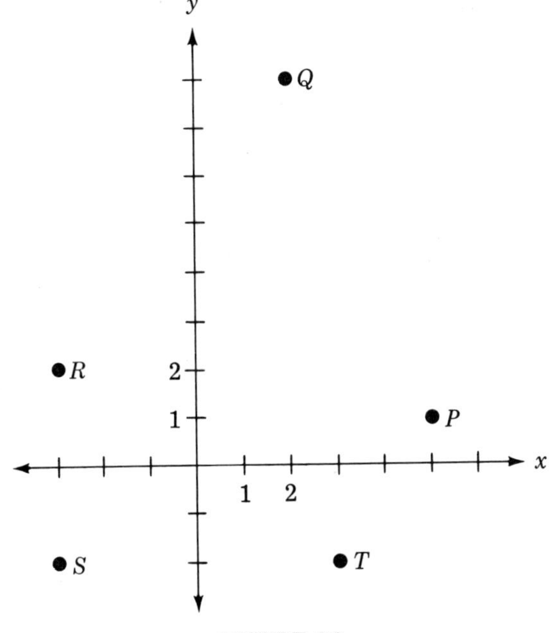

FIGURE 9.9

4. For each point in Figure 9.10, find
   (a) its x-coordinate,
   (b) its y-coordinate.
   (c) Identify the point with its pair of coordinates.

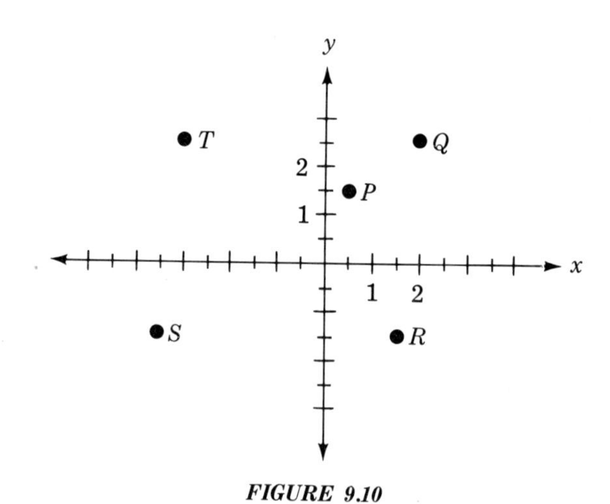

FIGURE 9.10

Locate the following ordered pairs on a rectangular coordinate system. (You may wish to draw a coordinate system for the odd-numbered exercises first, and then a separate coordinate system for the even-numbered exercises.)

5. $(5, 0)$
6. $(0, 3)$
7. $(-2, 0)$
8. $(0, -3)$
9. $\left(\frac{1}{2}, 0\right)$
10. $\left(0, \frac{-1}{2}\right)$
11. $\left(\frac{7}{2}, 0\right)$
12. $\left(0, \frac{-3}{2}\right)$
13. $(2, 2)$
14. $(-1, -1)$
15. $(4, 2)$
16. $(2, 4)$
17. $(-4, 2)$
18. $(-2, 4)$
19. $(-2, -4)$
20. $(-4, -2)$
21. $\left(6, \frac{1}{2}\right)$
22. $\left(\frac{-1}{2}, 4\right)$
23. $(2, -5)$
24. $(-6, -6)$
25. $(6, -1)$
26. $(1, -10)$
27. $(.5, 1.5)$
28. $(-.5, -2.5)$

9.2 WHAT IS A FUNCTION?

Definition and Notation

Suppose a car travels at the constant rate of 50 miles per hour. When you say that the distance it travels *is a function of* time, you mean that distance *depends upon* time. For instance, in 4 hours the car goes $50 \cdot 4$ (or 200) miles, in 6 hours, $50 \cdot 6$ (or 300) miles, etc.

Similarly, if you are given the equation

$$y = x + 5,$$

the value of y *depends upon that of* x.

$$\begin{aligned} &\text{If } x = 3, \text{ then } y = 3 + 5, \text{ or } 8; \\ &\text{if } x = 5, \text{ then } y = 5 + 5, \text{ or } 10; \\ &\text{if } x = 10, \text{ then } y = 10 + 5, \text{ or } 15. \end{aligned}$$

To indicate this dependence of y on x, you can say that

y is a function of x.

Definition

> *A **function** is a rule that indicates how one variable depends upon another variable. If y depends upon x, then y is called the **dependent variable** and x the **independent variable.** For every value of x, there must be exactly one corresponding value of y.*

The notation

$$f(x) = x + 5$$

is used to indicate how y depends on x by means of the rule f. Here

$$y = f(x)$$

For this function, each value of y is 5 more than that of the corresponding x. Thus when $x = 1$, replace x by 1 on both sides of the equation,

$$f(x) = x + 5,$$

and obtain $\quad f(1) = 1 + 5 = 6.$

Similarly, $\quad f(3) = 8$

and $\quad f(5) = 10$

EXAMPLE 1

Let $f(x) = x + 2$. Find:
(a) $f(0)$ (b) $f(2)$ (c) $f(7)$ (d) $f(100)$

Solution. Add 2 to each given value of x. Several corresponding values of x and $f(x)$ are given in the following table:

x	$f(x) = x + 2$
0	2
2	4
7	9
100	102

Letters other than x and y can be used for the variables of a function.

EXAMPLE 2

Let $t = F(s)$, and suppose

$$F(s) = 5s.$$

Find t when

(a) $s = -2$, (b) $s = 0$, (c) $s = \frac{2}{5}$, (d) $s = 4$.

Solution. Here s is the independent variable and t the dependent variable. Multiply each value of s by 5 to obtain t:

s	$t = f(s) = 5s$
-2	-10
0	0
$\frac{2}{5}$	2
4	20

A function can be thought of as a machine. The independent variables are then the inputs and the dependent variables the outputs. For each input there is exactly one output. Thus the function given by

$$f(x) = x + 2$$

of Example 1 can be pictured as follows:

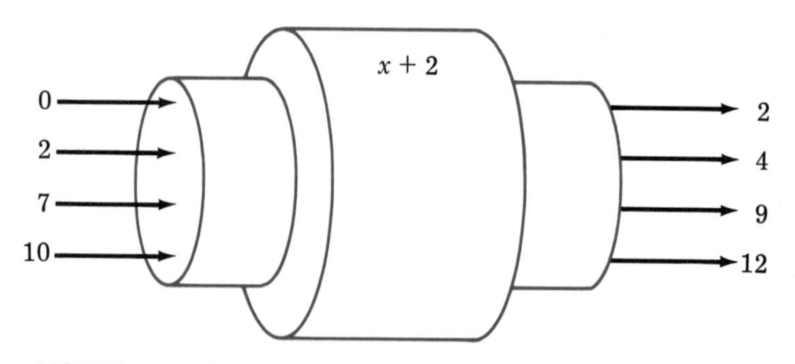

FIGURE 9.11. The $x + 2$ machine

The function given by

$$F(s) = 5s$$

of Example 2 would look like this:

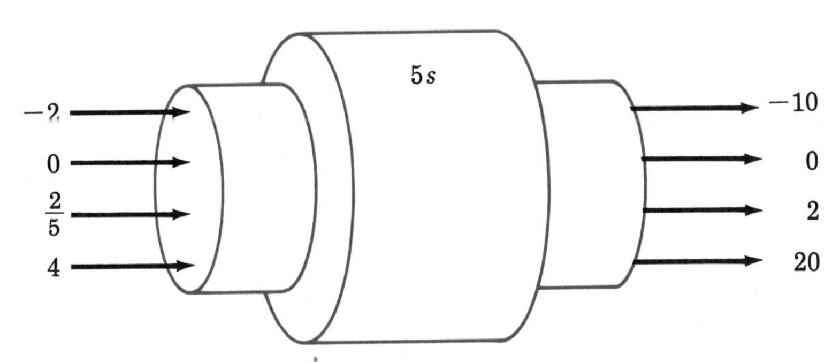

FIGURE 9.12. ***The 5s machine***

EXAMPLE 3

Let $g(x) = 2x - 7$. Find:

(a) $g(-3)$ (b) $g\left(\frac{1}{2}\right)$ (c) $g(4)$ (d) $g(10)$

Solution. To find each value of $g(x)$, substitute the given value of x in the polynomial $2x - 7$.

(a) $g(-3) = 2(-3) - 7 = -6 - 7 = -13$

(b) $g\left(\frac{1}{2}\right) = 2 \cdot \frac{1}{2} - 7 = 1 - 7 = -6$

(c) $g(4) = 2 \cdot 4 - 7 = 1$

(d) $g(10) = 2 \cdot 10 - 7 = 13$

x	$g(x) = 2x - 7$
-3	-13
$\frac{1}{2}$	-6
4	1
10	13

The definition of a function requires that *to each value of the independent variable,* say x, *there corresponds exactly one value of the dependent variable,* say y. However, several different values of x may correspond to the same value of y, as the next example illustrates.

EXAMPLE 4

Let

$$G(x) = 2$$

for all x. Find:

(a) $G(-1)$ (b) $G(2)$ (c) $G(10)$ (d) $G(50)$

Solution. The function assigns 2 to every value of x, large or small:

x	$G(x) = 2$
-1	2
2	2
10	2
50	2

A function in which the dependent variable has only one value, as in Example 4, is known as a **constant function.**

A function may not be defined for every number. For example, the function given by

$$f(x) = \frac{1}{x}$$

is not defined at $x = 0$. (Division by 0 is undefined.)

Curves and Functions

A curve can represent a function. To find the y-value corresponding to a specific x-value, draw perpendiculars to the coordinate axes, as in Example 5.

EXAMPLE 5

The following curve represents a function, say $f(x)$. Find:
(a) $f(1)$ (b) $f(2)$ (c) $f(3)$ (d) $f(4)$

Solution. Read off the function values from Figure 9.13.
(a) $f(1) = 3$ (b) $f(2) = 0$ (c) $f(3) = -1$ (d) $f(4) = 1$

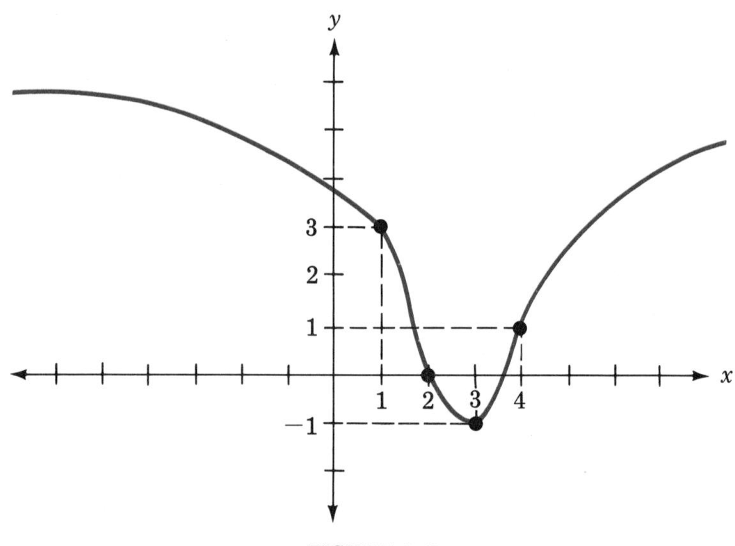

FIGURE 9.13

Not every curve represents a function. For, the definition of a function requires that for every x-value, there must be *exactly one* corresponding y-value. In the case of the circle of Figure 9.14(a), two values of y correspond to $x = 0$ and two values of y correspond to $x = 1$. Note that there are vertical lines that intersect this curve more than once. Whenever a vertical line intersects a curve, a y-value corresponds to a fixed x-value. Thus if there are two or more intersections, at least two y-values correspond to the same x-value. This violates the definition of a function.

A curve represents a function if no vertical line intersects the curve more than once; otherwise, the curve does not represent a function.

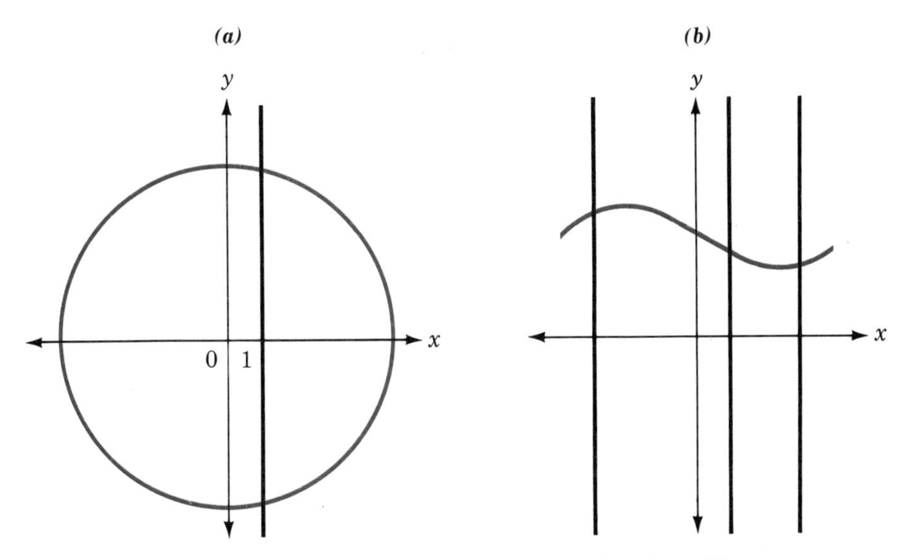

FIGURE 9.14(a). ***A circle does not represent a function. There are vertical lines that intersect the circle more than once.***

FIGURE 9.14(b). ***This curve represents a function. No vertical line intersects the curve more than once.***

Verbal Descriptions

A function can be defined verbally. This is often how functions are given in applications of mathematics. When a function is described verbally, you must express the function algebraically, as well.

EXAMPLE 6

An automobile cruises along a highway at the constant rate of 50 miles per hour.

(a) What function expresses the distance the automobile travels in t hours?
(b) Find the distance it travels in 3 hours.

Solution.

(a)
$$r \cdot t = d$$

Here, r is given as 50. The distance, d, can be described by means of the function

$$F(t) = 50t.$$

(b) Let $t = 3$. Then

$$F(3) = 50 \cdot 3$$
$$= 150$$

The car travels 150 miles in 3 hours.

EXAMPLE 7

A manufacturer finds that it costs him \$3 to produce a novelty item. There is also an overhead cost of \$500.
(a) What function expresses the cost of manufacturing n novelty items?
(b) Find the cost of manufacturing 100 items.
(c) Find the cost of manufacturing 1000 items.
(d) If the manufacturer has \$5000 to spend, how many items can be produced?

Solution.

(a) Here n is the number of novelty items manufactured. Each item costs \$3 to produce. Thus n items cost \$$3n$ to produce. Add \$500 for overhead. Therefore

$$f(n) = 3n + 500$$

expresses the cost (in dollars) of manufacturing n novelty items. Because n is always a positive integer or 0, *the function is only defined for positive integers and* 0.

(b)
$$f(100) = 3 \cdot 100 + 500$$
$$= 800$$

The cost of manufacturing 100 items is \$800.

(c)
$$f(1000) = 3 \cdot 1000 + 500$$
$$= 3500$$

The cost of manufacturing 1000 items is \$3500.

(d) $f(n)$, or $3n + 500$, is the cost of producing n items. Therefore, solve the equation

$$3n + 500 = 5000$$

to find how many items can be produced for \$5000.

$$3n = 4500$$
$$n = 1500$$

Thus 1500 items can be manufactured for \$5000.

EXERCISES

1. Let $f(x) = x + 4$. Find
   (a) $f(2)$, (b) $f(6)$, (c) $f(0)$, (d) $f(-4)$.

2. Let $g(x) = x - 2$. Find
   (a) $g(1)$, (b) $g(2)$, (c) $g(-2)$, (d) $g(10)$.

3. Let $F(x) = 3x$. Find
   (a) $F(1)$, (b) $F(5)$, (c) $F(-5)$, (d) $F(0)$.

4. Let $G(x) = -x$. Find
   (a) $G(1)$, (b) $G(-1)$, (c) $G(0)$, (d) $G(\pi)$.

5. Let $f(t) = \frac{t}{2}$. Find
   (a) $f(2)$, (b) $f(10)$, (c) $f(-4)$, (d) $f(1)$.

6. Let $g(r) = \frac{r}{3}$. Find
   (a) $g(6)$, (b) $g(24)$, (c) $g(2)$, (d) $g(-9)$.

7. Let $F(u) = -10u$. Find
   (a) $F(0)$, (b) $F(10)$, (c) $F(100)$, (d) $F(-2000)$.

8. Let $G(z) = \frac{-z}{4}$. Find
   (a) $G(4)$, (b) $G(-4)$, (c) $G(-2)$, (d) $G(6)$.

9. Let $f(u) = 2u + 1$. Find
   (a) $f(1)$, (b) $f(-1)$, (c) $f(5)$, (d) $f(-5)$.

10. Let $g(t) = 3t - 1$. Find
    (a) $g(0)$, (b) $g(2)$, (c) $g(-3)$, (d) $g\left(\frac{1}{3}\right)$.

11. Let $F(x) = -x - 2$. Find
    (a) $F(-3)$, (b) $F\left(\frac{-1}{2}\right)$, (c) $F(0)$, (d) $F(2)$.

12. Let $G(x) = 1 - x$. Find
    (a) $G(1)$, (b) $G(4)$, (c) $G\left(\frac{-1}{4}\right)$, (d) $G(-30)$.

13. Let $f(s) = 0$ for all s. Find
    (a) $f(3)$, (b) $f(0)$, (c) $f\left(\frac{2}{27}\right)$, (d) $f(-465\,381)$.

14. Let $g(x) = \pi$ for all x. Find
    (a) $g(-1)$, (b) $g(0)$, (c) $g\left(\frac{\pi}{2}\right)$, (d) $g(-\pi)$.

15. Let $F(x) = 1$, if $x > 0$, and let $F(x) = -1$, otherwise. Find
    (a) $F(1)$, (b) $F\left(\frac{1}{2}\right)$, (c) $F(0)$, (d) $F(-4)$.

16. Let $G(t) = t$, if $t > 0$, and let $G(t) = 0$, otherwise. Find
    (a) $G(1)$, (b) $G\left(\frac{3}{4}\right)$, (c) $G(0)$, (d) $G(-2)$.

17. Let $f(x) = x^2$. Find
(a) $f(1)$, (b) $f(5)$, (c) $f(-5)$, (d) $f(-10)$.
18. Let $g(t) = 2t^2$. Find
(a) $g(0)$, (b) $g(1)$, (c) $g(-2)$, (d) $g\left(\frac{1}{2}\right)$.
19. Let $F(u) = u^2 + 1$. Find
(a) $F(0)$, (b) $F(3)$, (c) $F(-2)$, (d) $F(-10)$.
20. Let $G(z) = 2 - z^2$. Find
(a) $G(0)$, (b) $G(-1)$, (c) $G(2)$, (d) $G\left(\frac{1}{2}\right)$.
21. Let $f(x) = \frac{1}{x}$. Find
(a) $f(1)$, (b) $f(3)$, (c) $f(10)$, (d) $f(-2)$.
22. Let $g(x) = \frac{2}{x}$. Find
(a) $g(1)$, (b) $g(5)$, (c) $g(6)$, (d) $g(-10)$.
23. Let $F(s) = \frac{-1}{s}$. Find
(a) $F(1)$, (b) $F(-1)$, (c) $F(2)$, (d) $F(-4)$.

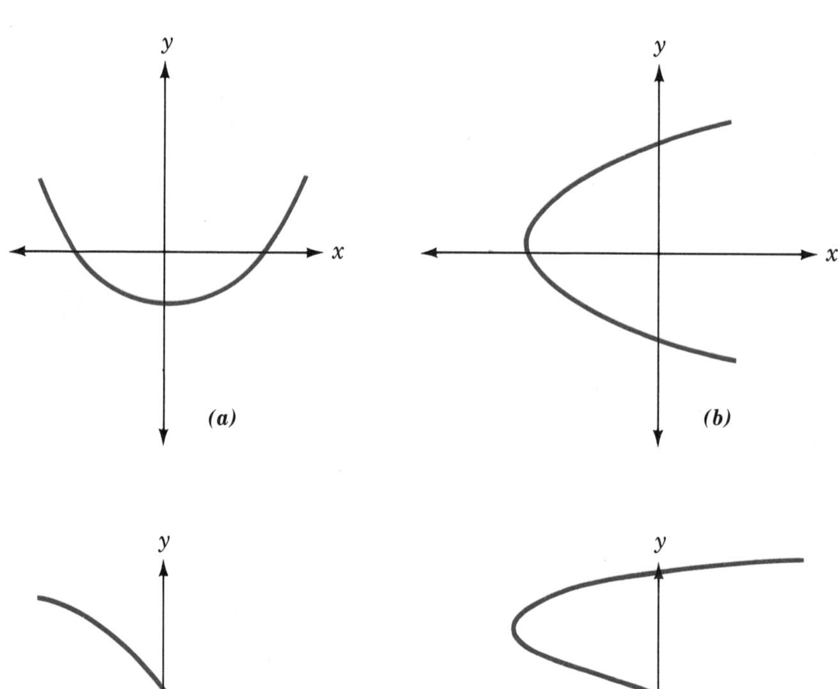

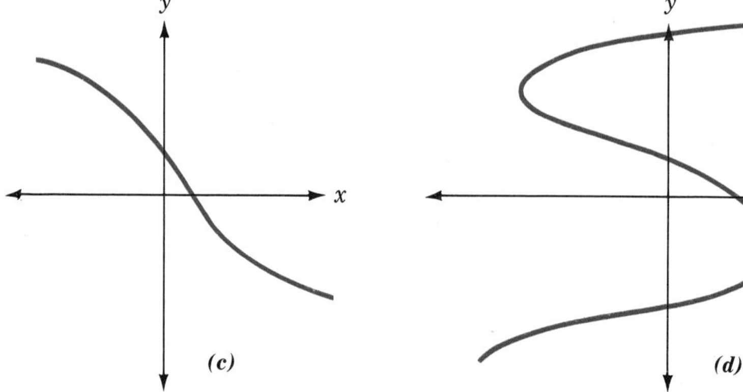

FIGURE 9.15

24. Let $G(t) = \frac{1}{t-1}$. Find
(a) $G(2)$, (b) $G(9)$, (c) $G(-2)$. (d) For which value of t is this function undefined?

25. Let $f(x) = \frac{1}{x}$. Find
(a) $f\left(\frac{1}{2}\right)$, (b) $f\left(\frac{1}{4}\right)$, (c) $f\left(\frac{3}{4}\right)$, (d) $f\left(\frac{-1}{2}\right)$.

26. Let $g(x) = \frac{1}{x+1}$. Find
(a) $g(-2)$, (b) $g\left(\frac{-1}{2}\right)$, (c) $g\left(\frac{1}{2}\right)$. (d) For which value of x is this function undefined?

27. Which of the curves in Figure 9.15 (page 244) represent functions?

28. Which of the lines in Figure 9.16 represent functions?

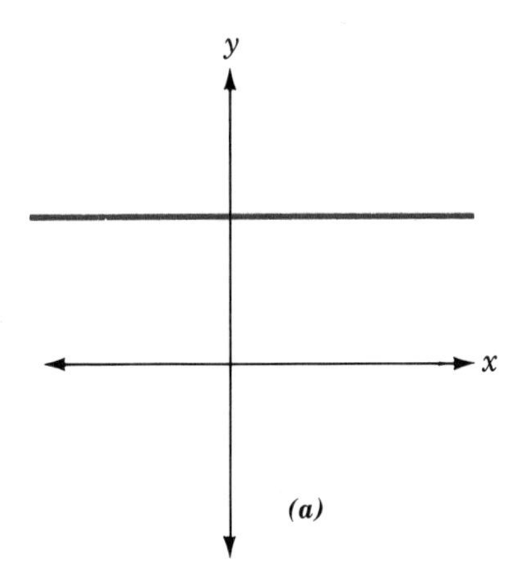

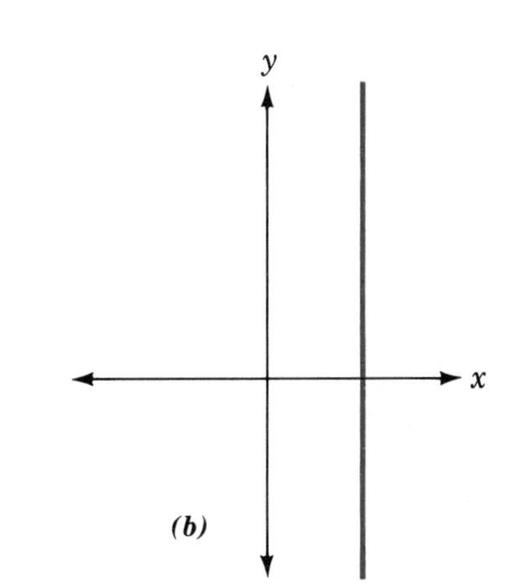

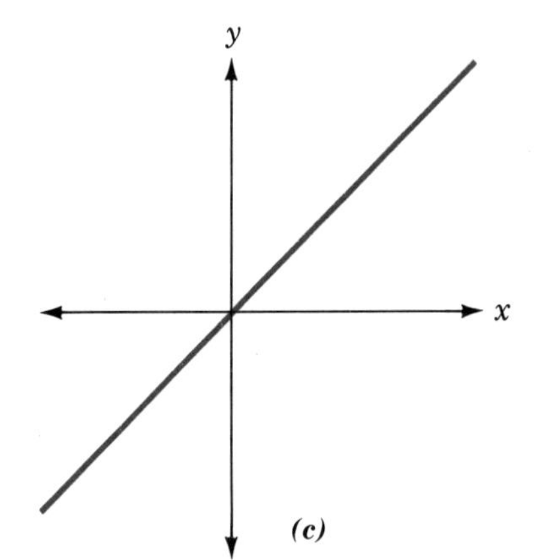

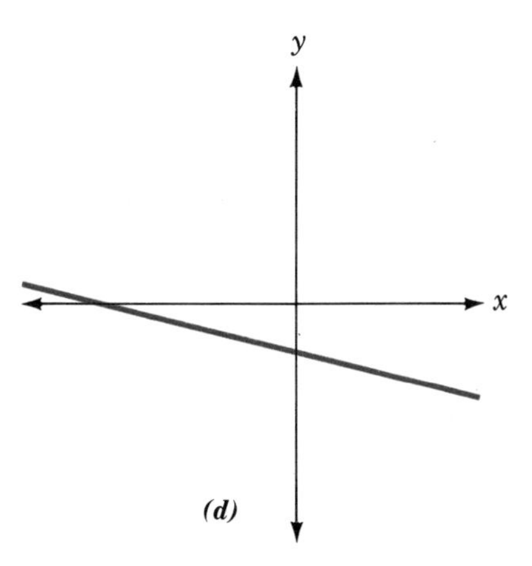

FIGURE 9.16

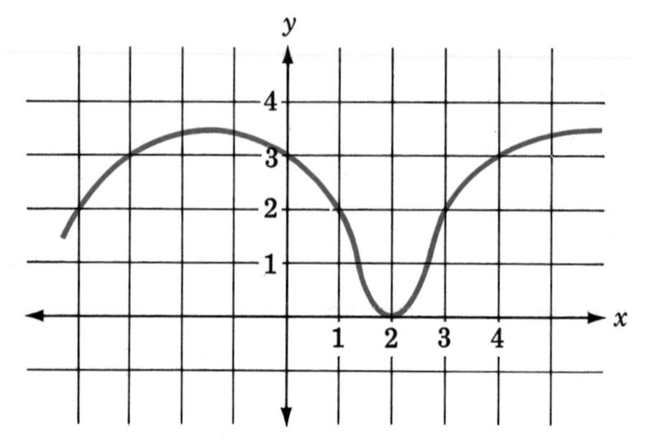

FIGURE 9.17

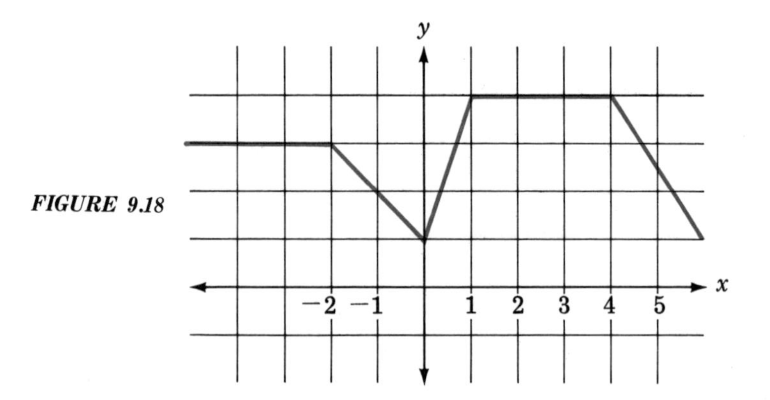

FIGURE 9.18

FIGURE 9.19

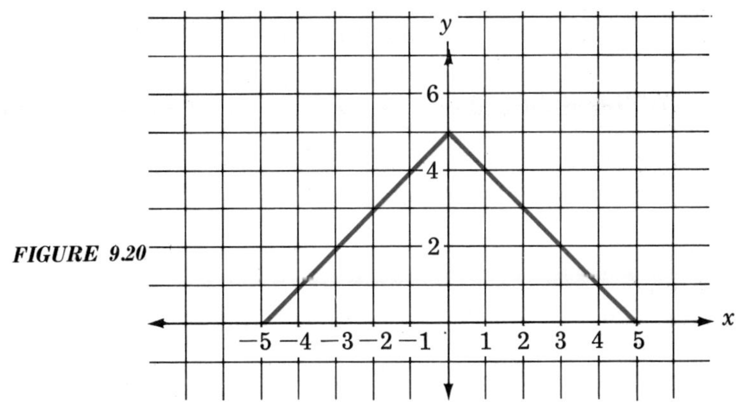

FIGURE 9.20

29. For the function f represented in Figure 9.17, find
    (a) $f(1)$, (b) $f(2)$, (c) $f(3)$, (d) $f(4)$.

30. For the function g represented in Figure 9.18, find
    (a) $g(-2)$, (b) $g(0)$, (c) $g(3)$, (d) $g(5)$.

31. For the function F represented in Figure 9.19, find
    (a) $F\left(\frac{1}{2}\right)$, (b) $F\left(\frac{3}{2}\right)$, (c) $F\left(\frac{-1}{2}\right)$, (d) $F\left(\frac{-5}{2}\right)$.

32. For the function G represented in Figure 9.20, above, find
    (a) $G(2)$, (b) $G(-2)$, (c) $G(5)$, (d) $G(-5)$.

33. (a) Express the distance a car travels at 45 miles an hour as a function of t, the number of hours it travels.
    (b) How far does the car go in 3 hours?
    (c) How far does it go in 4 hours and 20 minutes?

34. (a) Express the area of a square as a function of s, the length of a side.
    (b) Find the area if $s = 12$.

35. (a) Express the circumference of a circle as a function of r, the length of the radius.
    (b) Find the circumference when $r = 8$.

36. The charge for renting a car for a week is \$90 plus 10¢ a mile.
    (a) Express the charge for a week (in dollars) as a function of the number of miles driven.
    (b) What is the weekly rental charge for a 2000 mile trip?
    (c) How far can you travel for \$500?

37. A man is 24 when his son is born.
    (a) Express the man's age as a function of his son's age.
    (b) How old will the man be on his son's 24th birthday?
    (c) How old will the son be when the man is 60?

38. A salesman receives \$150 in wages per week plus a 5% rate of commission on sales.
    (a) Express his total weekly earnings as a function of his sales.
    (b) How much does he earn in a week in which his sales total \$1250?
    (c) What must his sales total in order to earn \$600 for a week's work?

9.3 GRAPHS OF FUNCTIONS

The Graphing Procedure

You have already seen how a curve can indicate a function. If the function is given algebraically, you can often draw the figure that represents it.

Definition

> *The **graph of a function** f is its pictorial representation on the plane. The graph consists of all points (x, y) such that*
>
> $$y = f(x)$$

To graph a function given by a "fairly simple" equation

$$y = f(x):$$

1. Consider several values of x.
2. Find the corresponding values of y.
3. Locate the points (x, y).
4. Connect these points by a curve. (This "curve" may be a straight line.)

Lines

The first few graphs you draw will be lines. The graph of any function of the form

$$y = \underbrace{mx + b}_{f(x)}$$

is always a line. For example, the graph of each of the functions

$$\begin{array}{ll} y = x + 4 & (m = 1,\ b = 4) \\ y = 3x - 5 & (m = 3,\ b = -5) \\ y = -2x & (m = -2,\ b = 0) \\ y = 6 & (m = 0,\ b = 6) \end{array}$$

is a line. In Chapter 10 you will learn that m expresses the "slope" (or steepness) of a line, and that the line intersects the y-axis at $(0, b)$.

Because *a line is determined by two points*, to graph the line, you need only locate two points. Then draw the line connecting them. You may wish to locate a third point to check whether a (straight) line goes through all three points.

EXAMPLE 1

Graph the function given by

$$y = x + 2.$$

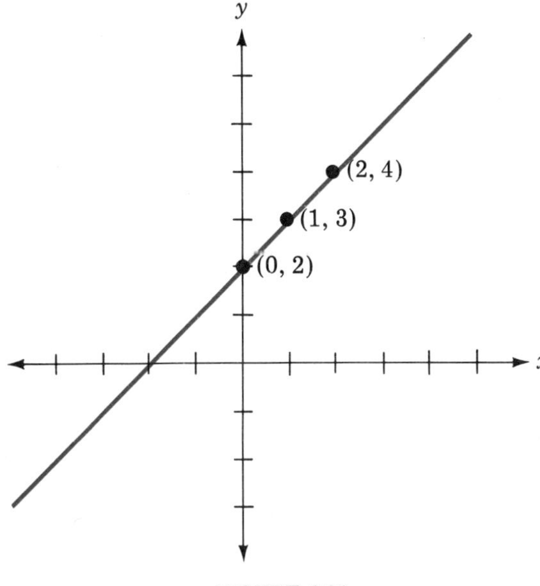

FIGURE 9.21

Solution. The function is of the form

$$y = mx + b$$

with $m = 1$, $b = 2$. Corresponding values of x and y are given in the following table:

x	$y = x + 2$
0	2
1	3
2	4

The graph of the function is drawn in Figure 9.21. Note that this line with equation

$$y = x + \underset{\substack{\uparrow \\ b}}{2}$$

intersects the y-axis at $(0, 2)$.

EXAMPLE 2

Graph the function given by

$$y = -2x.$$

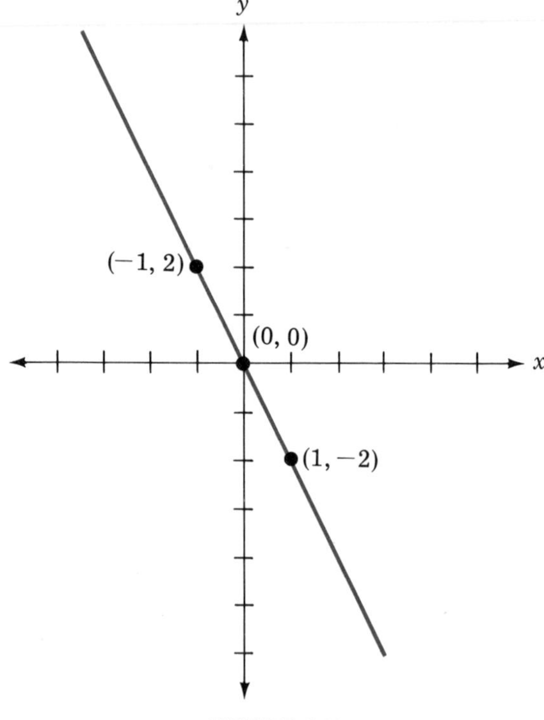

FIGURE 9.22

Solution. This is of the form

$$y = mx + b,$$

with $m = -2$, $b = 0$.

x	$y = -2x$
-1	2
0	0
1	-2

EXAMPLE 3

Graph the function given by

$$y = 3x - 4. \quad \text{(See Figure 9.23)}$$

Solution. This is of the form

$$y = mx + b$$

with $m = 3$, $b = -4$.

x	$y = 3x - 4$
0	-4
1	-1
2	2

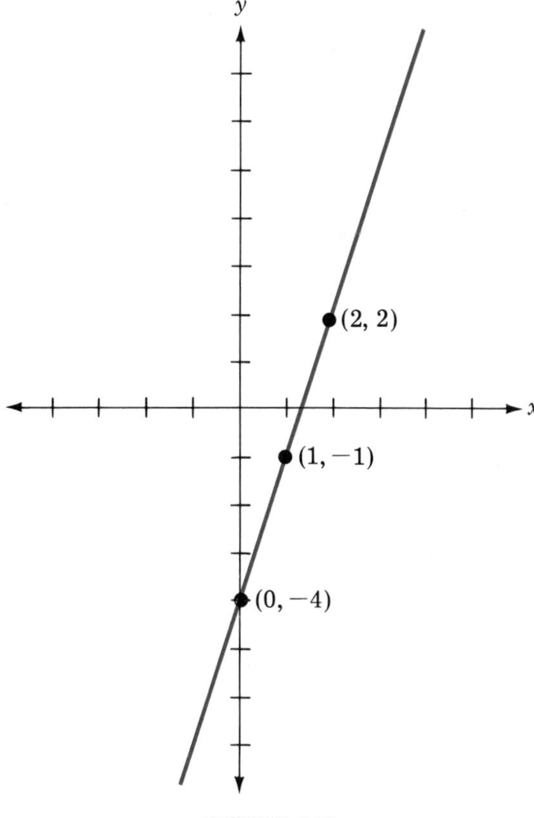

FIGURE 9.23

Absolute Value Function

Recall that $|x|$, the absolute value of x, is the distance from x to the origin. Thus

$$|x| = x, \text{ if } x > 0 \text{ or } x = 0$$

and

$$|x| = -x, \text{ if } x < 0$$

For example, $|9| = 9$, $|-9| = -(-9) = 9$, $|0| = 0$

The function defined by

$$f(x) = |x|$$

is called the **absolute value function.**

EXAMPLE 4

Graph the absolute value function,

$$f(x) = |x|.$$

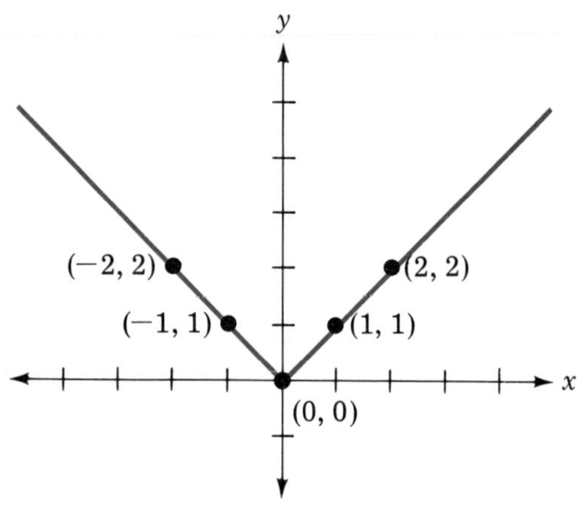

FIGURE 9.24

Solution. Use the conditions

$$|x| = x, \text{ if } x > 0 \text{ or } x = 0;$$
$$|x| = -x, \text{ if } x < 0.$$

x	$y = f(x) = \|x\|$
−2	2
−1	1
0	0
1	1
2	2

Note that $|-2| = -(-2) = 2$. The graph of the absolute value function is V-shaped. It consists of 2 lines meeting at the origin at a right angle. The graph is symmetric with respect to the y-axis, that is, the y-axis acts like a mirror. Whenever (x, y) is on the graph, so is $(-x, y)$.

There are variations of the absolute value function, two of which will be presented. The graphs of these are again V-shaped.

EXAMPLE 5

Graph the functions
(a) $F(x) = |x| + 1$, (b) $G(x) = |x + 1|$.

Solution.
(a) First find $|x|$. Then add 1.

x	$\|x\|$	$y = F(x) = \|x\| + 1$
−2	2	3
−1	1	2
0	0	1
1	1	2
2	2	3

The graph of F is given in Figure 9.25(a).

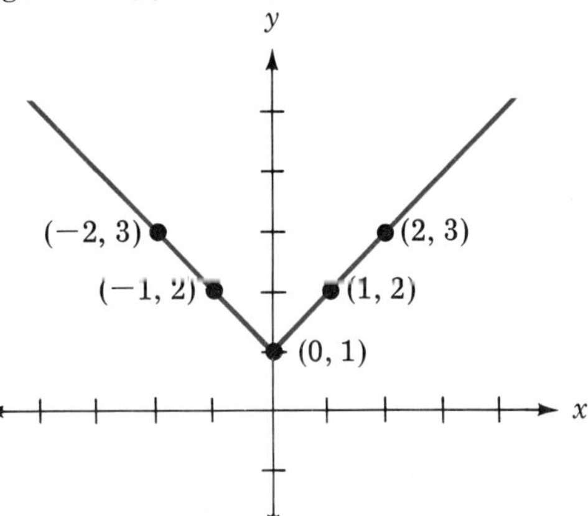

FIGURE 9.25(a) $F(x) = |x| + 1$

(b) First obtain $x + 1$. Then find the absolute value, $|x + 1|$.

x	$x + 1$	$y = G(x)$ $= \|x + 1\|$
−2	−1	1
−1	0	0
0	1	1
1	2	2
2	3	3

The graph of G is given in Figure 9.25(b).

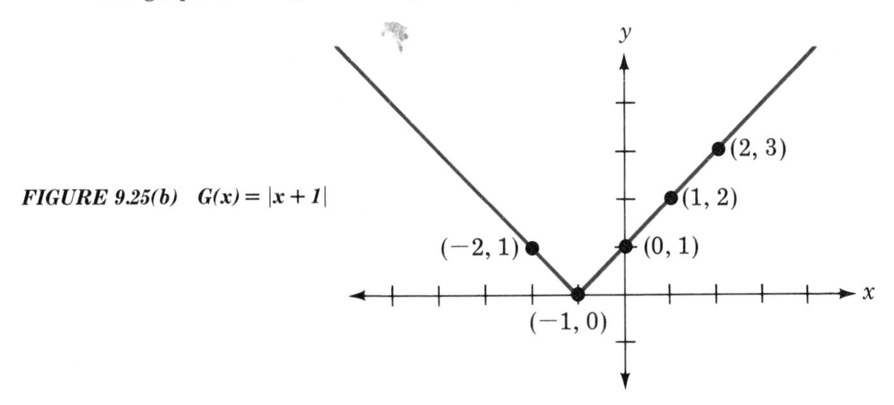

FIGURE 9.25(b) $G(x) = |x + 1|$

Parabolas

The function defined by

$$f(x) = x^2$$

is called the **squaring function.**

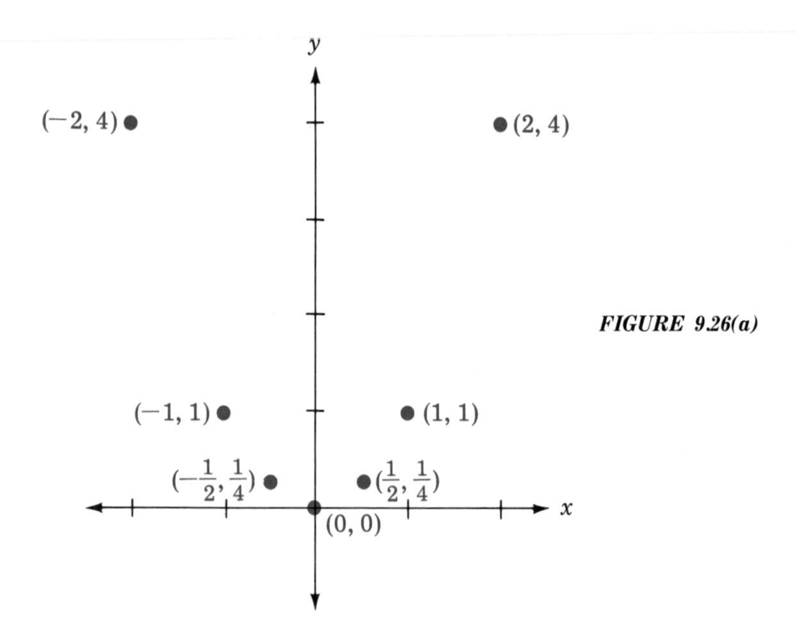

FIGURE 9.26(a)

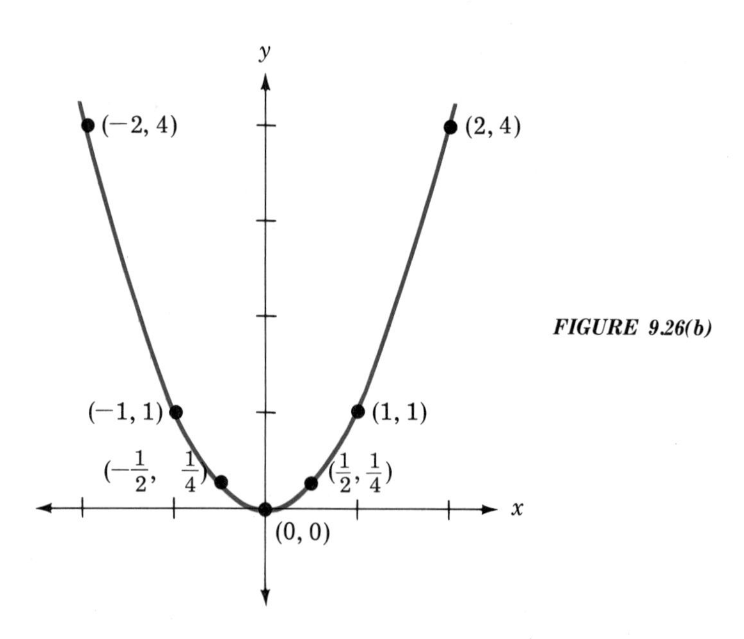

FIGURE 9.26(b)

EXAMPLE 6

Graph the squaring function, given by

$$f(x) = x^2.$$

Solution. First consider several values of x and the corresponding values of $f(x)$.

x	$f(x) = x^2$
-2	4
-1	1
$\frac{-1}{2}$	$\frac{1}{4}$
0	0
$\frac{1}{2}$	$\frac{1}{4}$
1	1
2	4

These points are located in Figure 9.26(a). They are connected by a curve in Figure 9.26(b).

Note that $x^2 = (-x)^2$. Because

$$y = x^2,$$

whenever (x, y) is one the graph, so is $(-x, y)$. The graph is symmetric with respect to the y-axis. The graph is known as a **parabola.** The turning point, $(0, 0)$, is known as the **vertex** of the parabola. The curve turns "smoothly" around the vertex. (In contrast, the graph of the absolute value function had a "sharp point".)

Other parabolas are given in Example 7 and in the Exercises.

EXAMPLE 7

Graph the functions given by
(a) $F(x) = x^2 - 1$, (b) $G(x) = 2x^2$.

Solution.

(a) First square x. Then subtract 1.

x	x^2	$F(x) = x^2 - 1$
-2	4	3
-1	1	0
$\frac{1}{2}$	$\frac{1}{4}$	$\frac{-3}{4}$
0	0	-1
$\frac{1}{2}$	$\frac{1}{4}$	$\frac{-3}{4}$
1	1	0
2	4	3

The graph of F is given in Figure 9.27(a) on page 257. The vertex of this parabola is at $(0, -1)$.

(b) First square x. Then multiply this by 2.

x	x^2	$G(x) = 2x^2$
-2	4	8
-1	1	2
$\frac{-1}{2}$	$\frac{1}{4}$	$\frac{1}{2}$
0	0	0
$\frac{1}{2}$	$\frac{1}{4}$	$\frac{1}{2}$
1	1	2
2	4	8

The graph of G is given in Figure 9.27(b). The vertex of this parabola is at the origin.

EXERCISES

Graph each function.

1. $y = x + 1$
2. $y = x + 4$
3. $y = x - 1$
4. $y = x - 2$
5. $y = x$
6. $y = 2x$
7. $y = -x$
8. $y = \frac{-3x}{2}$
9. $y = 2x + 1$
10. $y = 2x - 3$
11. $y = 3x - 1$
12. $y = 4x - 6$
13. $y = 1 - x$
14. $y = 4 - x$
15. $y = 6 - 2x$
16. $y = 1 - \frac{x}{2}$
17. $f(x) = |x| + 2$
18. $g(x) = |x + 2|$
19. $F(x) = |x| - 1$
20. $G(x) = |x - 1|$
21. $f(x) = 2|x|$
22. $g(x) = |2x|$
23. $F(x) = -|x|$
24. $G(x) = |-x|$
25. $f(x) = x^2 + 1$
26. $g(x) = x^2 + 3$
27. $F(x) = x^2 - 2$
28. $G(x) = (x - 2)^2$
29. $f(x) = 4x^2$
30. $g(x) = \frac{x^2}{2}$
31. $F(x) = -x^2$
32. $G(x) = 1 - x^2$

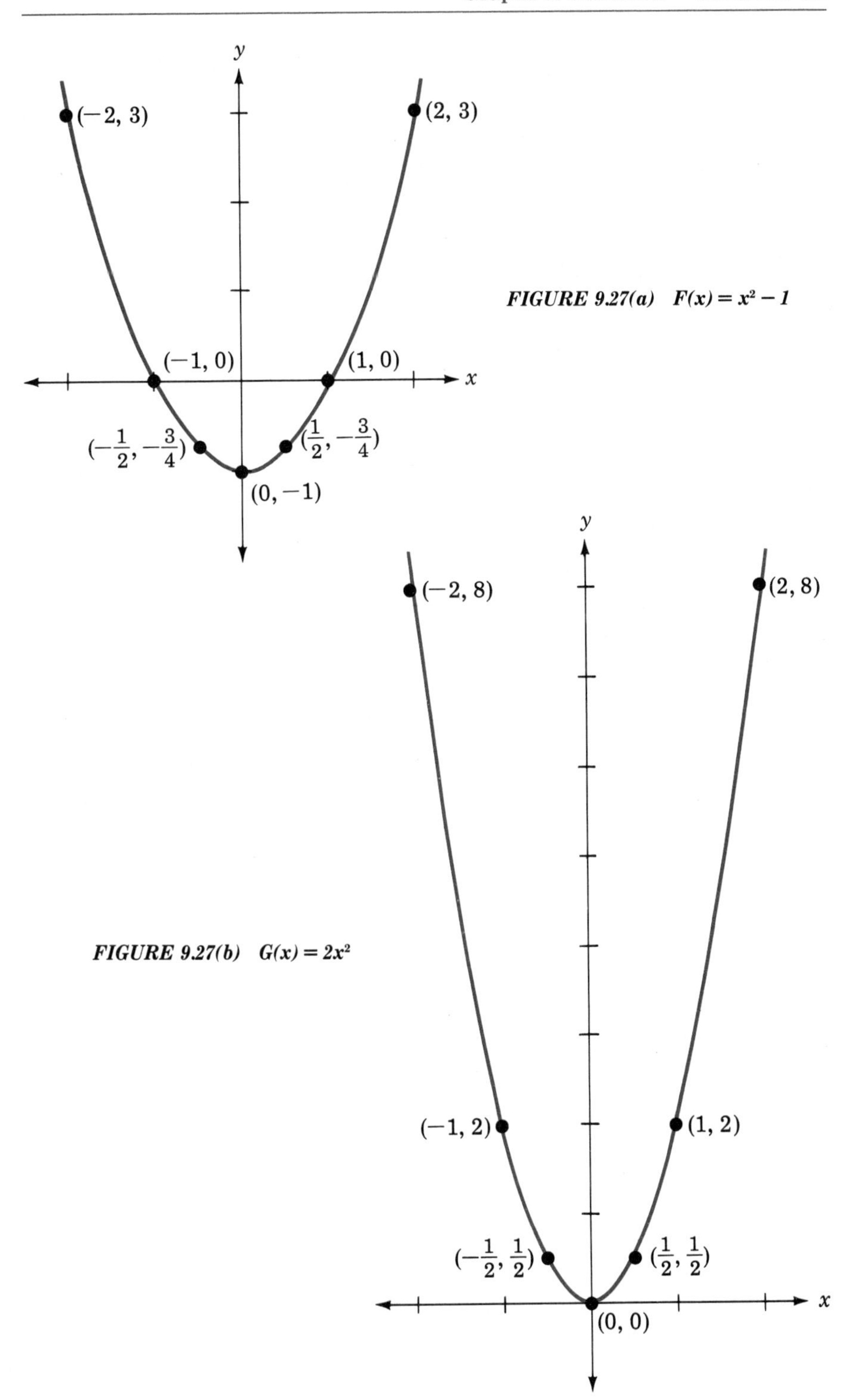

FIGURE 9.27(a) $F(x) = x^2 - 1$

FIGURE 9.27(b) $G(x) = 2x^2$

What Have You Learned in Chapter 9?

You can locate points on a rectangular coordinate system.

You know that a function is a rule that indicates how the dependent variable, say y, depends on the independent variable, say x. For each value of x, there is exactly one corresponding value of y.

And you know that the graph of a function is its pictorial representation on the plane.

Let's Review Chapter 9.

9.1 Rectangular Coordinates

1. For each point in Figure 9.28, find
(a) its x-coordinate, (b) its y-coordinate. (c) Identify the point with its ordered pair of coordinates.

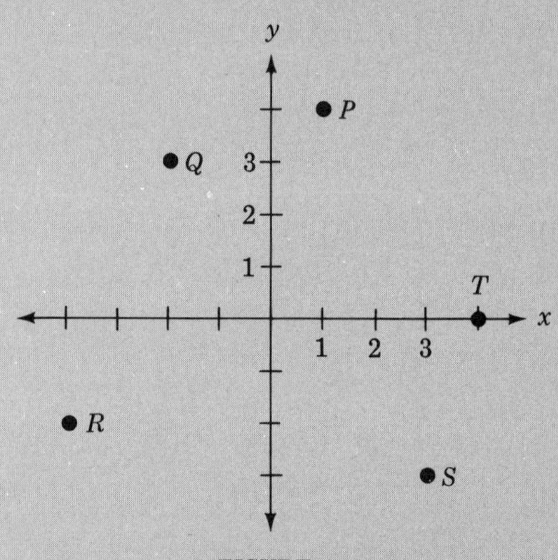

FIGURE 9.28

In exercises 2–6, locate the following points on a rectangular coordinate system:

2. (4, 0) 3. (0, −2) 4. (5, 3) 5. (−3, 4) 6. $\left(\frac{-1}{2}, \frac{-3}{2}\right)$

9.2 What is a Function?

7. Let $f(x) = x + 3$. Find

(a) $f(0)$, (b) $f(3)$, (c) $f\left(\frac{1}{2}\right)$, (d) $f(-3)$.

8. Let $g(x) = -x$. Find
   (a) $g(1)$, (b) $g(0)$, (c) $g(-1)$, (d) $g\left(\frac{-1}{2}\right)$.

9. Let $F(x) = 5x - 4$. Find
   (a) $F(2)$, (b) $F(-1)$, (c) $F\left(\frac{2}{5}\right)$, (d) $F\left(\frac{-1}{5}\right)$.

10. Let $G(x) = 6$ for all x. Find
    (a) $G(0)$, (b) $G(6)$, (c) $G(-6)$, (d) $G(100)$.

11. For which values of x are the following functions undefined?
    (a) $f(x) = \frac{1}{2x}$, (b) $g(x) = \frac{1}{x-2}$, (c) $F(x) = \frac{1}{x+2}$

12. (a) Express the distance a car travels at 60 miles an hour as a function of t, the number of hours it travels.
    (b) How far does it travel in 45 minutes?
    (c) How far does it go in an hour and a half?

9.3 Graphs of Functions

Graph the functions in exercises 13–18.

13. $y = x + 2$
14. $y = 2x$
15. $y = 2 - x$
16. $y = \frac{x}{2}$
17. $y = |x| - 2$
18. $y = x^2 + 2$

And these from Chapters 1–8:

19. Find (a) $|-8|$, (b) $-|8|$, (c) $-|-8|$.
20. Evaluate $x^2 + 100$ when
    (a) $x = -5$, (b) $x = 10$, (c) $x = -20$.
21. Solve the equation:

$$8(x + 5) - 5(x - 8) = 8$$

22. A man is two years older than his wife. Express his age as a function of hers.
23. Let $f(x) = |x + 2|$, $g(x) = |x| + 2$. Fill in "<", "=", or ">".
    (a) $f(-1)$ ☐ $g(-1)$ (b) $f(0)$ ☐ $g(0)$
    (c) $f(1)$ ☐ $g(1)$ (d) $f(10)$ ☐ $g(10)$
24. Let $f(x) = x^2$, $g(x) = 2x$. Fill in "<", "=", or ">".
    (a) $f\left(\frac{1}{2}\right)$ ☐ $g\left(\frac{1}{2}\right)$ (b) $f\left(\frac{3}{2}\right)$ ☐ $g\left(\frac{3}{2}\right)$ (c) $f(2)$ ☐ $g(2)$
    (d) $f(3)$ ☐ $g(3)$

Try These Exam Questions for Practice.

1. For each point in Figure 9.29, find
   (a) its x-coordinate, (b) its y-coordinate. (c) Identify the point with its ordered pair of coordinates.

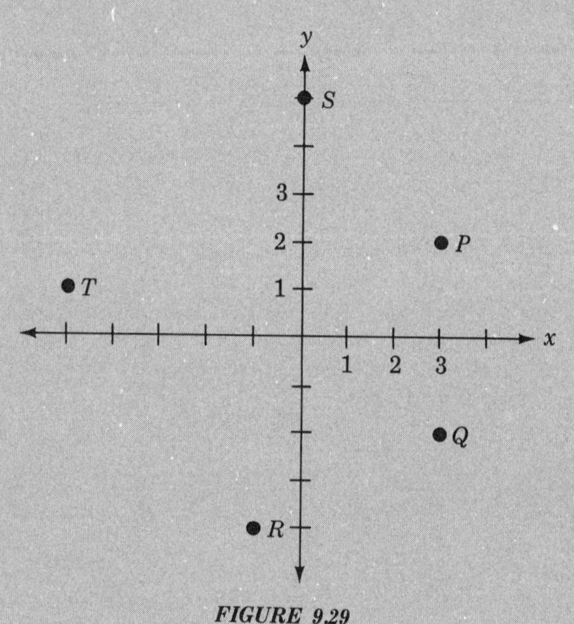

FIGURE 9.29

2. Let $f(x) = 3x - 1$. Find
   (a) $f(2)$, (b) $f\left(\frac{2}{3}\right)$, (c) $f(-2)$, (d) $f\left(\frac{1}{2}\right)$.
3. Let $g(x) = \frac{1}{x+3}$.
   (a) Find $g(1)$, (b) Find $g(-1)$, (c) For which value of x is the function undefined.
4. A doctor charges $25 for the first visit and $20 for each subsequent visit. Describe a patient's fee as a function of the number of visits.
5. Graph the function given by

$$y = 2x + 3.$$

10

Lines and Their Equations

10.1 SLOPE

Rise and Run

In Chapter 9 you learned that the graph of a function of the form

$$y = mx + b$$

is always a line. In this chapter you will study lines from an algebraic point of view.

The first matter of importance is the change in the y-coordinates along the line as the x-coordinates change. Do the y-coordinates increase or decrease as the x-coordinates increase?

Because you will be considering several x-coordinates as well as several y-coordinates, it is useful to use *letters with subscripts,* such as

$$x_1, x_2, \quad y_1, y_2$$

Thus let x_1 and x_2 be the x-coordinates of two different points on a line, and let y_1 and y_2 be the corresponding y-coordinates. Also, let

$$P_1 = (x_1, y_1) \text{ and } P_2 = (x_2, y_2)$$

represent these two points. (See Figure 10.1 on page 262.)

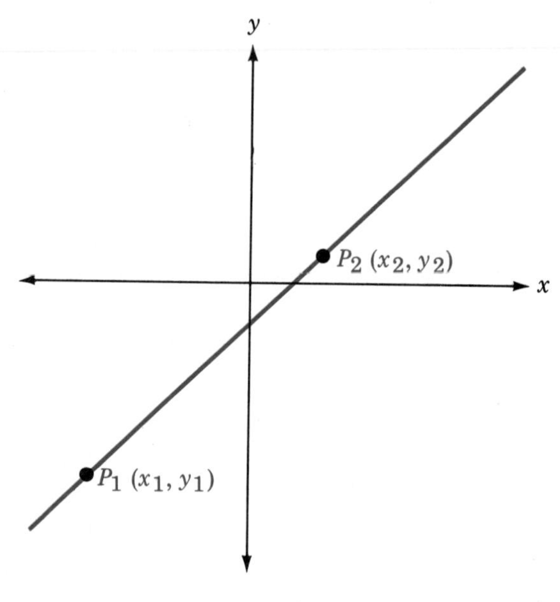

FIGURE 10.1

Definition

> *SLOPE. The* ***rise*** *or the* ***change in y along the line from P_1 to P_2*** *is defined to be*
>
> $$y_2 - y_1$$
>
> *The* ***run*** *or the* ***change in x along the line from P_1 to P_2*** *is defined to be*
>
> $$x_2 - x_1$$

Thus

$$\text{rise} = y_2 - y_1$$
$$\text{run} = x_2 - x_1$$

The **slope of the line** is defined to be

$$\frac{y_2 - y_1}{x_2 - x_1}$$

provided $x_2 \neq x_1$. (See Figure 10.2.)

Thus, if $x_2 \neq x_1$ (so that $x_2 - x_1 \neq 0$),

$$\text{slope of the line} = \frac{\text{rise}}{\text{run}} = \frac{y_2 - y_1}{x_2 - x_1}$$

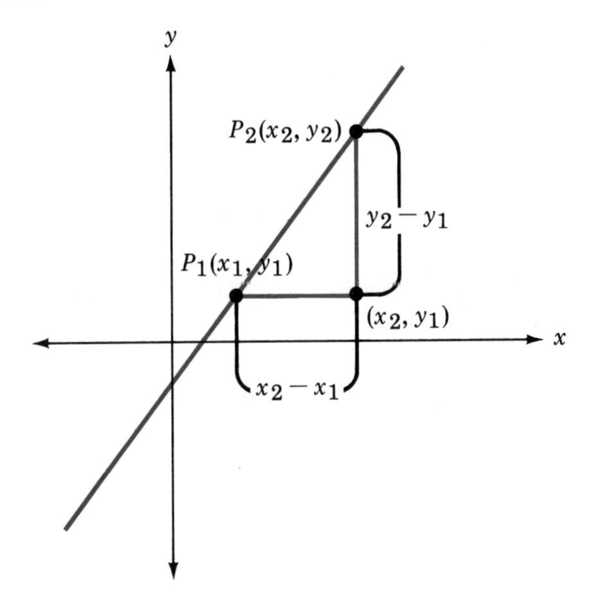

FIGURE 10.2. ***rise*** $= y_2 - y_1$, ***run*** $= x_2 - x_1$, ***slope*** $= \dfrac{y_2 - y_1}{x_2 - x_1}$

EXAMPLE 1

Suppose

$$P_1 = (4, 2) \text{ and } P_2 = (6, 3)$$

are two points on a line. (See Figure 10.3 on page 264.)

(a) Find the rise from P_1 to P_2.
(b) Find the run from P_1 to P_2.
(c) Find the slope of the line.

Solution. Here $x_1 = 4$, $y_1 = 2$, $x_2 = 6$, $y_2 = 3$

(a) $\text{rise} = y_2 - y_1$
$= 3 - 2$
$= 1$

(b) $\text{run} = x_2 - x_1$
$= 6 - 4$
$= 2$

(c) $\text{slope} = \dfrac{\text{rise}}{\text{run}}$
$= \dfrac{1}{2}$

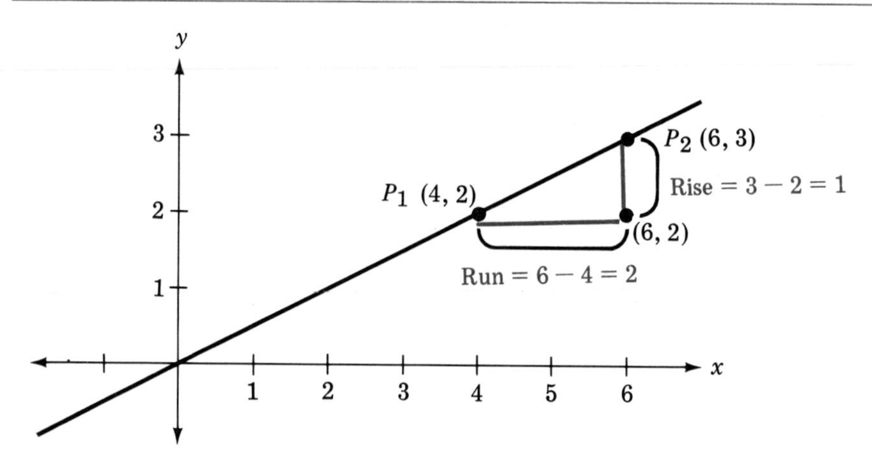

FIGURE 10.3

EXAMPLE 2

Suppose $P_1 = (5, 3)$ and $P_2 = (1, 5)$ are two points on a line. (Fig. 10.4)
(a) Find the rise from P_1 to P_2.
(b) Find the run from P_1 to P_2.
(c) Find the slope of the line.

Solution. Here $x_1 = 5$, $y_1 = 3$, $x_2 = 1$, $y_2 = 5$

(a) $\text{rise} = y_2 - y_1$
$= 5 - 3$
$= 2$

(b) $\text{run} = x_2 - x_1$
$= 1 - 5$
$= -4$

(c) $\text{slope} = \dfrac{\text{rise}}{\text{run}}$
$= \dfrac{2}{-4}$
$= \dfrac{-1}{2}$

Note that in Example 1 the slope was positive; in Example 2 the slope was negative. *When the rise and run are both of the same sign, the slope* $\left(= \dfrac{\text{rise}}{\text{run}}\right)$ *is positive. In this case the line slopes upward to the right.* [See Figure 10.5(a).] *When the rise and run are of different signs, the slope is negative. In this case the line slopes upward to the left.* [See Figure 10.5(b).]

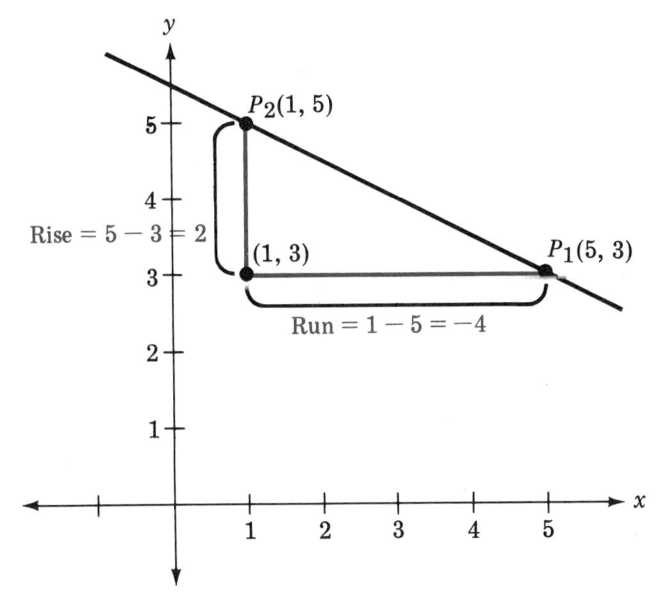

FIGURE 10.4

The slope of a line does not depend upon which points P_1 and P_2 are considered. It can be shown that no matter which points of the line are chosen, the slope of the line is the same. Also, in choosing two points it makes no difference which is called P_1 and which P_2. In fact,

$$\frac{y_1 - y_2}{x_1 - x_2} = \frac{-(y_2 - y_1)}{-(x_2 - x_1)} = \frac{y_2 - y_1}{x_2 - x_1}$$

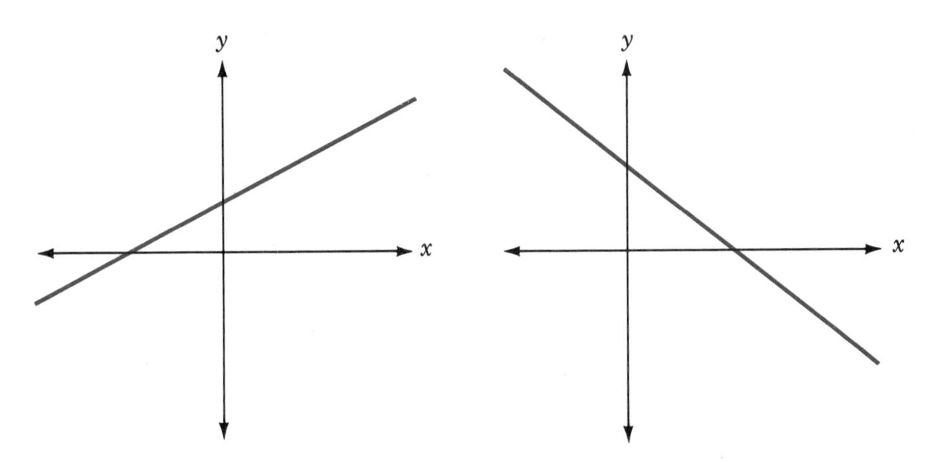

FIGURE 10.5(a). Line slopes upward to the right. Slope is positive.

FIGURE 10.5(b). Line slopes upward to the left. Slope is negative.

EXAMPLE 3

Find the slope of the line through (1, 3) and (0, −2)
(a) by letting $P_1 = (1, 3)$ and $P_2 = (0, -2)$,
(b) by letting $P_1 = (0, -2)$ and $P_2 = (1, 3)$.
(c) Suppose you are given that (2, 8) is also on the line. Find the slope by letting $P_1 = (1, 3)$ and $P_3 = (2, 8)$.

Solution.
(a) [See Figure 10.6(a).] Here $x_1 = 1, y_1 = 3, x_2 = 0, y_2 = -2$

$$\begin{aligned} \text{slope} &= \frac{y_2 - y_1}{x_2 - x_1} \\ &= \frac{-2 - 3}{0 - 1} \\ &= \frac{-5}{-1} \\ &= 5 \end{aligned}$$

(b) [See Figure 10.6(b).] Here $x_1 = 0, y_1 = -2, x_2 = 1, y_2 = 3$

$$\begin{aligned} \text{slope} &= \frac{y_2 - y_1}{x_2 - x_1} \\ &= \frac{3 - (-2)}{1 - 0} \\ &= \frac{5}{1} \\ &= 5 \end{aligned}$$

(c) [See Figure 10.6(c).] Let $P_3 = (x_3, y_3)$. Here $x_1 = 1, y_1 = 3, x_3 = 2, y_3 = 8$

$$\begin{aligned} \text{slope} &= \frac{y_3 - y_1}{x_3 - x_1} \\ &= \frac{8 - 3}{2 - 1} \\ &= \frac{5}{1} \\ &= 5 \end{aligned}$$

Locating Points on a Line

EXAMPLE 4

Draw the line through (3, 5) with slope 2.

Solution. Let $P_1 = (3, 5)$. To find a second point on the line, increase x by 1 unit. The new x-coordinate is thus 3 + 1, or 4. In order to find the y-coordinate of this point, consider the change in y. Because the

$$\begin{aligned} \text{slope} &= 2, \text{ and} \\ \text{slope} &= \frac{\text{change in } y}{\text{change in } x}, \\ 2 &= \frac{\text{change in } y}{1}. \end{aligned}$$

The change in y is 2, and the new y-coordinate is 5 + 2, or 7.

Let $P_2 = (4, 7)$. Now draw the line through P_1 and P_2, as in Figure 10.7 on page 268.

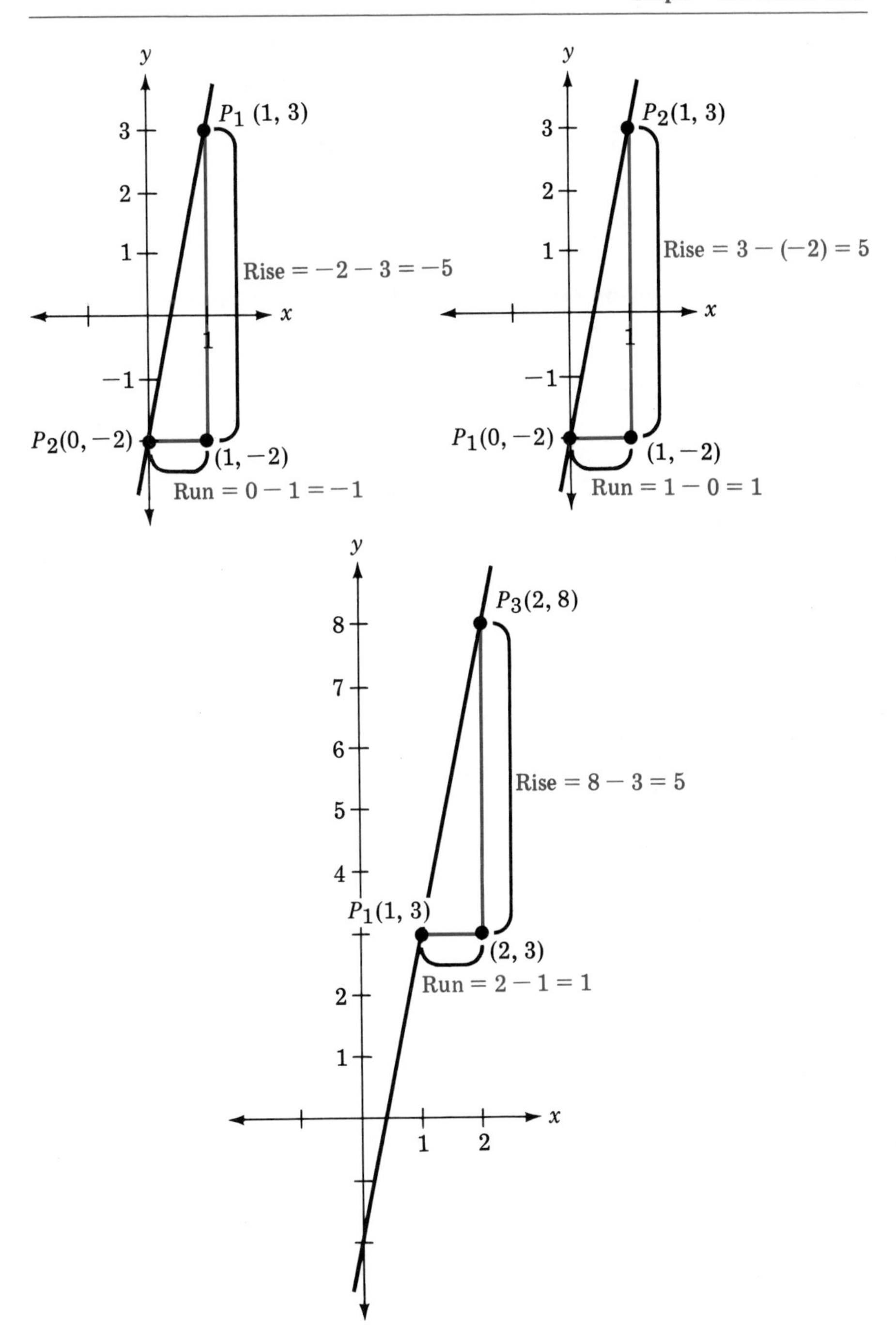

***FIGURE 10.6**(a). **slope** $=\frac{-5}{-1}=5$. (b). **slope** $=\frac{5}{1}=5$.*

*(c). **slope** $=\frac{8-3}{3-2}=5$.*

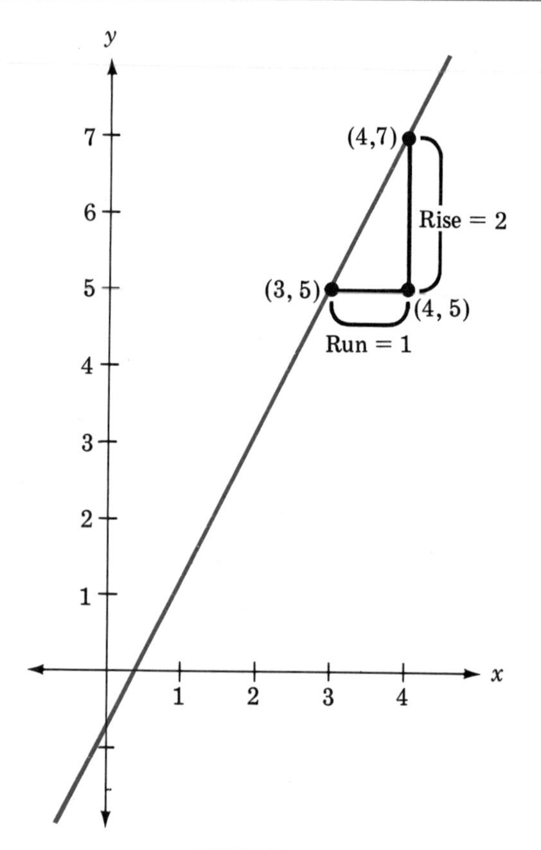

FIGURE 10.7

EXAMPLE 5

Consider the line through (0, 0) and (4, 3). Find the y-coordinate of the point on this line with x-coordinate 12.

Solution. First find the slope.
Let $(x_1, y_1) = (0, 0)$ and $(x_2, y_2) = (4, 3)$.

$$\begin{aligned} \text{slope} &= \frac{y_2 - y_1}{x_2 - x_1} \\ &= \frac{3 - 0}{4 - 0} \\ &= \frac{3}{4} \end{aligned}$$

Now consider the point P_3 on the line with x-coordinate 12. Let $P_3 = (12, y_3)$. The slope, $\frac{3}{4}$, can be written using P_1 and P_3. Thus

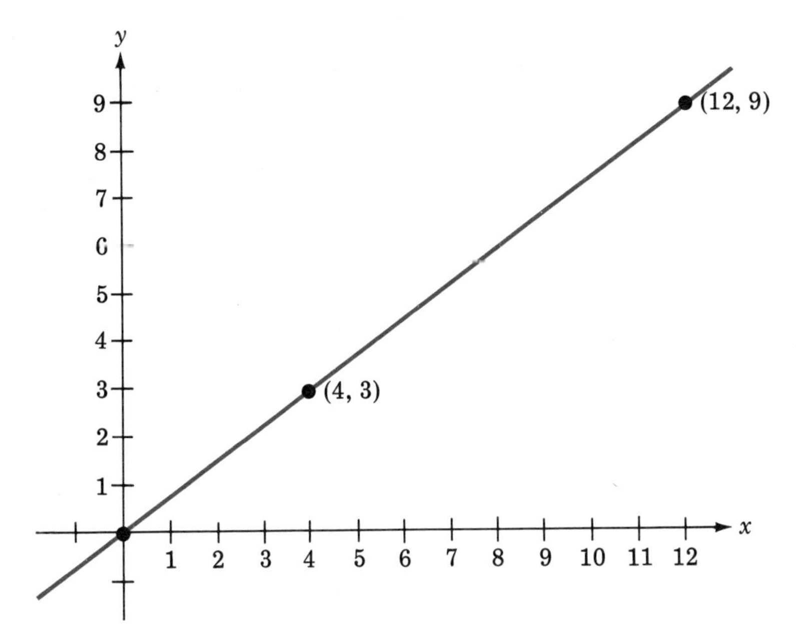

FIGURE 10.8

$$\text{slope} = \frac{y_3 - y_1}{x_3 - x_1}$$

$$\frac{3}{4} = \frac{y_3 - 0}{12 - 0}$$

$$\frac{3}{4} = \frac{y_3}{12} \qquad \text{Cross-multiply.}$$

$$36 = 4y_3$$

$$9 = y_3$$

See Figure 10.8.

EXAMPLE 6

Consider the line through (2, 6) and (−2, 8). Find the x-coordinate of the point on this line with y-coordinate 9.

Solution.

$$\text{slope} = \frac{y_2 - y_1}{x_2 - x_1}$$

$$= \frac{8 - 6}{-2 - 2}$$

$$= \frac{2}{-4}$$

$$= \frac{-1}{2}$$

Let $P_3 = (x_3, y_3)$ be the point on the line with y-coordinate 9. Thus

$$\text{slope} = \frac{y_3 - y_1}{x_3 - x_1}$$
$$\frac{-1}{2} = \frac{9 - 6}{x_3 - 2}$$
$$\frac{-1}{2} = \frac{3}{x_3 - 2} \qquad \text{Cross-multiply.}$$

$$-(x_3 - 2) = 6$$
$$2 - x_3 = 6 \qquad \text{Add } x_3 - 6 \text{ to both sides.}$$

See Figure 10.9.

$$-4 = x_3$$

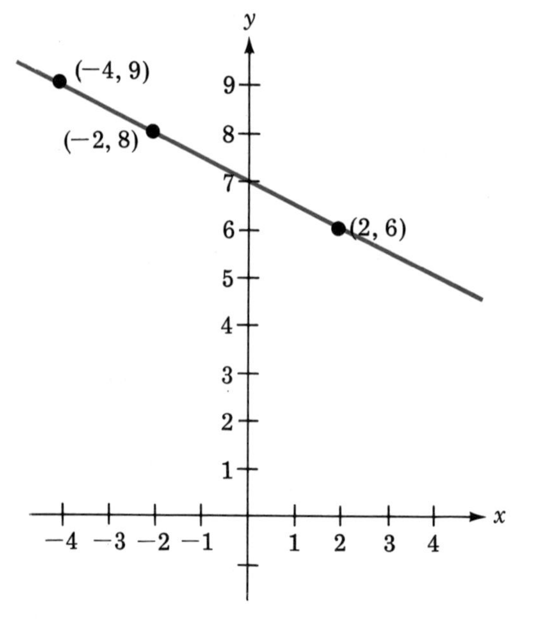

FIGURE 10.9

Horizontal and Vertical Lines

Every point on a *horizontal* line has the same y-coordinate. For example, if one point has y-coordinate 4, every point on the line has y-coordinate 4. Therefore $y = 4$ for every value of x. (See Figure 10.10) Thus the rise (or the change in y) from P_1 to P_2 is always 0. Also,

$$\text{slope} = \frac{\text{rise}}{\text{run}}$$
$$= \frac{0}{\text{run}}$$
$$= 0$$

The slope of a horizontal line is 0.

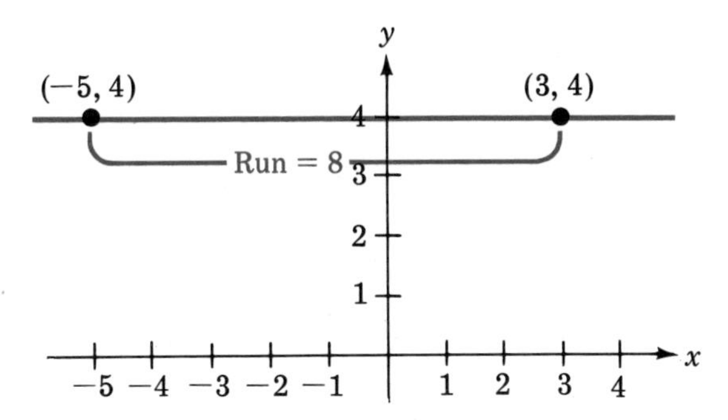

FIGURE 10.10. *slope* $=\frac{rise}{run}=\frac{0}{8}=0$. ***The slope of a horizontal line is 0.***

Every two points on a *vertical* line have the same x-coordinate. Thus if some point on the line has x-coordinate 5, every point has x-coordinate 5. Consequently, $x = 5$ for every value of y. (See Figure 10.11) Therefore, the run from P_1 to P_2 is always 0. The slope is defined as $\frac{\text{rise}}{\text{run}}$. Because the run is 0 and division by 0 is not defined, the slope is not defined. Thus *the slope of a vertical line is undefined.*

FIGURE 10.11. The run is 0. The slope of a vertical line is undefined.

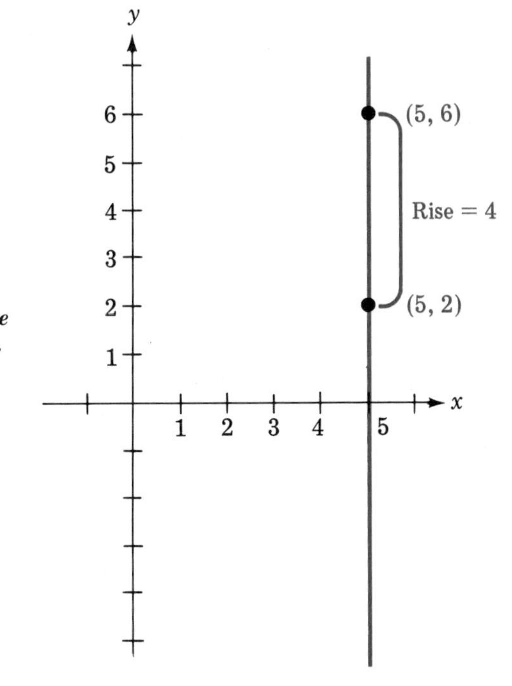

When two different points on a line have the same y-coordinate, the line is horizontal. When two different points have the same x-coordinate, the line is vertical.

EXERCISES

In exercises 1–20, suppose P_1 and P_2 are two points on a line. (a) Find the rise from P_1 to P_2, (b) find the run from P_1 to P_2, and (c) find the slope of the line.

1. $P_1 = (1, 1), P_2 = (2, 2)$
2. $P_1 = (1, 2), P_2 = (2, 5)$
3. $P_1 = (3, 3), P_2 = (4, 5)$
4. $P_1 = (0, 0), P_2 = (4, 1)$
5. $P_1 = (-1, -1), P_2 = (1, 0)$
6. $P_1 = (-2, -1), P_2 = (0, 0)$
7. $P_1 = (8, 3), P_2 = (9, 1)$
8. $P_1 = (2, 4), P_2 = (-2, 5)$
9. $P_1 = (3, 2), P_2 = (0, 1)$
10. $P_1 = (1, 0), P_2 = (0, 2)$
11. $P_1 = (4, 0), P_2 = (0, 2)$
12. $P_1 = (-3, 0), P_2 = (0, 3)$
13. $P_1 = (-7, -4), P_2 = (-3, -3)$
14. $P_1 = (-3, -2), P_2 = (-1, -1)$
15. $P_1 = \left(\frac{1}{2}, \frac{1}{2}\right), P_2 = (1, 2)$
16. $P_1 = \left(\frac{1}{4}, \frac{1}{2}\right), P_2 = \left(\frac{1}{2}, \frac{1}{4}\right)$
17. $P_1 = (2, 3), P_2 = (4, 3)$
18. $P_1 = (-1, -1), P_2 = (1, -1)$
19. $P_1 = (-7, 1), P_2 = (7, -1)$
20. $P_1 = \left(3, \frac{1}{2}\right), P_2 = \left(\frac{1}{2}, \frac{-1}{2}\right)$

In exercises 21–30, draw the line through the point P with the given slope.

SAMPLE. Draw the line through (1, 4) with slope 3.	***Solution.*** See Figure 10.12.

21. $P = (1, 1)$, slope $= 1$
22. $P = (0, 2)$, slope $= 2$
23. $P = (3, 1)$, slope $= -1$
24. $P = (5, -2)$, slope $= -2$
25. $P = (3, 3)$, slope $= 0$
26. $P = (1, -1)$, slope $= \frac{1}{2}$
27. $P = (4, 0)$, slope $= \frac{-1}{2}$
28. $P = (-1, -2)$, slope $= \frac{1}{4}$
29. $P = (1, 5)$, slope $= 5$
30. $P = (2, 4)$, slope undefined

In exercises 31–36, consider the line determined by P_1 and P_2. Find the y-coordinate, y_3, of the point (x_3, y_3) on this line.

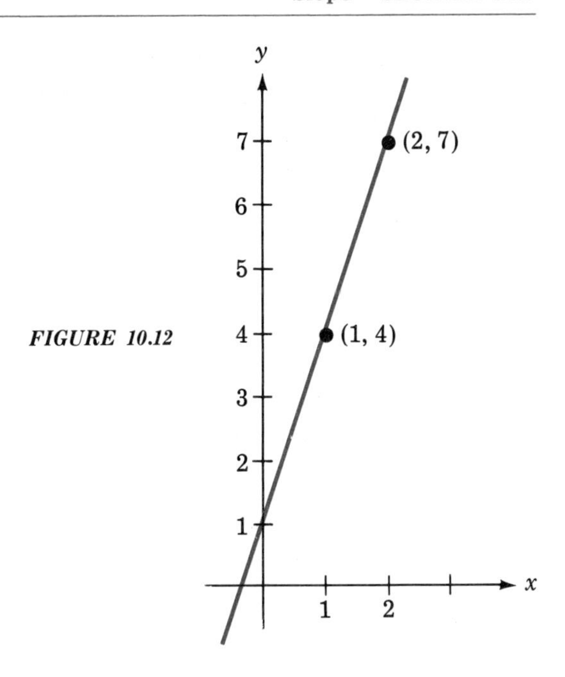

FIGURE 10.12

SAMPLE. $P_1 = (1, 2)$, $P_2 = (2, 5)$, $x_3 = 3$	***Solution.*** $\text{slope} = \dfrac{5-2}{2-1} = 3$ Also, $\text{slope} = \dfrac{y_3 - 2}{x_3 - 1}$; $3 = \dfrac{y_3 - 2}{3 - 1} = \dfrac{y_3 - 2}{2}$; $6 = y_3 - 2$; $8 = y_3$

31. $P_1 = (1, 3), P_2 = (2, 5), x_3 = 4$
32. $P_1 = (4, 0), P_2 = (0, 4), x_3 = 3$
33. $P_1 = (-2, 1), P_2 = (2, 2), x_3 = 0$
34. $P_1 = (8, 4), P_2 = (4, 6), x_3 = 0$
35. $P_1 = (-2, -1), P_2 = (1, 0), x_3 = -5$
36. $P_1 = (3, 1), P_2 = (-2, 1), x_3 = 4$

In exercises 37–42, consider the line determined by P_1 and P_2. Find the x-coordinate, x_3, of the point (x_3, y_3) on this line.

37. $P_1 = (4, 1), P_2 = (1, 4), y_3 = 5$
38. $P_1 = (3, 1), P_2 = (1, 4), y_3 = 7$
39. $P_1 = (0, 2), P_2 = (4, 0), y_3 = -4$
40. $P_1 = (8, 3), P_2 = (3, 13), y_3 = 8$
41. $P_1 = (1, 4), P_2 = (-1, -4), y_3 = 0$
42. $P_1 = (2, 5), P_2 = (1, 7), y_3 = 6$

In exercises 43 – 50, consider the line through P_1 and P_2. Indicate whether this line *(i)* is horizontal, *(ii)* is vertical, *(iii)* slopes upward to the right, *(iv)* slopes upward to the left.

SAMPLE. $P_1 = (4, 2)$, $P_2 = (4, 8)$	***Solution.*** *(ii)* [The line through (4, 2) and (4, 8) is vertical because two points on the line have the same x-coordinate.]

43. $P_1 = (1, 2), P_2 = (2, 6)$
44. $P_1 = (2, 4), P_2 = (2, 3)$
45. $P_1 = (3, 8), P_2 = (5, 8)$
46. $P_1 = (0, 6), P_2 = (-6, 0)$
47. $P_1 = (4, 0), P_2 = (-4, 0)$
48. $P_1 = (1, 10), P_2 = (1, -10)$
49. $P_1 = \left(\frac{1}{2}, \frac{1}{4}\right), P_2 = \left(\frac{1}{4}, \frac{1}{2}\right)$
50. $P_1 = \left(\frac{1}{2}, \frac{1}{3}\right), P_2 = \left(\frac{1}{4}, \frac{1}{3}\right)$

10.2 EQUATION OF A LINE

Slope and y-intercept

The graph of a function of the form

$$y = mx + b$$

is a line. To see the significance of m and b, first consider the line with equation

$$y = 2x + 5.$$

$$\text{When } x = 0, y = 2 \cdot 0 + 5 = 5$$
$$\text{When } x = 1, y = 2 \cdot 1 + 5 = 7$$

Thus the points $P_1 = (0, 5)$ and $P_2 = (1, 7)$ are on the line. The slope of the line is given by

$$\frac{\text{rise}}{\text{run}} = \frac{7-5}{1-0} = \frac{2}{1} = 2.$$

Therefore the line with equation

$$y = 2x + 5 \qquad \text{has slope 2.}$$

Similarly, it can be shown that the line with equation

$$y = \frac{x}{2} - 3 \qquad \text{has slope } \frac{1}{2}$$

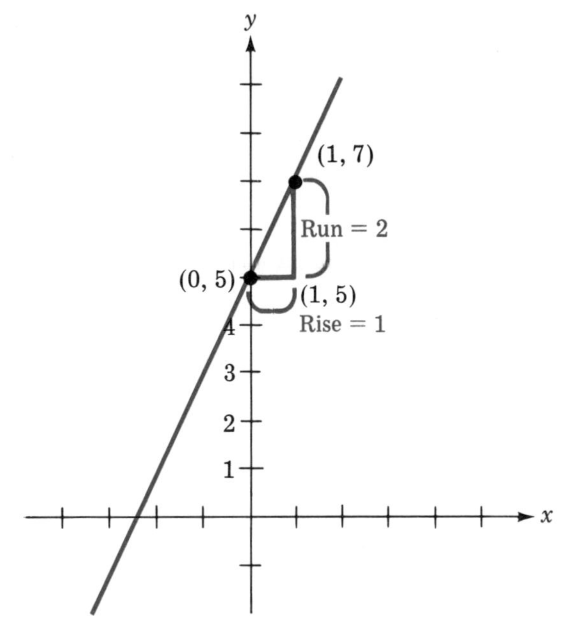

FIGURE 10.13. The line with equation $y = 2x + 5$ has slope 2 and y-intercept 5.

and that the line with equation

$$y = -4x + 5 \qquad \text{has slope } -4.$$

In general, *the line with equation*

$$y = mx + b \qquad \textit{has slope } m.$$

Let us return to the line given by

$$y = 2x + 5.$$

Replace x by 0 in this equation, and obtain

$$y = 2 \cdot 0 + 5$$

or

$$y = 5.$$

Thus the point (0, 5) is on the line because its coordinates satisfy the equation

$$y = 2x + 5.$$

Points with x-coordinate 0 are on the y-axis. Therefore (0, 5) is the point at which the line crosses the y-axis. The y-coordinate, 5, of this intersection point is known as the "y-intercept" of the line. (See Figure 10.13.)

In general, the ***y*-intercept** of a line is the y-coordinate of its intersection with the y-axis. To find the y-intercept of the line given by

$$y = mx + b,$$

replace x by 0 in this equation, and obtain

$$y = m \cdot 0 + b$$

or

$$y = b.$$

Thus *the line with equation*

$$y = mx + b$$

has y-intercept b.

To sum up, for the line with equation

$$y = mx + b,$$
$$m = \text{slope},$$
$$b = y\text{-intercept}.$$

EXAMPLE 1

Find (a) the slope and (b) the y-intercept of the line with equation

$$y = 4x - 1.$$

Solution.
(a) $m = 4$. Thus the slope is 4.
(b) $b = -1$. Thus the y-intercept is -1.

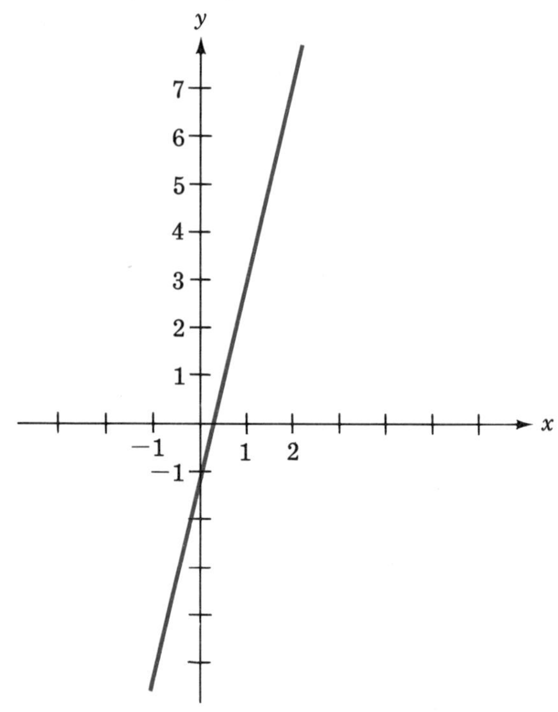

FIGURE 10.14

When you know the equation of a line in the form

$$y = mx + b,$$

you already know one point on the line—namely $(0, b)$. This is because b is the y-intercept. To graph the line, find another point. If you let

$$x = 1, \text{ then } y = m \cdot 1 + b = m + b.$$

Thus in Example 1, $m + b = 4 - 1 = 3$, and the point $(1, 3)$ is on the line.

EXAMPLE 2

Find

(a) the slope and

(b) the y-intercept of the line with equation

$$y = \frac{3x - 8}{4}.$$

(c) Graph the line.

Solution. Rewrite the equation of the line in the form

$$y = \frac{3}{4}x - \frac{8}{4}$$

or

$$y = \frac{3}{4}x - 2.$$

(a) $m = \text{slope} = \frac{3}{4}$

(b) $b = y\text{-intercept} = -2$

(c) When $x = 0$, $y = b = -2$. When $x = 1$, $y = \frac{3}{4} - 2 = \frac{3 - 8}{4} = \frac{-5}{4}$

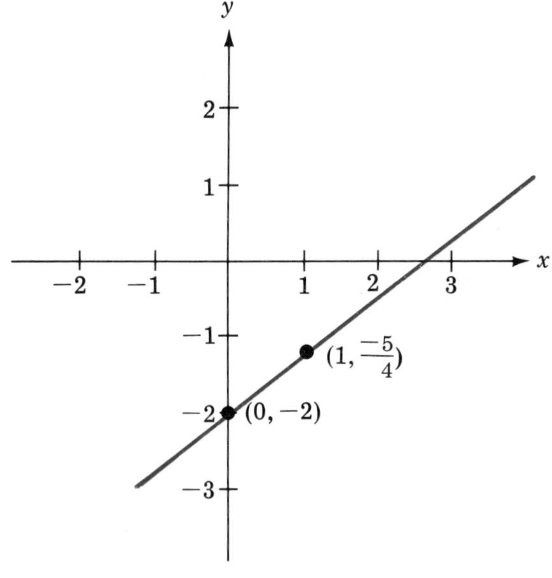

FIGURE 10.15

EXAMPLE 3

(a) Find the equation of the line with slope 4 and y-intercept 3.
(b) Graph this line.

Solution.
(a) $m = 4$, $b = 3$. Thus the equation of the line is

$$y = 4x + 3.$$

(b) When $x = 0$, $y = 3$. When $x = 1$, $y = 4 + 3 = 7$

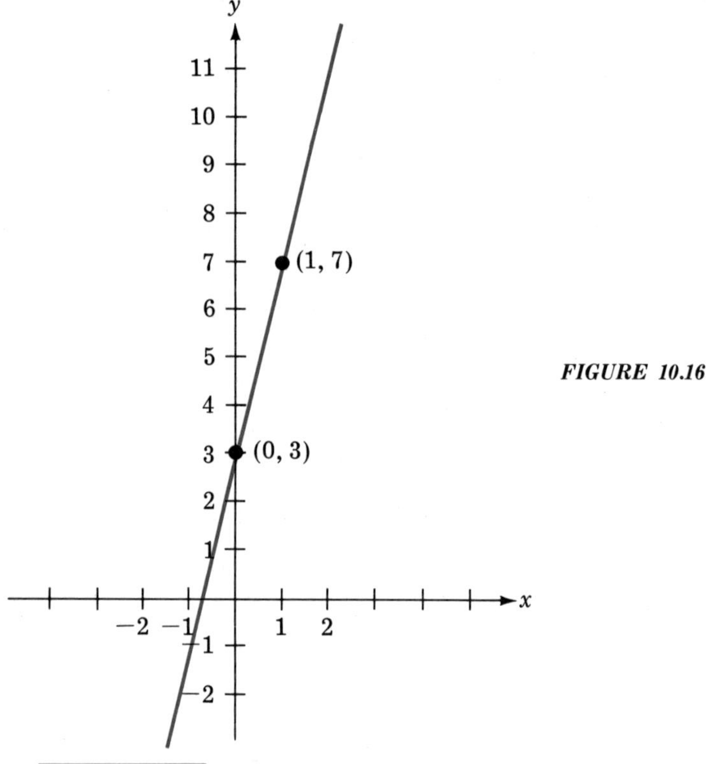

FIGURE 10.16

EXAMPLE 4

(a) Find the equation of the line with slope $\frac{-1}{2}$ and y-intercept 0.
(b) Graph this line.

Solution.
(a) $m = \frac{-1}{2}$, $b = 0$. The equation of this line is

$$y = \frac{-1}{2}x$$

or

$$y = \frac{-x}{2}.$$

(b) When $x = 0$, $y = 0$. When $x = 1$, $y = \frac{-1}{2}$

FIGURE 10.17

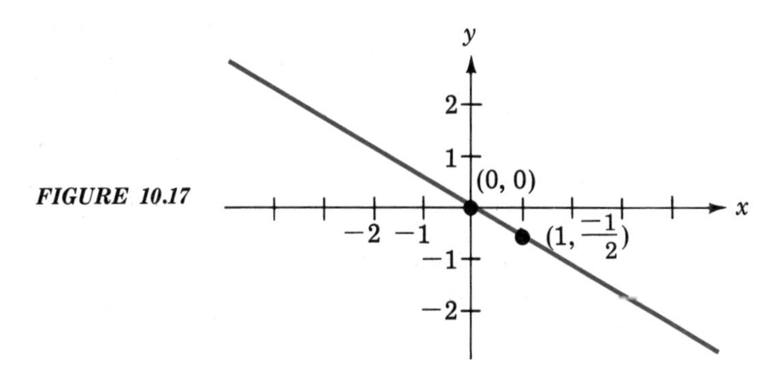

Even if you are not given *both* the slope and the y-intercept, you may still be able to determine the equation of the line if you are given other information.

EXAMPLE 5

(a) Find the equation of the line that crosses the x-axis at (6, 0) and the y-axis at (0, 3).
(b) Graph this line.

Solution.

(a) The y-intercept, b, is 3 because the line crosses the y-axis at (0, 3). You can find the slope, m, by letting $P_1 = (6, 0)$ and $P_2 = (0, 3)$. Thus

$$m = \frac{3-0}{0-6}$$
$$= \frac{-1}{2}$$

and

$$y = mx + b$$
$$y = \frac{-x}{2} + 3$$

(b) You know two points, (6, 0) and (0, 3), on the line. Draw the line through these two points. See Figure 10.18.

FIGURE 10.18

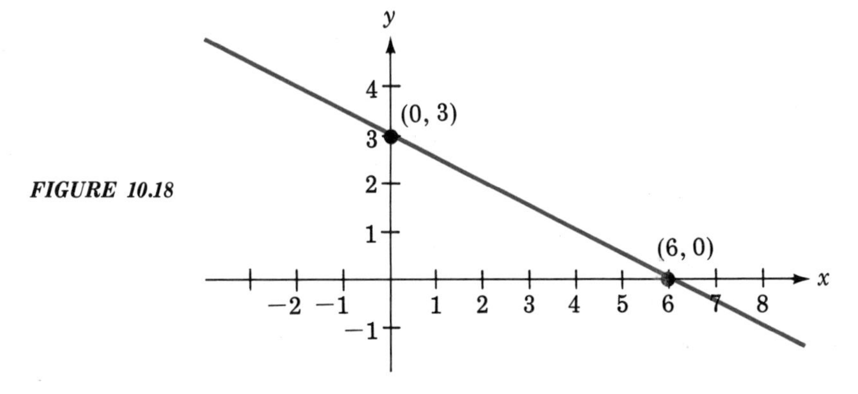

Horizontal and Vertical Lines

Horizontal lines have slope 0. Thus $m = 0$, and, instead of

$$y = mx + b,$$

the equation of a horizontal line is simply

$$y = b,$$

where b is the y-intercept.

EXAMPLE 6

(a) Find the equation of the horizontal line with y-intercept 5.
(b) Graph this line.

Solution.
(a) $b = 5$. Thus the equation of the line is

$$y = 5.$$

(b) When $x = 0$, $y = 5$. When $x = 1$, $y = 5$. See Figure 10.19.

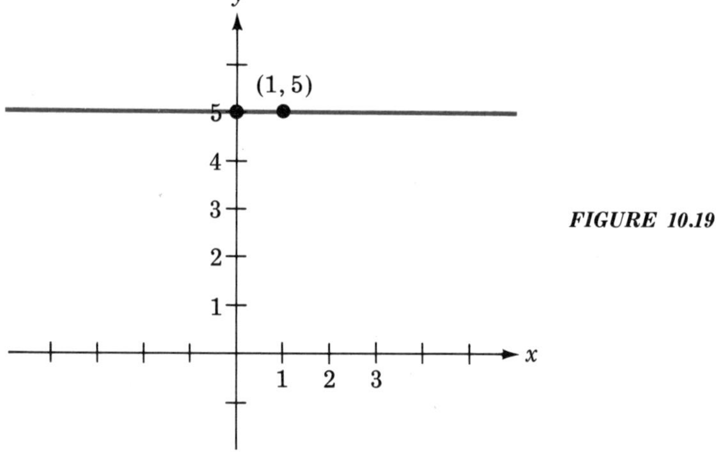

FIGURE 10.19

Actually, if you know *any* point (x_1, y_1) on a *horizontal* line, you can find its equation. For every point on a horizontal line has the same y-coordinate, y_1. Therefore, the equation of the line is

$$y = y_1.$$

EXAMPLE 7

Find the equation of the horizontal line that passes through $(2, -2)$.

Solution. Here $y_1 = -2$, the y-coordinate of $(2, -2)$. Thus the equation of the line is

$$y = -2.$$

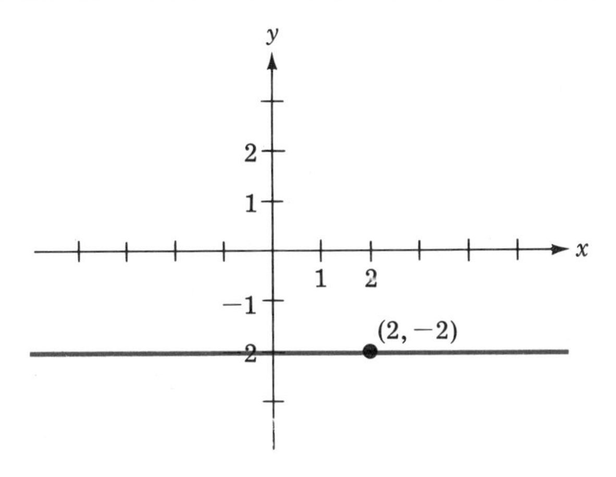

FIGURE 10.20

The x-axis is the horizontal line with y-intercept 0. Thus the equation of the x-axis is

$$y = 0.$$

For a vertical line the slope is undefined, and the equation

$$y = mx + b$$

no longer applies. Instead, observe that every point on a vertical line has the same x-coordinate. Thus, all you need to know is a single point, (x_1, y_1), on a vertical line, and you then know this x-coordinate, x_1. The equation of the line is then

$$x = x_1.$$

EXAMPLE 8

Find the equation of the vertical line that passes through $(4, -5)$.

Solution. The x-coordinate of the given point is 4. Thus the equation of the vertical line is

$$x = 4.$$ See Figure 10.21 on page 282.

The y-axis is the *vertical* line that passes through $(0, 0)$. Thus the equation of the y-axis is

$$x = 0.$$

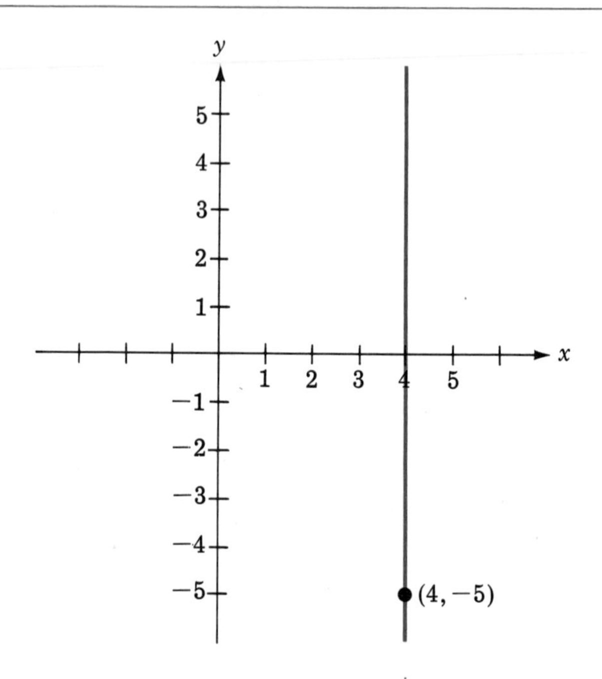

FIGURE 10.21

EXERCISES

In exercises 1–16, find (a) the slope, and (b) the y-intercept of the line with the given equation.

1. $y = 2x + 4$
2. $y = 3x - 2$
3. $y = \frac{x}{2} + 5$
4. $y = -x - 1$
5. $y = 2 - x$
6. $y = \frac{3x + 1}{4}$
7. $y = \frac{x + 2}{2}$
8. $y = \frac{4 - x}{2}$
9. $y = 3(x + 1)$
10. $y = -2(x + 2)$
11. $y = -(2 - 3x)$
12. $x + y = 1$
13. $2y = x + 8$
14. $3y = x - 6$
15. $\frac{y}{2} = x + 5$
16. $-y = x + \frac{1}{2}$

In exercises 17–24, find the equation of the line with the given slope and y-intercept.

17. slope 4, y-intercept 1
18. slope 2, y-intercept 2
19. slope -1, y-intercept 5
20. slope 10, y-intercept 10
21. slope $\frac{1}{2}$, y-intercept $\frac{1}{4}$
22. slope $\frac{-1}{2}$, y-intercept -3
23. slope -2, y-intercept $\frac{1}{2}$
24. slope 0, y-intercept 3

In exercises 25–30, find (a) the slope, and (b) the y-intercept of the line with the given equation. (c) Graph this line.

25. $y = 2x + 1$
26. $y = -x + 4$
27. $y = -2x + 1$
28. $y = 3x$
29. $y = \frac{x+1}{2}$
30. $y = \frac{-3-x}{2}$

In exercises 31–42, find the equation of the indicated line.

31. x-intercept 3, y-intercept 3. (The **x-intercept** of a line is the x-coordinate of its intersection with the x-axis.)
32. x-intercept 1, y-intercept 2
33. x-intercept 2, y-intercept 1
34. x-intercept 2, y-intercept -1
35. the horizontal line with y-intercept 3
36. the horizontal line that passes through $(1, -2)$
37. the horizontal line that passes through $(-1, 2)$
38. the vertical line with x-intercept 1
39. the vertical line that passes through $(2, -3)$
40. the vertical line that passes through $(-2, 3)$
41. the line that passes through the origin and $(2, 4)$
42. the line that passes through the origin and $(4, 2)$

In exercises 43–48, (a) find the equation of the indicated line, and (b) graph this line.

43. x-intercept 2, y-intercept 3. (See Exercise 31.)
44. x-intercept -4, y-intercept 2
45. the horizontal line with y-intercept -3
46. the horizontal line that passes through $(4, -1)$
47. the vertical line that passes through $(4, -1)$
48. the line that passes through the origin and $(2, 6)$

10.3 INTERSECTION OF LINES

Two different lines (on the plane) that are not parallel intersect (or meet) at a single point.

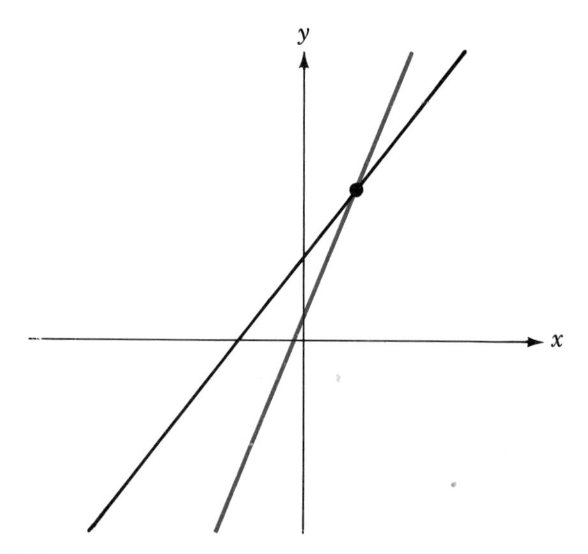

FIGURE 10.22(a). Two different lines (on the plane) that are not parallel intersect at a single point.

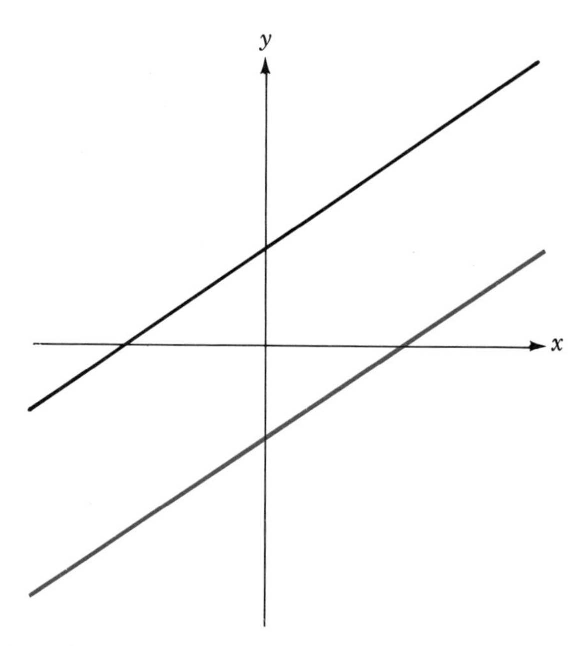

FIGURE 10.22(b). Parallel lines are lines (on the same plane) that do not intersect.

Now you will find the intersection point by graphing both lines on the same coordinate system. In Section 10.4, you will solve this problem algebraically.

Intersecting Lines

EXAMPLE 1

Find the intersection of the lines L_1, given by

$$y = x + 1,$$

and L_2, given by

$$y = 2x - 1.$$

Solution.

L_1:

x	y
0	1
1	2

L_2:

x	y
0	−1
1	1

Graph both lines on the same coordinate system. As you see in Figure 10.23, the intersection is the point (2, 3).

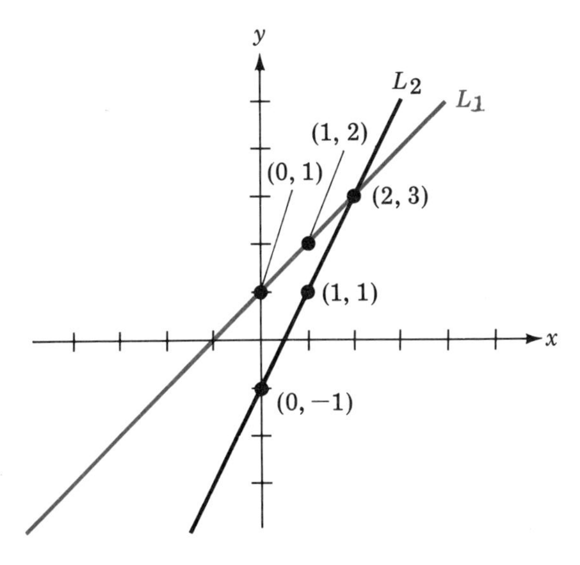

FIGURE 10.23

CHECK.
You can check your graphical solution by replacing x by 2 and $\boldsymbol{y}$ by **3** in each of the given equations:

$$\begin{array}{ll} \boldsymbol{y} = x + 1 & \boldsymbol{y} = 2x - 1 \\ \mathbf{3} \stackrel{?}{=} 2 + 1 & \mathbf{3} \stackrel{?}{=} 2 \cdot 2 - 1 \\ 3 \stackrel{\checkmark}{=} 3 & 3 \stackrel{\checkmark}{=} 3 \end{array}$$

In Example 1, the check showed that you found the *exact* point by graphing. However, in general, the intersection point can only be *approximated* by graphing. It is often difficult to approximate closely an intersection point, such as $\left(\frac{3}{5}, \frac{4}{7}\right)$, with fractional coordinates.

EXAMPLE 2

Find the intersection of the lines L_1, given by

$$y - 2 = 3(x - 5),$$

and L_2, given by

$$2y = x - 6.$$

Solution. Add 2 to both sides of the first equation:

$$\begin{array}{rl} y - 2 &= 3(x - 5) \\ y &= 3(x - 5) + 2 \\ \text{or} \quad y &= 3x - 13 \end{array}$$

L_1:

x	y
5	2
6	5

Divide both sides of the second equation by 2:

$$\begin{array}{rl} 2y &= x - 6 \\ y &= \dfrac{x - 6}{2} \end{array}$$

L_2:

x	y
0	−3
2	−2

As you see in Figure 10.24, the intersection is the point $(4, -1)$.

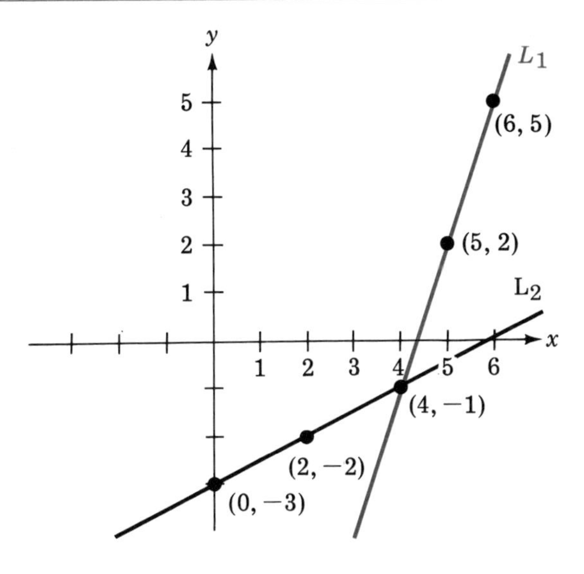

FIGURE 10.24

EXAMPLE 3

Find the intersection of the lines L_1, given by

$$y = -4x,$$

and L_2, given by

$$y + 3 = 2x.$$

Solution.

L_1:

x	y
0	0
1	−4

Subtract 3 from both sides of the second equation:

$$y + 3 = 2x$$
$$y \quad = 2x - 3$$

L_2:

x	y
0	−3
1	−1

As you see in Figure 10.25 on page 288, the intersection is the point $\left(\frac{1}{2}, -2\right)$.

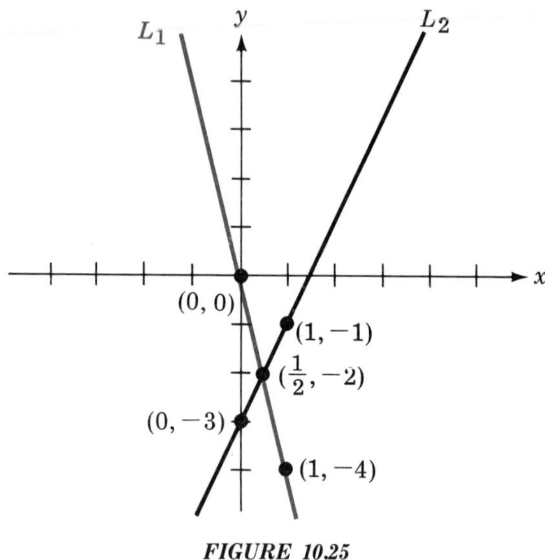

FIGURE 10.25

Parallel Lines

Definition

> ***Parallel lines*** *are lines (in the same plane) that do not intersect.*

EXAMPLE 4

Show that the lines L_1, given by

$$y = 5 - 2x,$$

and L_2, given by

$$y = \frac{1 - 4x}{2},$$

are parallel.

Solution.

L_1:

x	y
0	5
1	3

L_2:

x	y
0	$\frac{1}{2}$
1	$\frac{-3}{2}$

For any value of x, the y-coordinates of the corresponding points on L_1 and L_2

are $4\frac{1}{2}$ units apart. For example, for each x, let y_1 be the y-coordinate of the point on L_1, and let y_2 be the y-coordinate of the corresponding point on L_2. (See Figure 10.26.)

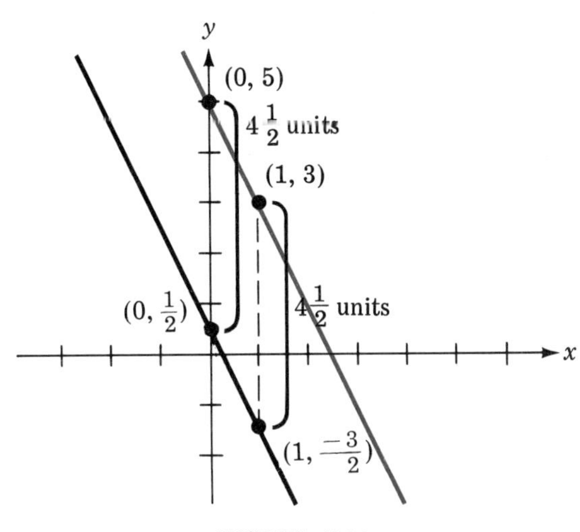

FIGURE 10.26

$$\text{When } x = 0,\ y_2 - y_1 = 5 - \tfrac{1}{2} = 4\tfrac{1}{2}$$
$$\text{When } x = 1,\ y_2 - y_1 = 3 - \left(\tfrac{-3}{2}\right) = 4\tfrac{1}{2}$$

In Example 4, observe that both lines have slope -2. *Two different* (nonvertical) *lines are parallel if they have the same slope and intersect if they have different slopes.* Thus the lines given by the equation

$$y = 2x + 3 \quad \text{and} \quad y = 2x - 1$$

are parallel because each of these lines has slope 2. But the lines given by

$$y = 3x \quad \text{and} \quad y = x + 2$$

intersect because the first line has slope 3, whereas the second has slope 1.

EXERCISES

In exercises 1–12, find the intersection of the given lines, graphically. In exercises 1–4, check your solution, as in Example 1, p. 286.

1. L_1: $y = x$
   L_2: $y = 4 - x$

2. L_1: $y = 2x$
   L_2: $y - 3 = \frac{x}{2}$

3. L_1: $y - 4 = x$
   L_2: $y - 2 = 2(x + 1)$

4. L_1: $y + 2 = \frac{x + 4}{2}$
   L_2: $y = 2x$

5. L_1: $2y = x + 3$
   L_2: $y = 2x$

6. L_1: $y = 10x - 7$
   L_2: $y + 5 = 2(x - 1)$

7. L_1: $y - 4 = 3x$
   L_2: $y = x$

8. L_1: $y = 2x + 5$
   L_2: $y = x + 1$

9. L_1: $y = \frac{x}{4}$
   L_2: $y + 2 = 2(x - 6)$

10. L_1: $3y + 1 = 2x$
    L_2: $y + 1 = x$

11. L_1: $y + 3 = 2(x - 5)$
    L_2: $2y = \frac{-x}{3}$

12. L_1: $y + 8 = 3(x - 1)$
    L_2: $3y = x - 1$

In exercises 13–20, show, graphically, that the indicated lines are parallel.

13. L_1: $y = x + 3$
    L_2: $2y = 2x - 4$

14. L_1: $2y = x - 1$
    L_2: $y = \frac{x + 1}{2}$

15. L_1: $y = 5x - 3$
    L_2: $2y = 5(2x - 1)$

16. L_1: $y = 3x$
    L_2: $\frac{y}{3} = x + 4$

17. L_1: $y - 4 = 2(x - 1)$
    L_2: $y = 2x + 1$

18. L_1: $y - 1 = 4(x - 2)$
    L_2: $\frac{y}{2} = 2(x - 1)$

19. L_1: $y + 3 = 3(x + 2)$
    L_2: $1 - y = 3(2 - x)$

20. L_1: $x + 3y = 4$
    L_2: $2 + y = \frac{1 - x}{3}$

In exercises 21–28, graph the indicated lines to determine whether or not they intersect. If they do, find the intersection point.

21. L_1: $y = x - 1$
    L_2: $2y = x + 1$

22. L_1: $y = 2x - 1$
    L_2: $2y = 2x + 1$

23. L_1: $y = x - 7$
    L_2: $y = x + 3$

24. L_1: $2y = x + 5$
    L_2: $4y = 2(x - 1)$

25. L_1: $y = x + 6$
    L_2: $y - 3 = 2x + 2$

26. L_1: $y = 1 + x$
    L_2: $y = 1 - x$

27. L_1: $\frac{x}{2} + \frac{y}{4} = 0$
    L_2: $y = 1 - 2x$

28. L_1: $y = 4$
    L_2: $x = 5$

29. Without graphing, indicate which two of these lines are parallel
    L_1: $y = x$
    L_2: $y = 2x$
    L_3: $y = 3x + 1$
    L_4: $y = x + 1$

30. Without graphing, indicate which two of these lines are parallel
    L_1: $x + y = 0$
    L_2: $y = 1 - x$
    L_3: $y = x + 1$
    L_4: $y = 0$

31. Without graphing, indicate which two of these lines are parallel

L_1: $y = 2x + 3$ $\qquad$ L_2: $2y = x + 3$

L_3: $\frac{y}{2} = x + 3$ $\qquad$ L_4: $\frac{x}{2} + \frac{y}{2} = 3$

32. Without graphing, indicate which two of these lines are parallel

L_1: $y = 4(x + 1)$ $\qquad$ L_2: $\frac{y}{4} = \frac{x}{4} + 1$

L_3: $y = x + 1$ $\qquad$ L_4: $y = \frac{x}{4} + 1$

10.4 SYSTEMS OF LINEAR EQUATIONS

Systems and Their Solutions

The graphical method that you used to find the intersection of two lines tends to be inaccurate, particularly when the coordinates of the intersection point are fractions. There are also algebraic methods of determining this intersection.

An equation of a line is called a **linear equation.** Thus

$$y = 2x + 3$$

is a linear equation.

Definition

> *An ordered pair (x_1, y_1) is a* ***solution*** *of a linear equation in the variables x and y if a true statement results when x_1 replaces x and y_1 replaces y.*

Thus (0, 3) is a solution of the equation

$$y = 2x + 3$$

because when you replace x by 0 and y by **3,** a true statement,

$$\mathbf{3} = 2 \cdot 0 + 3$$

results. Note that (1, 5) is also a solution because when you replace x by 1 and y by **5,** a true statement,

$$\mathbf{5} = 2 \cdot 1 + 3$$

results.

Geometrically, a solution represents a point on the line given by this equation. There are "infinitely many" solutions of a linear equation because every point on the line corresponds to a solution.

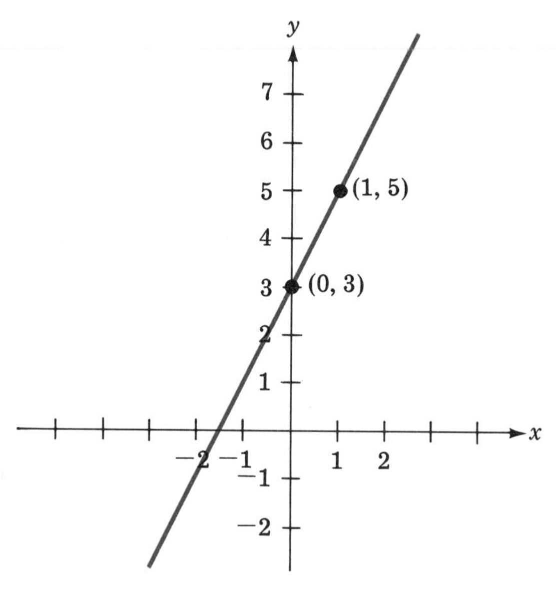

FIGURE 10.27. ***Every point on a line corresponds to a solution of an equation of the line. Thus, (0, 3) and (1, 5) are both solutions of the equation, $y = 2x + 3$, for the above line.***

Two linear equations, taken together, form a **system of linear equations (in 2 variables).** For example, the two equations

$$y = 2x + 3$$

$$y + 1 = 3(x + 1)$$

together form such a system.

Definition

> *An ordered pair (x_1, y_1) is a* ***solution of a system of linear equations*** *if (x_1, y_1) is a solution of both equations of the system.*

The ordered pair (1, 5) is a solution of

$$y + 1 = 3(x + 1)$$

because

$$\mathbf{5} + 1 = 3(1 + 1)$$

$$[\text{or } 6 = 6].$$

Recall that (1, 5) is also a solution of

$$y = 2x + 3.$$

Thus (1, 5) is a solution of the system:

$$y = 2x + 3$$

$$y + 1 = 3(x + 1)$$

Geometrically, a solution of a system of linear equations represents the intersection of the lines whose equations comprise the system.

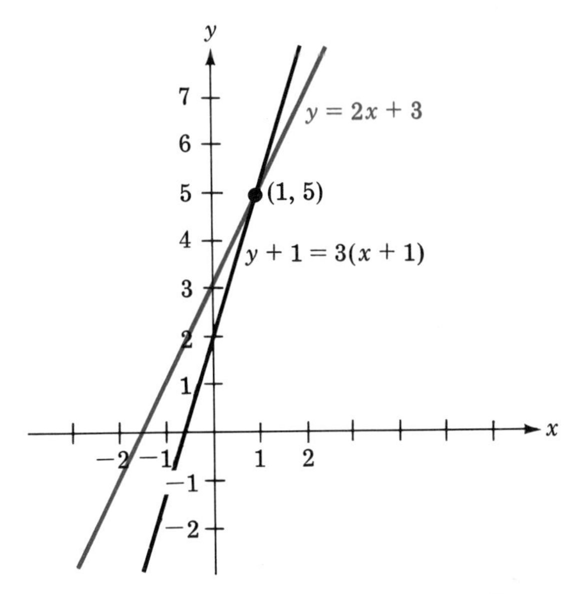

FIGURE 10.28. A solution of a system of linear equations represents the intersection of the corresponding lines. Thus, (1, 5) is the solution of the system $y=2x+3$, $y+1=3(x+1)$.

Substituting

You can solve a system of equations by substituting for one of the variables, as in Example 1.

EXAMPLE 1
Solve the system:

$$y = 2x + 3$$

$$y + 2 = 4x + 1$$

Check the solution.

Solution. Substitute

$$2x + 3 \text{ for } y$$

in the second equation.

$$2x + 3 + 2 = 4x + 1$$
$$2x + 5 = 4x + 1$$
$$4 = 2x$$
$$2 = x$$

Now replace x by **2** in the first equation, and find y:

$$y = 2 \cdot \mathbf{2} + 3$$
$$y = 7$$

The solution of the system is (2, 7).

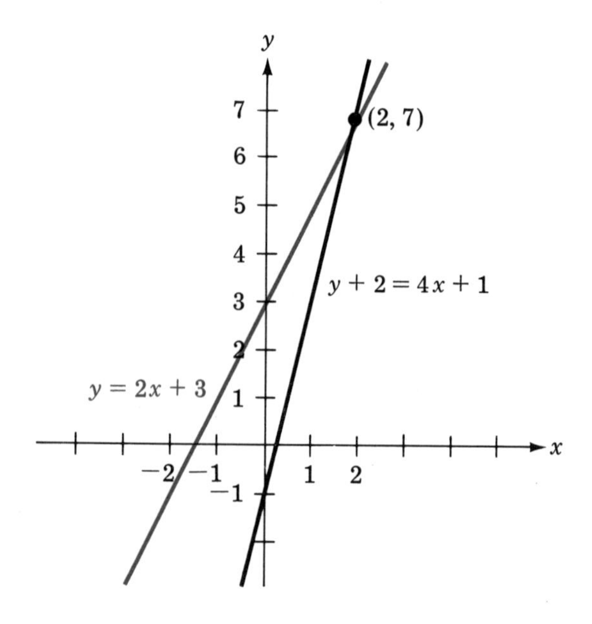

FIGURE 10.29

CHECK. You can check the solution by replacing x by **2** and y by 7 in each of the given equations.

$$7 \stackrel{?}{=} 2 \cdot \mathbf{2} + 3 \qquad 7 + 2 \stackrel{?}{=} 4 \cdot \mathbf{2} + 1$$
$$7 \stackrel{\checkmark}{=} 7 \qquad 9 \stackrel{\checkmark}{=} 9$$

In the **substitution method:**

1. Solve one equation for one of the variables, say y, in terms of the other, x.
2. Replace y by this expression in the other equation.
3. Solve this equation for x.
4. To find y, replace x by this value in the equation for y in Step (1).

EXAMPLE 2

Solve the system:

$$y + 3 = 4(x - 1)$$
$$y - 2 = -5x$$

Solution.

(1) From the second equation:

$$y = 2 - 5x$$

(2) Replace y by the expression $2 - 5x$ in the first equation:

$$2 - 5x + 3 = 4(x - 1)$$

(3) Solve this equation for x:

$$5 - 5x = 4x - 4$$
$$9 = 9x$$
$$1 = x$$

(4) Replace $\boldsymbol{x}$ by **1** in the equation $y = 2 - 5\boldsymbol{x}$ in Step (1):

$$y = 2 - 5 \cdot \mathbf{1}$$
$$y = -3$$

The solution is $(1, -3)$.

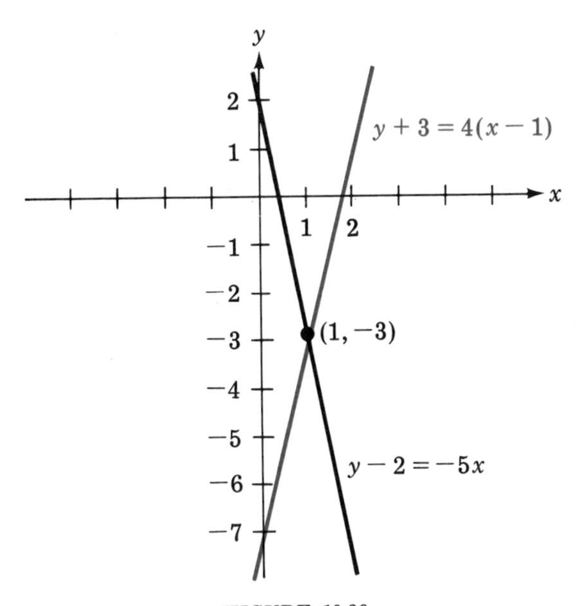

FIGURE 10.30

A linear equation in the form $y = mx + b$ can be written in the form

$$Ax + By = C,$$

where A, B, and C are numbers, and where A and B are not both 0. For example,

$$y = 2x + 5$$

becomes

$$-2x + y = 5.$$

Here $A = -2$, $B = 1$, $C = 5$. Also, the linear equation

$$y - 3 = \frac{1 - x}{2},$$

which is not in either form, becomes

$$2y - 6 = 1 - x,$$

and then

$$x + 2y = 7.$$

Here $A = 1$, $B = 2$, $C = 7$.

Often the equations of a system are given in the form

$$Ax + By = C$$

$$ax + by = c$$

EXAMPLE 3

Solve the system:

$$x + 2y = 10$$

$$2x - 3y = 6$$

Solution.

(1) Because x appears in the first equation with coefficient 1, solve this equation for x in terms of y:

$$x = 10 - 2y$$

(2) Replace x by the expression $10 - 2y$ in the second equation:

$$2(10 - 2y) - 3y = 6$$

(3) Solve this for y:

$$20 - 4y - 3y = 6$$

$$20 - 7y = 6 \qquad \text{Add } 7y - 6 \text{ to both sides.}$$

$$14 = 7y$$

$$2 = y$$

(4) Replace $\boldsymbol{y}$ by **2** in the equation $x = 10 - 2\boldsymbol{y}$ in Step (1):

$$x = 10 - 2 \cdot \mathbf{2}$$

$$x = 6$$

The solution is (6, 2). [See Figure 10.31.]

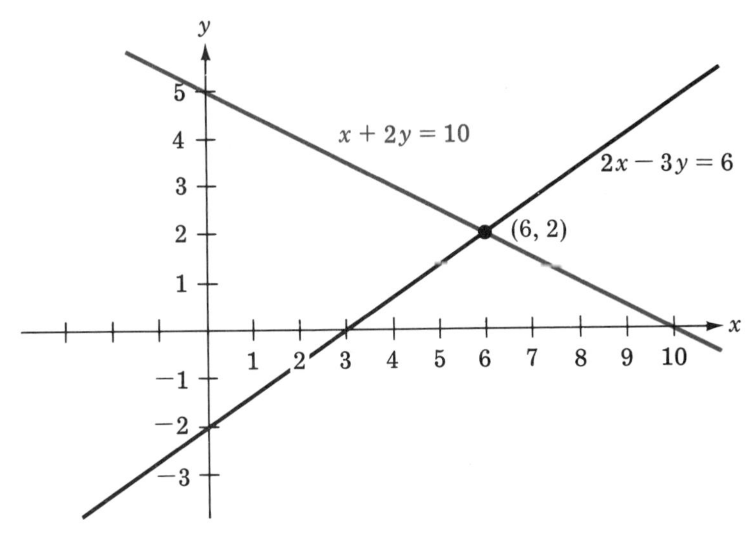

FIGURE 10.31

Adding

You can also solve a system of equations by first adding or subtracting the corresponding sides of the equations to eliminate one of the variables.

EXAMPLE 4

Solve the system:

$$2x + 3y = 5$$
$$4x - 3y = 1$$

Solution. Add the corresponding sides of the equations to eliminate y:

$$\begin{array}{l} 2x + 3y = 5 \\ \underline{4x - 3y = 1} \\ 6x \qquad\ \ = 6 \\ \ \ x \qquad\ \ = 1 \end{array}$$

Now replace x by 1 in either equation. (For example, use the first equation.)

$$\begin{aligned} 2 \cdot 1 + 3y &= 5 \\ 3y &= 3 \\ y &= 1 \end{aligned}$$

The solution is (1, 1). [See Figure 10.32 on page 298.]

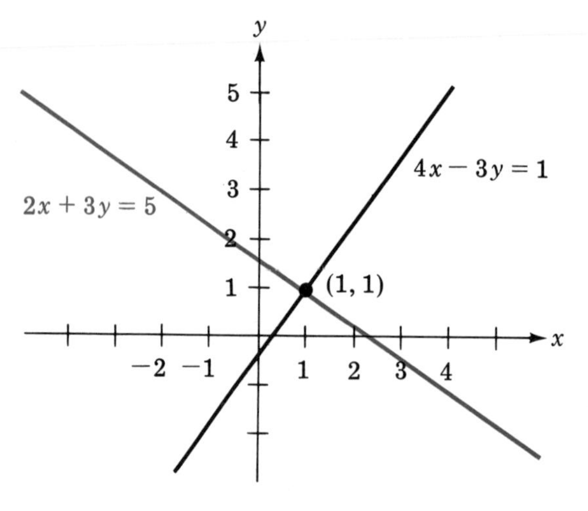

FIGURE 10.32

EXAMPLE 5

Solve the system:

$$2x + y = 10$$
$$x + y = 6$$

Solution. Subtract each side of the second equation from the correspond- side of the first equation to eliminate y:

$$\begin{array}{r} 2x + y = 10 \\ \underline{\mp x \mp y = \mp 6} \\ x \qquad = 4 \end{array}$$

Replace x by 4 in the second equation.

$$\begin{aligned} 4 + y &= 6 \\ y &= 2 \end{aligned}$$

The solution is (4, 2).

These algebraic methods apply to systems in variables other than x and y.

In Example 6, both equations are first transformed, so that when the resulting equations are added, a variable is eliminated.

EXAMPLE 6

Solve the system:

$$3s + 2t = 10$$
$$2s + 3t = 5$$

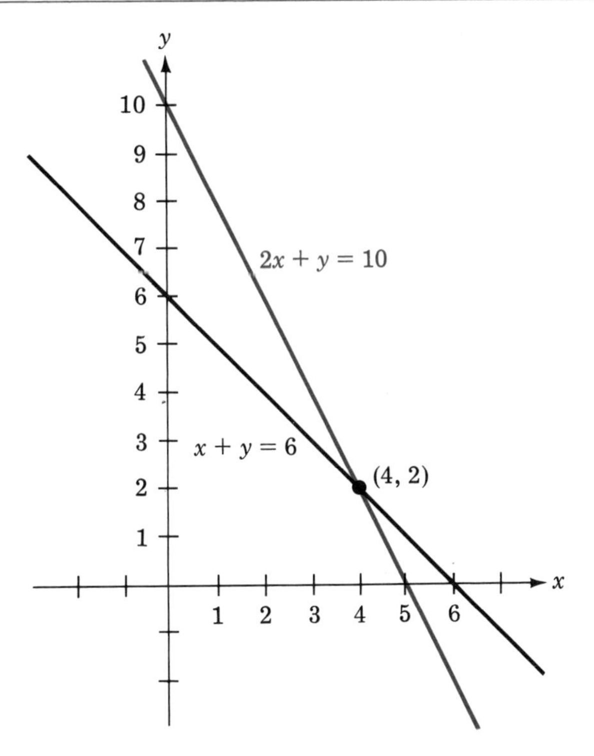

FIGURE 10.33

Solution. Multiply both sides of the first equation by 2, and both sides of the second equation by −3. You will then eliminate s.

$$\begin{array}{rlr} 2(3s+2t) = 2 \cdot 10 & \text{or} & 6s + 4t = 20 \\ -3(2s+3t) = -3 \cdot 5 & \text{or} & \underline{-6s - 9t = -15} \\ & & -5t = 5 \\ & & t = -1 \end{array}$$

Replace t by −1 in the (given) first equation:

$$\begin{aligned} 3s + 2(-1) &= 10 \\ 3s &= 12 \\ s &= 4 \end{aligned}$$

The solution is (4, −1).

You can graph this system by letting the horizontal axis be the s-axis and the vertical axis be the t-axis, as in Figure 10.34 on page 300.

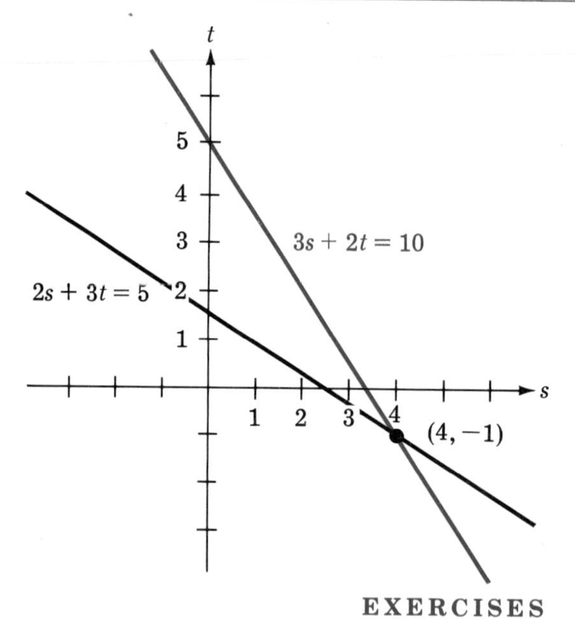

FIGURE 10.34

EXERCISES

In exercises 1–10, solve by substituting. In exercises 1–4, check your solution.

1. $y = x + 1$
   $y = 3x - 5$

2. $y = x - 2$
   $y = 2x - 8$

3. $y = 2x - 3$
   $y = 3x - 7$

4. $y = 4x - 1$
   $y = 2x + 3$

5. $y = 4 - 3x$
   $y - 3 = 2(x - 2)$

6. $y - 8 = 5x$
   $y + 1 = 4(x + 2)$

7. $v + 2 = 4(u - 3)$
   $v - 4 = \dfrac{u - 1}{2}$

8. $t + 5 = s$
   $t + 3 = 4(s - 8)$

9. $b - 2 = -3(a - 1)$
   $b = -2a$

10. $n + 7 = 4(m - 1)$
    $n + 5 = \dfrac{3(m + 1)}{2}$

In exercises 11–20, solve by the "adding" method. In exercises 11–14, check your solution.

11. $x + y = 4$
    $x - y = 2$

12. $x - y = 3$
    $-x - y = 1$

13. $a + 2b = -2$
    $3a - 2b = 6$

14. $2x + y = 7$
    $3x + y = 12$

15. $c + 5d = 2$
    $5c + 5d = 6$

16. $x + 3y = -1$
    $5x + 3y = 7$

17. $r + 2s = 8$
    $4r - s = 5$

18. $3m - n = 2$
    $2m - 3n = -1$

19. $6x - 5y = 7$
    $2x + 3y = 7$

20. $2x + 3y = 9$
    $3x + 2y = 11$

In exercises 21–30, solve by either substituting or adding.

21. $x + 2y = 6$
    $3x - y = 4$

22. $u + v = 8$
    $5v = 3u$

23. $3y = 4x$
    $y - 4 = x + 2$

24. $x + y = 3$
    $2x - 3y = 6$

25. $a + 2b = 5$
    $4a - 6b = 6$

26. $6x - 3y = 3$
    $3x + y = 9$

27. $3y = 2x$
    $y - 4 = 2x$

28. $2x + 5y = 9$
    $4x + 3y = 11$

29. $x - 3y = 2$
    $4x - 10y = 10$

30. $2s + 3t = 10$
    $3s + 2t = 10$

10.5 VARIATION

Direct Variation

A linear equation of the form

$$y = kx$$

describes many important relationships in science and everyday affairs. For example, if a car is traveling at the constant rate of 60 miles per hour, the distance, y, traveled in x hours is given by the linear equation

$$y = 60x.$$

Thus in half an hour, the car travels $60 \cdot \frac{1}{2}$ or 30 miles; in 3 hours the car travels $60 \cdot 3$ or 180 miles.

Definition

> *Let x and y be variables. Then y **varies directly as** x if*
>
> $$y = kx$$
>
> *for some nonzero number k. This number k is called the **constant of variation.***

EXAMPLE 1

Suppose

$$y = 3x.$$

Then y varies directly as x. The constant of variation is 3. Some of the corresponding values of x and y are given in the following table.

x	y
1	3
2	6
5	15
−1	−3
−3	−9

The relationship between x and y is described by a line through the origin with slope 3. (See Figure 10.35.)

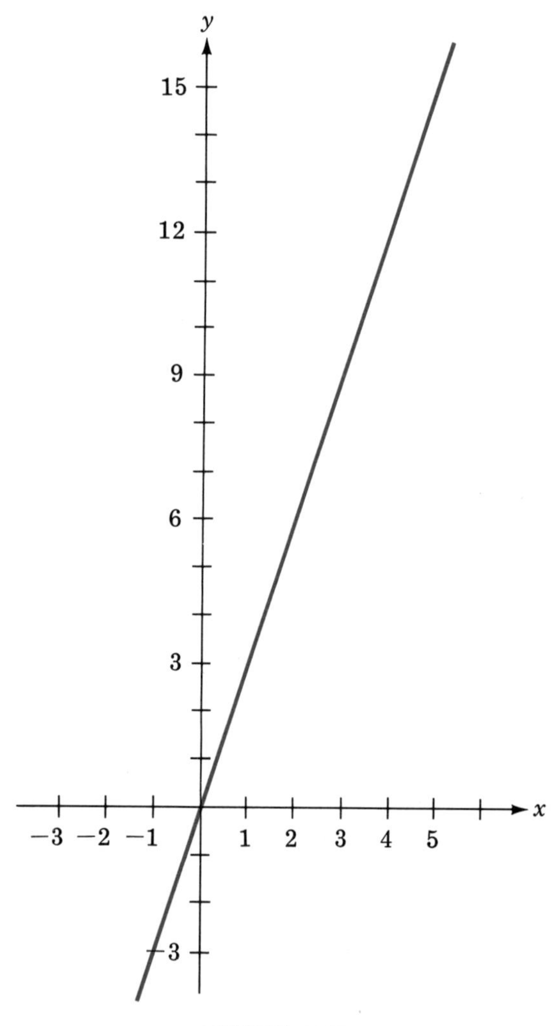

FIGURE 10.35

EXAMPLE 2

The circumference C of a circle is given by

$$C = 2\pi r,$$

where r is the length of the radius. Thus C varies directly as r. Here, the constant of variation is 2π. As r increases, C increases. For example,

if $r = 5$, $C = 10\pi$
if $r = 10$, $C = 20\pi$
if $r = 50$, $C = 100\pi$

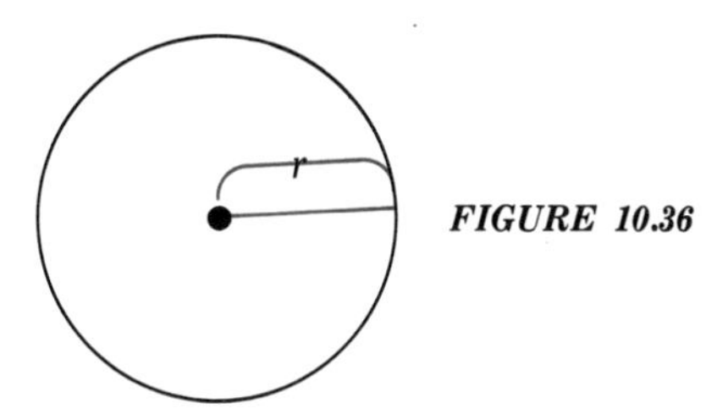

FIGURE 10.36

When y varies directly as x, then

$$y = kx,\ k \neq 0.$$

It follows that *x also varies directly as y* because if you divide both sides of the equation by k, you obtain

$$\frac{1}{k}y = x.$$

The constant of variation is now $\frac{1}{k}$. Also, if y varies directly as x, then

$$\frac{y}{x} = k,\ x \neq 0.$$

Thus *the quotient, $\frac{y}{x}$, remains fixed.* Let x_1 and x_2 be values of x and let y_1 and y_2 be the corresponding values of y. Then

$$\frac{y_1}{x_1} = \frac{y_2}{x_2} = k$$

Thus *corresponding values of y and x are proportional. When one variable increases, the other variable also increases.*

EXAMPLE 3

Suppose

$$y = 5x.$$

The constant of variation is 5. Then you can also write

$$x = \frac{1}{5}y.$$

Here the constant of variation is $\frac{1}{5}$. Also,

$$\frac{y}{x} = 5$$

Here are some corresponding values of x and y.

x	y
1	5
2	10
3	15
4	20

EXAMPLE 4

Suppose y varies directly as x, and $y = 12$ when $x = 3$.
(a) Find the constant of variation.
(b) Find y when $x = 5$.
(c) Find x when $y = 8$.

Solution.
(a)

$$y = kx$$

Replace y by 12 and $\boldsymbol{x}$ by **3.**

$$12 = k \cdot \mathbf{3}$$
$$4 = k$$

Thus the constant of variation is 4, and $y = 4x$.
(b) Replace x by 5.

$$y = 4 \cdot 5$$
$$y = 20$$

(c) Replace y by 8.

$$8 = 4x$$
$$2 = x$$

Inverse Variation

Definition

> *Let x and y be variables. Then y* ***varies inversely as*** *x if*
>
> $$y = \frac{k}{x}, x \neq 0,$$
>
> *for some nonzero number k. Again, k is called the* ***constant of variation.***

EXAMPLE 5

$$\text{Let } y = \frac{2}{x},\ x \neq 0.$$

Then y varies inversely as x. The constant of variation is 2. Some corresponding values of x and y are given in the following table.

x	y
1	2
2	1
4	$\frac{1}{2}$
6	$\frac{1}{3}$
8	$\frac{1}{4}$

Recall that

$$\frac{2}{\frac{1}{2}} = 2 \div \frac{1}{2} = 2 \cdot \frac{2}{1} = 4$$

and that

$$\frac{2}{\frac{1}{3}} = 2 \div \frac{1}{3} = 2 \cdot 3 = 6.$$

Thus when x takes on the fractional values $\frac{1}{2}, \frac{1}{3}, \frac{1}{4}$, the above values of x and y are interchanged.

x	y
$\frac{1}{2}$	4
$\frac{1}{3}$	6
$\frac{1}{4}$	8

Note that x can also take on negative values. The corresponding values of y are also negative.

x	y
-1	-2
-2	-1
-4	$\frac{-1}{2}$
$\frac{-1}{2}$	-4

However, neither x nor y can take on the value 0 in the equation

$$y = \frac{2}{x}.$$

First, x cannot be 0 because you cannot divide by 0. Also, y cannot be 0 because y equals the fraction $\frac{2}{x}$, and a fraction is 0 only if the numerator is 0.

The graph is given in Figure 10.37, and is known as a **hyperbola.** Each of the curves is called a **branch** of the hyperbola.

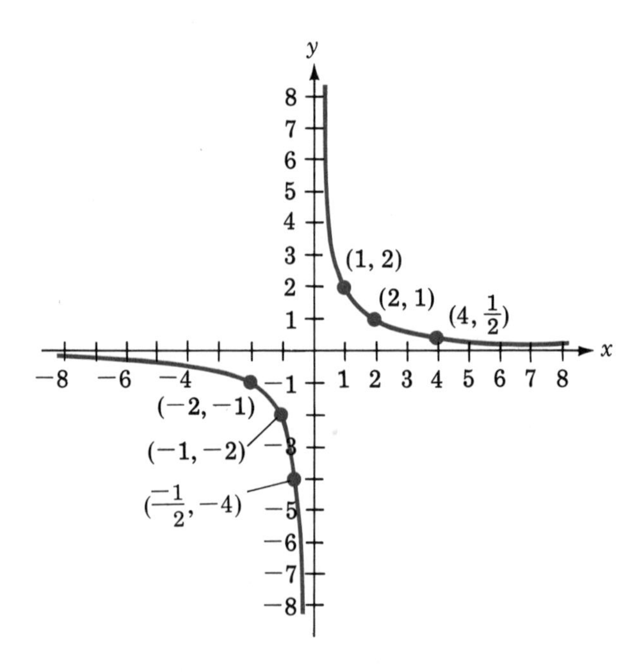

FIGURE 10.37. A hyperbola with two branches.

If y varies inversely as x, then

$$y = \frac{k}{x}, \; k \neq 0.$$

Cross-multiply, and obtain

$$x = \frac{k}{y}, \; y \neq 0.$$

Thus x *varies inversely as* y, *with the same constant of variation*, k. Also, *if* y *varies inversely as* x, *then*

$$xy = k.$$

Thus *the product of the two variables is constant. When one variable increases, the other variable decreases.*

EXAMPLE 6

Suppose

$$y = \frac{-6}{x}, \; x \neq 0.$$

Here the constant of variation is −6. Cross-multiply to obtain

$$x = \frac{-6}{y}, \; y \neq 0.$$

Again the constant of variation is −6. Finally,

$$xy = -6$$

Some of the corresponding values of x and y are:

x	y
1	−6
2	−3
3	−2
6	−1
12	$\frac{-1}{2}$
18	$\frac{-1}{3}$
$\frac{1}{2}$	−12
$\frac{1}{3}$	−18

EXAMPLE 7

Suppose

$$y = \frac{10}{x}, \; x \neq 0.$$

(a) What is the constant of variation?
(b) Find y when $x = 2$.
(c) Find x when $y = -2$.

Solution.
(a) The constant of variation is 10.
(b) Replace x by 2:

$$y = \frac{10}{2}$$
$$y = 5$$

(c) Replace y by −2:

$$-2 = \frac{10}{x} \qquad \text{Cross-multiply}$$

$$-2x = 10$$
$$x = \frac{10}{-2}$$
$$x = -5$$

EXAMPLE 8

Boyle's Law states that the pressure, p, of a compressed gas varies inversely as the volume, v, of gas. Suppose the pressure is 40 pounds per square inch when the volume is 100 cubic inches. Find the pressure when the gas is compressed to 80 cubic inches.

Solution. Because p varies inversely as v,

$$p = \frac{k}{v}, \; v \neq 0.$$

To find k, let $p = 40$ and $\mathbf{v = 100}$.

$$40 = \frac{k}{\mathbf{100}}$$
$$4000 = k$$

Thus the relationship between pressure and volume is given by the equation

$$p = \frac{4000}{v}.$$

Now replace $\mathbf{v}$ by **80** to find the corresponding value of p:

$$p = \frac{4000}{\mathbf{80}}$$
$$p = 50$$

EXERCISES

In exercises 1–8, suppose y varies directly as x. Find the constant of variation.

1. $y = 4$ when $x = 2$
2. $y = 10$ when $x = 2$
3. $y = 9$ when $x = 3$
4. $y = 36$ when $x = 6$
5. $y = 4$ when $x = -2$
6. $y = 10$ when $x = 4$
7. $y = \frac{1}{2}$ when $x = \frac{1}{4}$
8. $y = \frac{1}{2}$ when $x = \frac{1}{3}$

In exercises 9–16, assume y varies directly as x.

9. Suppose $y = 6$ when $x = 3$. Find y when $x = 5$.
10. Suppose $y = 12$ when $x = 2$. Find y when $x = 6$.
11. Suppose $y = 15$ when $x = 3$. Find y when $x = 2$.
12. Suppose $y = 20$ when $x = 4$. Find y when $x = -2$.
13. Suppose $y = 100$ when $x = 10$. Find x when $y = 400$.
14. Suppose $y = 49$ when $x = 7$. Find x when $y = 7$.

15. Suppose $y = -12$ when $x = 36$. Find x when $y = 48$.

16. Suppose $y = -64$ when $x = -16$. Find x when $y = 16$.

In exercises 17–24, suppose y varies inversely as x. Find the constant of variation.

17. $y = 4$ when $x = 2$

18. $y = 3$ when $x = 4$

19. $y = -2$ when $x = -1$

20. $y = 12$ when $x = 2$

21. $y = \frac{1}{2}$ when $x = 4$

22. $y = -3$ when $x = 2$

23. $y = -8$ when $x = -4$

24. $y = \frac{1}{3}$ when $x = \frac{1}{2}$

In exercises 25–32, assume y varies inversely as x.

25. Suppose $y = 2$ when $x = 2$. Find y when $x = 4$.

26. Suppose $y = 5$ when $x = 2$. Find y when $x = 1$.

27. Suppose $y = 10$ when $x = 5$. Find y when $x = 25$.

28. Suppose $y = 3$ when $x = 4$. Find y when $x = 3$.

29. Suppose $y = 12$ when $x = 5$. Find x when $y = -5$.

30. Suppose $y = 3$ when $x = 10$. Find x when $y = 15$.

31. Suppose $y = 7$ when $x = 4$. Find x when $y = 28$.

32. Suppose $y = 5$ when $x = 3$. Find x when $y = 10$.

In exercises 33–36, fill in "directly" or "inversely."

33. When width is held constant, the area of a rectangle varies ____________ as the length.

34. For rectangles of a fixed area, the length of the rectangle varies ____________ as the width.

35. A car travels at a constant rate along the throughway from Albany to Buffalo. The time it spends traveling varies ____________ as this rate.

36. If all seats are $1.50, the gross of a movie theatre varies ____________ as the number of tickets sold.

37. The cost of a slice of cheese varies directly as its weight. If a 12 ounce slice costs $1.80, how much does a 14 ounce slice cost?

38. The tension in a spring varies directly as the distance it is stretched. If the tension is 36 pounds when the distance stretched is 9 inches, what is the tension when the distance stretched is 1 foot?

39. The boll weevil population of a cotton field varies inversely as the amount of insecticide used. If 5000 boll weevils are in a field when 60 pounds of insecticide are used, how many boll weevils will be left when 300 pounds are used?

40. The weight of a body varies inversely as its distance from the *center* of earth. The earth's radius is (approximately) 4000 miles. How much does a 200-pound man weigh 16000 miles above the *surface* of the earth?

What Have You Learned in Chapter 10?

You have learned how to find the slope and y-intercept of a line.

You can find the equation of a line with a given slope and y-intercept.

You can find the intersection of two lines graphically.

Also, you can solve the corresponding system of linear equations algebraically.

And you have learned that:

$$y \text{ varies directly as } x, \text{ if } \frac{y}{x} = k, \quad x \neq 0,$$

$$y \text{ varies inversely as } x, \text{ if } xy = k, \text{ where } k \text{ is a nonzero number.}$$

Let's Review Chapter 10.

10.1 Slope

1. Let $P_1 = (3, 1)$ and $P_2 = (5, 6)$.
   (a) Find the rise from P_1 to P_2.
   (b) Find the run from P_1 to P_2.
   (c) Find the slope of the line through P_1 and P_2.
2. Draw the line through $(2, 1)$ with slope $\frac{1}{2}$.
3. Consider the line determined by $(0, 4)$ and $(3, 0)$. Find the y-coordinate, y_3, of the point $(-3, y_3)$ on this line.
4. Indicate whether each of the following lines through P_1 and P_2 (i) is horizontal, (ii) is vertical, (iii) slopes upward to the right, (iv) slopes upward to the left.
   (a) $P_1 = (4, 4)$, $P_2 = (6, 0)$
   (b) $P_1 = (3, -3)$, $P_2 = (6, -3)$

10.2 Equation of a Line

5. Find (a) the slope, and (b) the y-intercept of the line with equation

$$y = \frac{-x}{2} + 7.$$

6. Find the equation of the line with slope 4 and y-intercept 0.
7. Find the equation of the line with x-intercept 3 and y-intercept -2.
8. Find the equation of the horizontal line through $(3, 5)$.

10.3 *Intersection of Lines*

9. Graph the intersection of the lines given by:

$$L_1\text{: } y = 2x + 4 \qquad L_2\text{: } y = 7 - x$$

10. Show, graphically, that the indicated lines are parallel.

$$L_1\text{: } y = 4x - 3 \qquad L_2\text{: } y - 1 = 4(x + 1)$$

11. Graph the following lines to determine whether or not they intersect. If they do, find the intersection point.

$$L_1\text{: } y = x - 3 \qquad L_2\text{: } y = 2x + 2$$

12. Without graphing, indicate which two of these lines are parallel:

L_1: $y = 5x$ $\qquad$ L_2: $y = -5x$

L_3: $y = x + 5$ $\qquad$ L_4: $y = x - 5$

10.4 *Systems of Linear Equations*

13. Solve by substituting:

$$y = 2x + 1, \qquad y = 6 - x$$

14. Solve by substituting. Check your solution.

$$y + 1 = 2x, \qquad \frac{y}{3} = 3 - x$$

15. Solve by the "adding" method.

$$6a - 5b = 7$$
$$3a - 2b = 4$$

16. Solve by either substituting or adding:

$$3x + 4y = 2$$
$$2x + 3y = -1$$

10.5 *Variation*

17. Suppose y varies directly as x. Find the constant of variation if $y = 4$ when $x = -2$.
18. Assume y varies directly as x. Suppose $y = 10$ when $x = 4$. Find y when $x = 6$.
19. Assume y varies inversely as x. Suppose $y = 8$ when $x = 2$. Find x when $y = \frac{1}{2}$.
20. Fill in "directly" or "inversely." If y varies inversely as x, then x varies ____________ as y.

And these from Chapters 1–9:

21. Simplify: $\dfrac{\frac{3}{4}}{\frac{1}{8}}$

22. Let $f(x) = 3x - 2$. Find
(a) $f(1)$, (b) $f(-1)$, (c) $f(3)$.
(d) For which value of x does $f(x) = 0$?

23. Let $f(x) = \frac{4}{x}$. Find:

(a) $f(2)$, (b) $f(-2)$, (c) $f\left(\frac{1}{4}\right)$.

(d) For which value of x is this function undefined?

24. Factor: $4x^4 - 9x^2$

Try These Exam Questions for Practice.

1. Let $P_1 = (6, 3)$ and $P_2 = (3, 6)$.
   (a) Find the rise from P_1 to P_2.
   (b) Find the run from P_1 to P_2.
   (c) Find the slope of the line through P_1 and P_2.
2. Find (a) the slope and (b) the y-intercept of the line with equation $y = 4x - 3$.
3. Find the equation of the line with slope -2 and y-intercept 3.
4. Solve by either substituting or adding. Check your solution.
$$2x + y = 8$$
$$x - y = 1$$
5. Solve by either substituting or adding.
$$5x + 3y = 4$$
$$10x + 7y = 6$$
6. Assume y varies directly as x. Suppose $y = 3$ when $x = 2$. Find y when $x = \frac{1}{2}$.

11

Roots

11.1 SQUARE ROOTS

Roots of Numbers

You know that

$$2^2 = 2 \cdot 2 = 4.$$

Suppose you are asked, "What *positive* number times itself is 4?" In symbols, you are asked to find x if

$$x \cdot x = 4.$$

Clearly, 2 is the positive root of this equation. 2 is called the "square root of 4." Note that the negative number -2 is also a root of this equation because

$$(-2)(-2) = 4.$$

Definition

> *SQUARE ROOT. Let a be positive. Then b is called the* ***square root of*** *a if b is positive and if*
>
> $$b^2 = a.$$
>
> *Also, the* ***square root of 0*** *is 0.*

Write

$$b = \sqrt{a}$$

[Read this: b equals the square root of a.]

if b is the square root of a. Here, the symbol

$$\sqrt{}$$

is known as a **radical sign.**

EXAMPLE 1

(a) $\sqrt{9} = 3$ because 3 is positive and $3^2 = 9$.
(b) $\sqrt{25} = 5$ because 5 is positive and $5^2 = 25$.

Note that the definition requires a square root to be positive or 0. Thus

$$\sqrt{4} = 2,$$

whereas

$$-\sqrt{4} = -2$$

Negative numbers do not have square roots (within the real number system). Thus

$$\sqrt{-4} \text{ is not defined.}$$

EXAMPLE 2

Find: (a) $\sqrt{100}$ (b) $\sqrt{144}$ (c) $\sqrt{400}$

Solution.
(a) $\sqrt{100} = 10$
(b) $\sqrt{144} = 12$
(c) Note that $20 \cdot 20 = 400$. Thus
$\sqrt{400} = 20$

EXAMPLE 3

Find: (a) $\sqrt{4^2}$ (b) $\sqrt{(-4)^2}$ (c) $-\sqrt{4^2}$

Solution.
(a)

$$4^2 = 16.$$

Therefore

$$\sqrt{4^2} = \sqrt{16}$$
$$= 4$$

(b)

$$(-4)^2 = (-4)(-4)$$
$$= 16$$

Therefore

$$\sqrt{(-4)^2} = \sqrt{16} = 4$$

(c)

$$-\sqrt{4^2} = -\sqrt{16} = -4$$

Roots of Algebraic Expressions

Just as you have considered squares such as

$$a^2, \quad (3x)^2, \quad \text{and } (st)^2,$$

so too, you will consider square roots of algebraic expressions. Square roots are positive or zero. *When working with algebraic expressions, you must consider the signs of the variables.*

EXAMPLE 4

Assume $a > 0$, $x > 0$, $s > 0$, $t > 0$. Find:
(a) $\sqrt{a^2}$ (b) $\sqrt{(3x)^2}$ (c) $\sqrt{(st)^2}$

Solution.

(a)
$$\sqrt{a^2} = a$$
because $a > 0$ and $a^2 = a \cdot a$

(b) Note that $3x \cdot 3x = (3x)^2$. Thus
$$\sqrt{(3x)^2} = 3x$$

(c)
$$\sqrt{(st)^2} = st$$

The product of positive numbers is positive. Thus in (b) and (c), $3x$ and st are each positive.

Before considering the next example, observe that

$$\sqrt{(-3)^2} = \sqrt{9} = 3.$$

Thus
$$\sqrt{(-3)^2} = -(-3)$$

EXAMPLE 5

Let $a > 0$ and $b < 0$.
Find: (a) $\sqrt{b^2}$, (b) $\sqrt{(ab)^2}$

Solution.

(a)
$$b < 0$$
Thus
$$-b > 0$$
Therefore
$$\sqrt{b^2} = -b,$$
because $-b$ is positive and

$$(-b)^2 = (-b)(-b) = b^2.$$

(b)
$$a > 0, b < 0$$
Therefore
$$ab < 0$$
and
$$-ab > 0.$$
Thus
$$\sqrt{(ab)^2} = -ab.$$

Roots of Even Powers

$$\sqrt{5^2} = \sqrt{25} = 5,$$

but
$$\sqrt{(-5)^2} = \sqrt{25} = 5 = -(-5)$$

In general, $\sqrt{a^2} = a$, if $a > 0$ or $a = 0$,

but $\sqrt{a^2} = -a$, if $a < 0$

Next observe that

$$10000 = 100 \cdot 100 \text{ and } 1000000 = 1000 \cdot 1000.$$

Thus $\sqrt{10^4} = \sqrt{10000} = 100 = 10^2$

and $\sqrt{10^6} = \sqrt{1000000} = 1000 = 10^3$

Therefore, $\sqrt{10^{2 \cdot 2}} = 10^2$

and $\sqrt{10^{2 \cdot 3}} = 10^3$

You can also consider square roots of *even* powers of variables.

EXAMPLE 6

Let $b > 0$. Find:
(a) $\sqrt{a^4}$ (b) $\sqrt{b^6}$ (c) $\sqrt{c^{20}}$

Solution.

(a) $$\sqrt{a^4} = a^2$$

because $a^4 = a^2 \cdot a^2$

and a^2 is *nonnegative* (positive or 0).

(b) $$\sqrt{b^6} = b^3$$

Note that $b^6 = b^3 \cdot b^3$.
Also, $b^3 > 0$ because $b > 0$

(c) $$\sqrt{c^{20}} = c^{10}$$

because $c^{20} = c^{10} \cdot c^{10}$
Also, $c^{10} = c^5 \cdot c^5 = (c^5)^2$
Thus c^{10} is nonnegative.

In general, when k is a positive integer,

$$\sqrt{a^{2k}} = a^k,$$

provided that a is nonnegative and hence a^k is nonnegative, when k is odd. Thus

$$\sqrt{a^8} = a^4 \quad \text{for all } a,$$

but $\sqrt{a^{10}} = a^5$ for nonnegative a.

Pythagorean Theorem

A **right triangle** is a triangle in which one of the angles is 90°. This angle is called a **right angle.** The side opposite the right angle is called the **hypotenuse.** (See Figure 11.1.) The ancient Greeks discovered that the lengths of the sides, a, b, and c, of a right triangle are related by the formula

$$c^2 = a^2 + b^2.$$

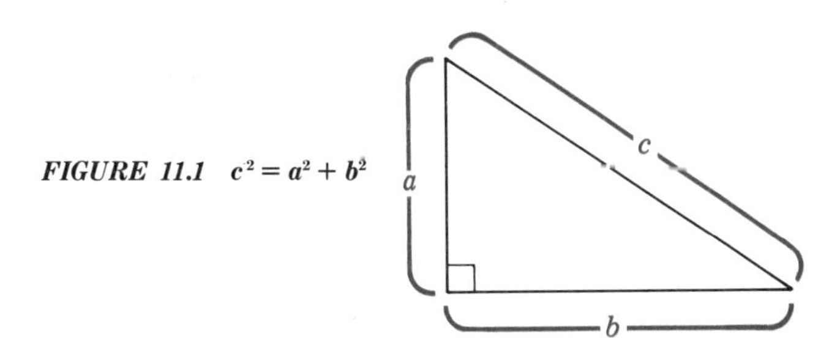

FIGURE 11.1 $c^2 = a^2 + b^2$

Here c is the length of the hypotenuse. This relationship is known as the **Pythagorean Theorem.** Obtain the square root of each side of the above equation. Because $c > 0$,

$$c = \sqrt{a^2 + b^2}.$$

EXAMPLE 7

In the right triangle of Figure 11.2, find c if $a = 3$ inches and $b = 4$ inches.

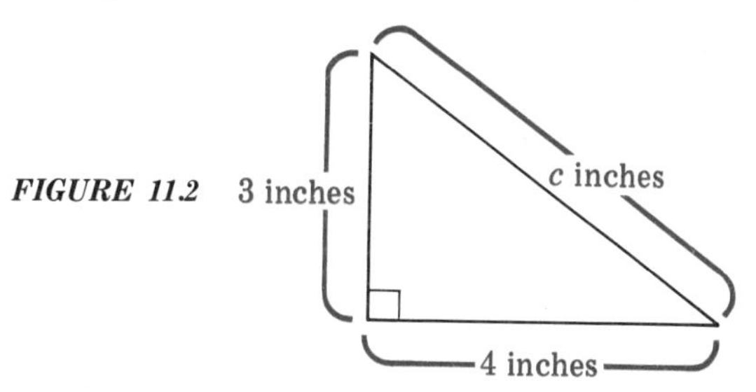

FIGURE 11.2

Solution.

$$c = \sqrt{a^2 + b^2}$$

Replace a by 3 and b by 4.

$$\begin{aligned} c &= \sqrt{3^2 + 4^2} \\ &= \sqrt{9 + 16} \\ &= \sqrt{25} \\ &= 5 \end{aligned}$$

The length of the hypotenuse is 5 inches.

EXAMPLE 8

In the right triangle of Figure 11.3, find b if $a = 5$ inches and $c = 13$ inches.

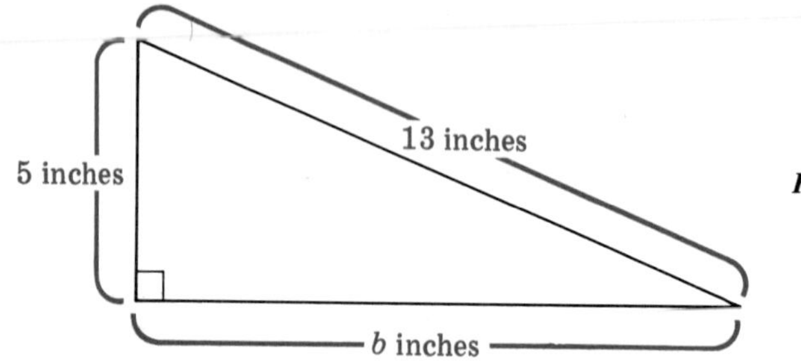

FIGURE 11.3

Solution.

$$a^2 + b^2 = c^2$$
$$b^2 = c^2 - a^2$$
$$b = \sqrt{c^2 - a^2}$$

Replace c by 13 and **a** by **5.**

$$b = \sqrt{13^2 - \mathbf{5}^2}$$
$$= \sqrt{169 - 25}$$
$$= \sqrt{144}$$
$$= 12 \text{ (inches)}$$

EXERCISES

In exercises 1–24, find the square root or its inverse, as indicated.

1. $\sqrt{16}$
2. $\sqrt{36}$
3. $\sqrt{49}$
4. $\sqrt{64}$
5. $-\sqrt{25}$
6. $-\sqrt{100}$
7. $\sqrt{81}$
8. $\sqrt{121}$
9. $-\sqrt{144}$
10. $\sqrt{900}$
11. $\sqrt{1600}$
12. $\sqrt{2500}$
13. $\sqrt{10000}$
14. $\sqrt{40000}$
15. $\sqrt{1000000}$
16. $-\sqrt{3600}$
17. $\sqrt{0}$
18. $\sqrt{(-3)^2}$
19. $-\sqrt{3^2}$
20. $-\sqrt{(-3)^2}$
21. $\sqrt{6^2}$
22. $\sqrt{(-6)^2}$
23. $\sqrt{(-9)^2}$
24. $-\sqrt{9}$

In exercises 25–38, find the square root or its inverse, as indicated. Assume $a, b, c, x, y,$ and z are each positive or 0.

25. $\sqrt{x^2}$
26. $-\sqrt{z^2}$
27. $\sqrt{(ab)^2}$
28. $\sqrt{(xyz)^2}$

29. $\sqrt{c^4}$

30. $\sqrt{x^8}$

31. $\sqrt{y^{10}}$

32. $\sqrt{a^{12}}$

33. $\sqrt{(xy)^4}$

34. $\sqrt{(bc)^6}$

35. $\sqrt{x^{18}}$

36. $\sqrt{z^{64}}$

37. $\sqrt{a^{20}}$

38. $\sqrt{b^{200}}$

In exercises 39–48, find the square root or its inverse, as indicated. Assume $m > 0$ and $n < 0$.

39. $\sqrt{m^2}$

40. $\sqrt{(-m)^2}$

41. $-\sqrt{m^2}$

42. $\sqrt{n^2}$

43. $\sqrt{(-n)^2}$

44. $-\sqrt{n^2}$

45. $\sqrt{(mn)^2}$

46. $\sqrt{(-mn)^2}$

47. $\sqrt{n^4}$

48. $\sqrt{n^6}$

In exercises 49–52, find the length of the indicated side of the right triangle.

49.

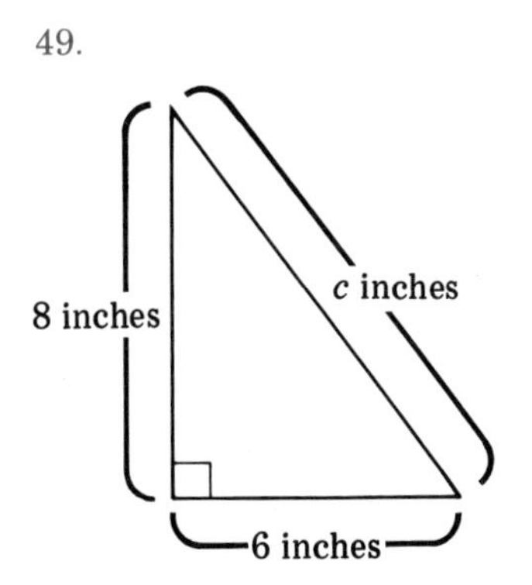

FIGURE 11.4

50.

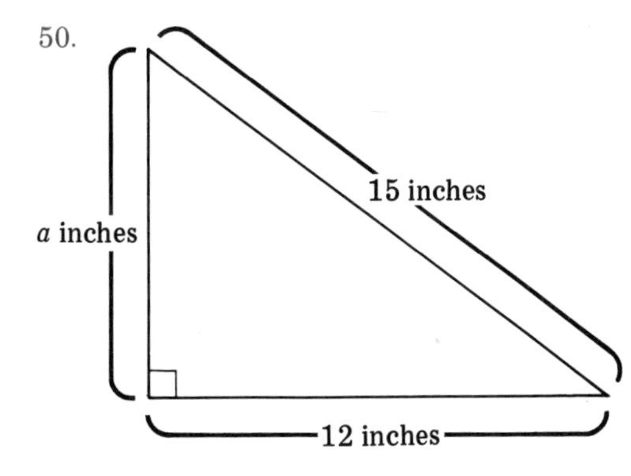

FIGURE 11.5

51.

52.

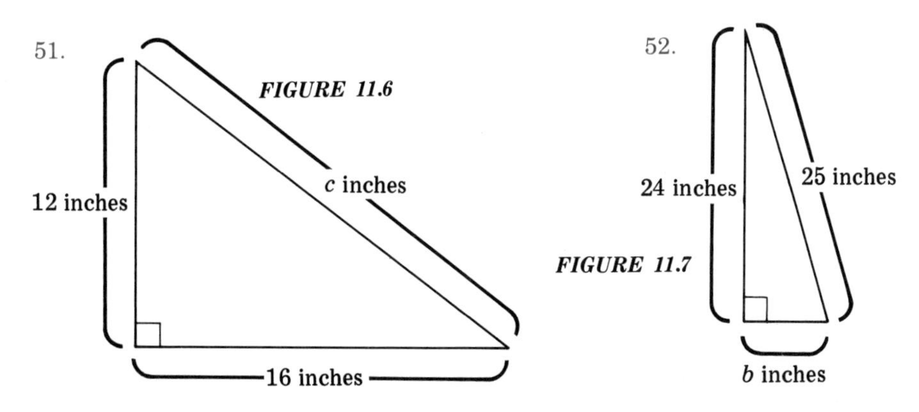

FIGURE 11.6

FIGURE 11.7

53. Find the length of the side of a square whose area is 64 square inches.
54. Find the length of the side of a square whose area is 121 square inches.

11.2 IRRATIONAL SQUARE ROOTS.

Approximating Square Roots

A rational number is a real number that can be written in the form $\frac{N}{D}$, where N and D are integers and $D \neq 0$. An *irrational number* is a real number that *cannot* be expressed in this form. Recall that π is irrational.

The integers

$$0, 1, 4, 9, 16, 25, \ldots$$

are squares of other integers. Their square roots are also integers. Thus,

$$\sqrt{0} = 0, \sqrt{1} = 1, \sqrt{4} = 2, \sqrt{9} = 3, \text{etc.}$$

These square roots are rational numbers, but the square roots of integers such as

$$2, 3, 5, 6, 7, \ldots$$

are irrational. For example, it can be shown that $\sqrt{2}$ is irrational. Therefore $\sqrt{2}$ cannot be written as a decimal, which is a rational number. But $\sqrt{2}$ can be *approximated* by a decimal to as many places as desired. Observe how this can be done.

$$\begin{array}{r} 1.4 \\ \underline{1.4} \\ 56 \\ \underline{14} \\ 1.96 \end{array} \qquad \begin{array}{r} 1.5 \\ \underline{1.5} \\ 75 \\ \underline{15} \\ 2.25 \end{array}$$

$(1.4)^2 = 1.96$ and $(1.5)^2 = 2.25$. Thus,

$$(1.4)^2 < 2 < (1.5)^2$$

[Read this: $(1.4)^2 < 2$ *and* $2 < (1.5)^2$]

Therefore, it seems reasonable that

$$1.4 < \sqrt{2} < 1.5.$$

Next, observe:

$$\begin{array}{rr} 1.41 & 1.42 \\ \underline{1.41} & \underline{1.42} \\ 1\ 41 & 2\ 84 \\ 56\ 4 & 56\ 8 \\ \underline{1\ 41} & \underline{1\ 42} \\ 1.98\ 81 & 2.01\ 64 \end{array}$$

$$(1.41)^2 = 1.9881 \text{ and } (1.42)^2 = 2.0164$$

Thus $$(1.41)^2 < 2 < (1.42)^2$$

Therefore

$$1.41 < \sqrt{2} < 1.42$$

(See Figure 11.8.) To three decimal places, it can be shown that

$$\sqrt{2} \approx 1.414$$ [Read: $\sqrt{2}$ *is approximately equal to* 1.414.]

Similarly, it can be shown that

$$\sqrt{3} \approx 1.732$$

(See Figure 11.8.)

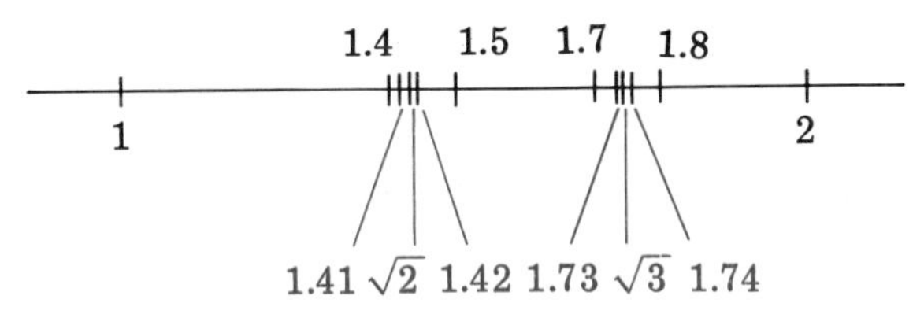

FIGURE 11.8. The irrational number $\sqrt{2}$ lies between the rational numbers 1.41 and 1.42. The irrational number $\sqrt{3}$ lies between the rational numbers 1.73 and 1.74.

EXAMPLE 1

Use $\sqrt{2} \approx 1.41$ to approximate

(a) $5\sqrt{2}$, (b) $5 + \sqrt{2}$, (c) $\dfrac{\sqrt{2}}{2}$

to 1 decimal place.

Solution.

(a) $$5\sqrt{2} \approx 5(1.41) = 7.05$$

When the last digit is 5 or more, round off by increasing the next to last digit by 1. Thus to 1 decimal place,

$$5\sqrt{2} \approx 7.1$$

(b) $$5 + \sqrt{2} \approx 5 + 1.41$$

Thus

$$5 + \sqrt{2} \approx 6.41$$

Because the last digit is less than 5, to 1 decimal place,

$$5 + \sqrt{2} \approx 6.4$$

(c) $$\frac{\sqrt{2}}{2} \approx \frac{1.410}{2} = \frac{1.205}{2}$$

To 1 decimal place,

$$\frac{\sqrt{2}}{2} \approx 1.2$$

EXAMPLE 2

Find consecutive integers N and $N + 1$ such that $N < \sqrt{92} < N + 1$.

Solution. 81 and 100 are squares and

$$81 < 92 < 100.$$

Thus

$$\sqrt{81} < \sqrt{92} < \sqrt{100}$$

or

$$9 < \sqrt{92} < 10$$

Therefore $N = 9$ and $N + 1 = 10$. (See Figure 11.9.)

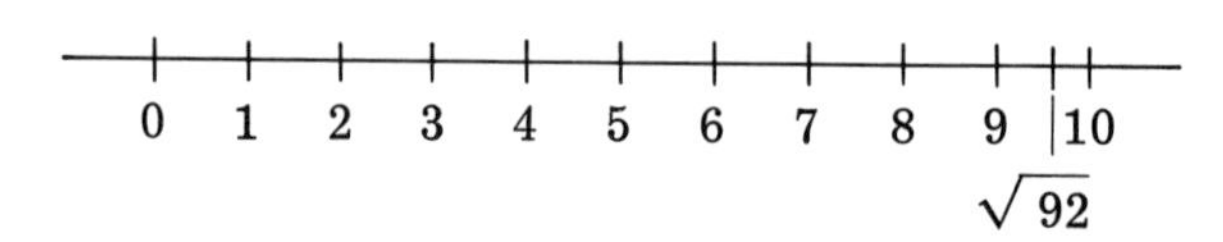

FIGURE 11.9

Decimal Representation

Recall that every rational number can be expressed as either a (regular) decimal or as an infinite repeating decimal (in which one or more digits repeat). Thus

$$\frac{1}{4} = .25, \quad \frac{2}{3} = .666666\ldots, \quad \frac{4}{11} = .363636\ldots$$

Every irrational number can be expressed as an infinite nonrepeating decimal. The digits do not repeat in any regular pattern. For example, the decimal representation of the irrational number $\sqrt{2}$ *to six decimal places* is 1.414 214. Thus $\sqrt{2}$ can be *approximated* by the rational number 1.414 214. Also, the decimal representation of the irrational number π to six decimal places is 3.141 593. Recall that π is often *approximated* by the rational number $\frac{22}{7}$. By dividing 22 by 7, you can see that

$$\frac{22}{7} = 3.1\underbrace{42857}\ \underbrace{142857}\ldots$$

Thus,

$$\pi < \frac{22}{7}$$

because

$$3.141593 < 3.142857$$

Square Roots in Geometry

What is the geometric significance of $\sqrt{2}$? Consider the squares in Figure 11.10.

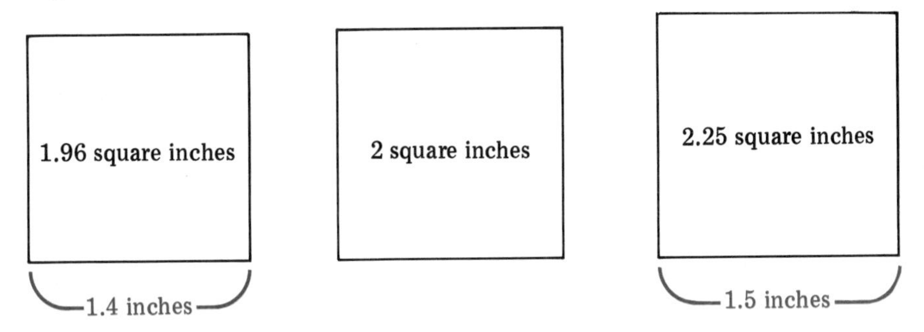

FIGURE 11.10

It is easy enough to draw the first and third squares, whose side lengths are rational. Whether or not you can draw the middle square, doesn't it seem reasonable that there should be a square whose area is 2 square inches? The length of a side is then $\sqrt{2}$ inches.

Next, an **isosceles triangle** is one in which two sides are of equal length. Consider the isosceles right triangle of Figure 11.11, in which the two equal sides are each of length 1 inch. What is the length of the hypotenuse? By the Pythagorean Theorem,

$$c^2 = a^2 + b^2$$

Thus

$$c^2 = 1^2 + 1^2$$
$$= 2$$

Therefore

$$c = \sqrt{2}$$

The *exact* length of the hypotenuse is $\sqrt{2}$ inches. (The *approximate* length is 1.414 inches.)

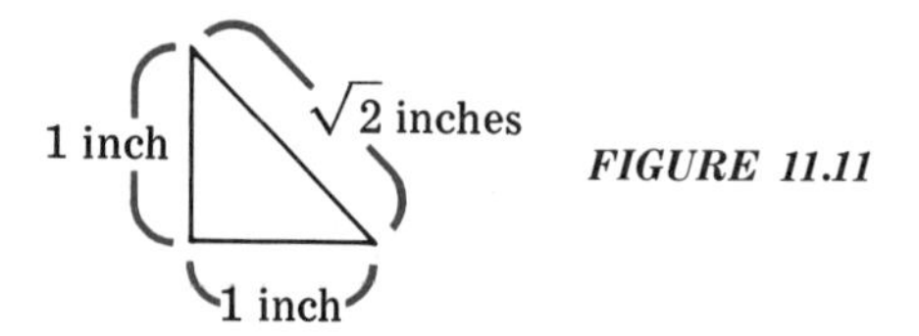

FIGURE 11.11

EXAMPLE 3

In the right triangle of Figure 11.12, find c if $a = 3$ inches and $b = 5$ inches.

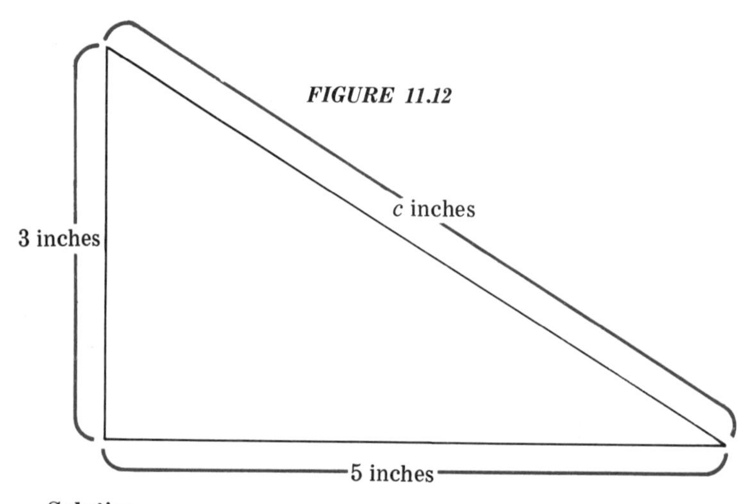

FIGURE 11.12

Solution.

$$c = \sqrt{a^2 + b^2}$$

Replace a by 3 and $\boldsymbol{b}$ by **5.**

$$\begin{aligned} c &= \sqrt{3^2 + \mathbf{5}^2} \\ &= \sqrt{9 + 25} \\ &= \sqrt{34} \end{aligned}$$

The length of the hypotenuse is slightly less than 6 inches (because $\sqrt{36} = 6$).

Square Root Function

In Section 11.3 you will consider the square roots of rational numbers such as $\frac{1}{9}$ and $\frac{4}{25}$. In fact, *every nonnegative number has a square root.* Thus you can consider the following **square root function:**

$$\text{Let } f(x) = \sqrt{x} \text{ for } x \text{ nonnegative.}$$

Then
$$f(0) = \sqrt{0} = 0$$
$$f(1) = \sqrt{1} = 1$$
$$f(4) = \sqrt{4} = 2$$
$$f(9) = \sqrt{9} = 3$$

EXAMPLE 4

Let $f(x) = \sqrt{x}$, for x nonnegative. Find:

(a) $f(36)$ (b) $f(144)$

(c) $f\left(\frac{1}{4}\right)$ (d) $f(2)$

Solution.

(a) $f(36) = \sqrt{36} = 6$

(b) $f(144) = \sqrt{144} = 12$

(c) Observe that

$$\frac{1}{2} \cdot \frac{1}{2} = \frac{1}{4}$$

Thus
$$f\left(\frac{1}{4}\right) = \sqrt{\frac{1}{4}} = \frac{1}{2}$$

(d) $f(2) = \sqrt{2}$

Leave this in radical form.

The graph of the square root function is given in Figure 11.13.

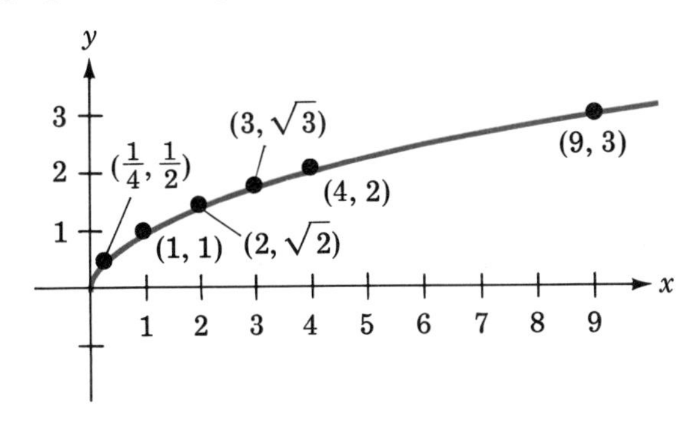

FIGURE 11.13

EXERCISES

In exercises 1–12, find consecutive integers N and $N + 1$ such that for the indicated square root $\sqrt{M}$, $N < \sqrt{M} < N + 1$.

1. $\sqrt{8}$
2. $\sqrt{14}$
3. $\sqrt{23}$
4. $\sqrt{32}$
5. $\sqrt{53}$
6. $\sqrt{65}$

7. $\sqrt{85}$

8. $\sqrt{99}$

9. $\sqrt{125}$

10. $\sqrt{145}$

11. $\sqrt{401}$

12. $\sqrt{9999}$

In exercises 13–18, use

$$\sqrt{2} \approx 1.41$$

to approximate each of the following to one decimal place:

13. $3\sqrt{2}$

14. $10 + \sqrt{2}$

15. $\sqrt{2} - 1$

16. $7 - \sqrt{2}$

17. $\frac{\sqrt{2}}{5}$

18. $1.5 + \sqrt{2}$

In exercises 19–24, use

$$\sqrt{3} \approx 1.73$$

to approximate each of the following to one decimal place:

19. $2\sqrt{3}$

20. $100\sqrt{3}$

21. $\frac{\sqrt{3}}{2}$

22. $3.9 - \sqrt{3}$

23. $\frac{1 + \sqrt{3}}{2}$

24. $\frac{\sqrt{3}}{3} - 1$

In exercises 25–30, use

$$\sqrt{2} \approx 1.41, \quad \sqrt{3} \approx 1.73$$

to approximate each of the following to one decimal place.

25. $\sqrt{2} + \sqrt{3}$

26. $\sqrt{3} - \sqrt{2}$

27. $\sqrt{2} + 3\sqrt{3}$

28. $\frac{\sqrt{2} - \sqrt{3}}{2}$

29. $\frac{\sqrt{3}}{5} + 10\sqrt{2}$

30. $\sqrt{2}\,\sqrt{3}$

31. Approximate the length of a side of a square whose area is 41 square inches. (Use the next *smallest* integer.)

32. Approximate the length of a side of a square whose area is 118 square feet. (Use the next *largest* integer.)

In exercises 33–36, (Figs. 11.14–11.17) approximate the length of the indicated side of the right triangle. (Use the next *largest* integer.)

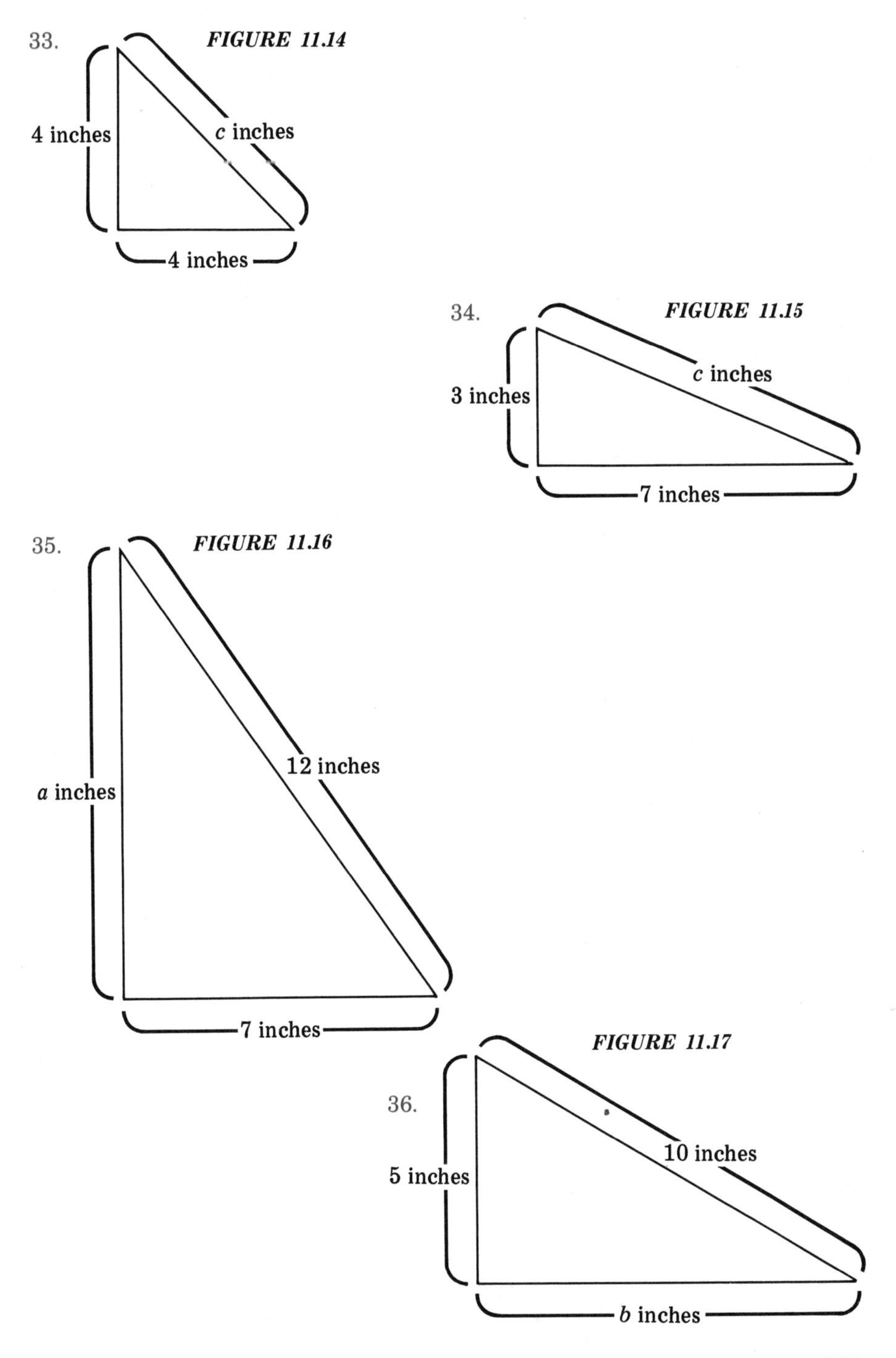

FIGURE 11.14

FIGURE 11.15

FIGURE 11.16

FIGURE 11.17

In exercises 37–44, let $f(x) = \sqrt{x}$, for x nonnegative. Find each value. (Leave irrational numbers in radical form.)

37. $f(16)$
38. $f(49)$
39. $f(81)$
40. $f(100)$
41. $f(10000)$
42. $f(3)$
43. $f(5)$
44. $f(19)$

11.3 ROOTS OF PRODUCTS AND QUOTIENTS

Roots of Products

Observe that

$$\sqrt{36} = 6 = 3 \cdot 2$$

and

$$36 = 9 \cdot 4.$$

Also,

$$\sqrt{9} = 3, \quad \sqrt{4} = 2$$

Because

$$\sqrt{36} = 3 \cdot 2$$

it follows that

$$\sqrt{9 \cdot 4} = \sqrt{9}\sqrt{4}.$$

In general, *the square root of a product equals the product of the square roots.*

$$\sqrt{ab} = \sqrt{a}\sqrt{b}, \quad \text{where } a > 0,\ b > 0$$

EXAMPLE 1

Let $x > 0$. Find:

$$\sqrt{9x^2}$$

Solution.

$$\begin{aligned}\sqrt{9x^2} &= \sqrt{9}\sqrt{x^2}\\ &= 3x\end{aligned}$$

EXAMPLE 2

Let $a > 0$. Find:

$$\sqrt{a^2b^4}$$

Solution.

$$\begin{aligned}\sqrt{a^2b^4} &= \sqrt{a^2}\sqrt{b^4}\\ &= ab^2\end{aligned}$$

The above product rule applies to three or more factors. For example,

$$\sqrt{abc} = \sqrt{a}\sqrt{b}\sqrt{c}, \quad a > 0, b > 0, c > 0$$

EXAMPLE 3

Let $x > 0$, $y > 0$. Find:

$$\sqrt{25x^6y^{10}}$$

Solution.

$$\begin{aligned} \sqrt{25x^6y^{10}} &= \sqrt{25}\sqrt{x^6}\sqrt{y^{10}} \\ &= 5x^3y^5 \end{aligned}$$

The product rule enables you to simplify some irrational square roots, and often to approximate them quickly.

EXAMPLE 4

(a) Simplify:

$$\sqrt{8}$$

(b) Estimate $\sqrt{8}$, roughly.

Solution.

(a)

$$\begin{aligned} 8 &= 4 \cdot 2 \\ \sqrt{8} &= \sqrt{4}\sqrt{2} \\ &= 2\sqrt{2} \end{aligned}$$

The equivalent form $2\sqrt{2}$ is simpler than $\sqrt{8}$ in that the radical sign applies to a *smaller* positive integer. *Leave the answer in radical form.* Note that $2\sqrt{2}$ is the *exact* answer. In part (b) you will *approximate* $\sqrt{8}$.

(b)

$$\sqrt{2} \approx 1.4$$

Thus

$$\begin{aligned} \sqrt{8} &= 2\sqrt{2} \\ &\approx 2(1.4) \\ \sqrt{8} &\approx 2.8 \end{aligned}$$

You can also simplify square roots of *odd* powers of a variable.

EXAMPLE 5

Let $x > 0$. Simplify:

$$\sqrt{x^9}$$

Solution. The "largest square factor" that can be removed from the radical symbol is x^8:

$$\begin{aligned} x^9 &= x^8 \cdot x \\ \sqrt{x^9} &= \sqrt{x^8 \cdot x} \\ &= \sqrt{x^8}\sqrt{x} \\ &= x^4\sqrt{x} \end{aligned}$$

Roots of Quotients

Observe that

$$\sqrt{4} = 2, \quad 4 = \frac{100}{25}, \quad \text{and } 2 = \frac{10}{5}.$$

Also,

$$\sqrt{100} = 10$$
$$\sqrt{25} = 5$$

Because

$$\sqrt{4} = \frac{10}{5},$$

it follows that

$$\sqrt{\frac{100}{25}} = \frac{\sqrt{100}}{\sqrt{25}}.$$

In general, *the square root of a quotient equals the quotient of the square roots.*

$$\sqrt{\frac{a}{b}} = \frac{\sqrt{a}}{\sqrt{b}}, \quad a > 0, b > 0$$

EXAMPLE 6

Find:

$$\sqrt{\frac{4}{9}}$$

Solution.

$$\sqrt{\frac{4}{9}} = \frac{\sqrt{4}}{\sqrt{9}}$$
$$= \frac{2}{3}$$

EXAMPLE 7

Find:

$$\sqrt{\frac{a^4}{b^8}}, \quad b \neq 0$$

Solution.

$$\sqrt{\frac{a^4}{b^8}} = \frac{\sqrt{a^4}}{\sqrt{b^8}} = \frac{a^2}{b^4}$$

EXAMPLE 8

Let $x > 0$ and $z \neq 0$. Find:

$$\sqrt{\frac{25x^2y^4}{81z^{12}}}$$

Solution.

$$\sqrt{\frac{25x^2y^4}{81z^{12}}} = \frac{\sqrt{25x^2y^4}}{\sqrt{81z^{12}}}$$
$$= \frac{\sqrt{25}\sqrt{x^2}\sqrt{y^4}}{\sqrt{81}\sqrt{z^{12}}}$$
$$= \frac{5xy^2}{9z^6}$$

EXAMPLE 9

Let $c > 0$. Simplify:

$$\sqrt{\frac{c^5}{32}}$$

Solution.

$$\sqrt{\frac{c^5}{32}} = \frac{\sqrt{c^5}}{\sqrt{32}}$$
$$= \frac{\sqrt{c^4}\sqrt{c}}{\sqrt{16}\sqrt{2}}$$
$$= \frac{c^2\sqrt{c}}{4\sqrt{2}}$$
$$= \frac{c^2}{4}\sqrt{\frac{c}{2}}$$

Let $a > 0$, $b > 0$. Although

$$\sqrt{ab} = \sqrt{a}\sqrt{b}$$

and

$$\sqrt{\frac{a}{b}} = \frac{\sqrt{a}}{\sqrt{b}},$$

make sure that you realize that, in general,

$$\sqrt{a+b} \neq \sqrt{a} + \sqrt{b}$$

and

$$\sqrt{a-b} \neq \sqrt{a} - \sqrt{b}.$$

For example, if $a = 25$ and $b = 4$, then

$$\sqrt{25+4} = \sqrt{29} < \sqrt{36} = 6,$$

whereas

$$\sqrt{25} + \sqrt{4} = 5 + 2 = 7$$

Also,

$$\sqrt{25-4} = \sqrt{21} > \sqrt{16} = 4,$$

whereas

$$\sqrt{25} - \sqrt{4} = 5 - 2 = 3$$

Roots of Decimals

Observe that

$$\sqrt{\frac{1}{100}} = \frac{\sqrt{1}}{\sqrt{100}} = \frac{1}{10}$$

and that

$$\sqrt{\frac{1}{10000}} = \frac{\sqrt{1}}{\sqrt{10000}} = \frac{1}{100}.$$

Thus, in decimal notation:

$$\sqrt{.01} = .1$$

and

$$\sqrt{.0001} = .01$$

EXAMPLE 10

Find:
(a) $\sqrt{.04}$ (b) $\sqrt{.0025}$ (c) $\sqrt{.36x^4}$

Solution.

(a)

$$\begin{aligned}\sqrt{.04} &= \sqrt{4(.01)}\\ &= \sqrt{4}\sqrt{.01}\\ &= 2(.1)\\ &= .2\end{aligned}$$

(b)

$$\begin{aligned}\sqrt{.0025} &= \sqrt{25(.0001)}\\ &= \sqrt{25}\sqrt{.0001}\\ &= 5(.01)\\ &= .05\end{aligned}$$

(c)

$$\begin{aligned}\sqrt{.36x^4} &= \sqrt{36(.01)x^4}\\ &= \sqrt{36}\sqrt{.01}\sqrt{x^4}\\ &= 6(.1)x^2\\ &= .6x^2\end{aligned}$$

EXERCISES

In exercises 1–34, assume $a > 0$, $b > 0$, $c > 0$, $d > 0$, $x > 0$, $y > 0$, $z > 0$. Find each square root.

1. $\sqrt{4x^2}$
2. $\sqrt{16a^4}$

3. $\sqrt{25b^8}$
4. $\sqrt{100a^{18}}$
5. $\sqrt{a^2b^2}$
6. $\sqrt{x^2y^4}$
7. $\sqrt{x^4y^4}$
8. $\sqrt{a^8x^{10}}$
9. $\sqrt{x^2y^2z^2}$
10. $\sqrt{a^4b^2c^6}$
11. $\sqrt{a^2x^{10}y^4}$
12. $\sqrt{a^2b^4c^6d^8}$
13. $\sqrt{4a^4b^8}$
14. $\sqrt{49x^2y^4}$
15. $\sqrt{36a^4x^2}$
16. $\sqrt{121y^2z^{12}}$
17. $\sqrt{\frac{1}{4}}$
18. $\sqrt{\frac{9}{25}}$
19. $\sqrt{\frac{36}{49}}$
20. $\sqrt{\frac{81}{4}}$
21. $\sqrt{\frac{25}{144}}$
22. $\sqrt{\frac{100}{121}}$
23. $\sqrt{\frac{a^2}{c^2}}$
24. $\sqrt{\frac{c^4}{d^2}}$
25. $\sqrt{\frac{x^2}{y^8}}$
26. $\sqrt{\frac{z^6}{a^{10}}}$
27. $\sqrt{\frac{x^2}{4}}$
28. $\sqrt{\frac{y^8}{9}}$
29. $\sqrt{\frac{c^{16}}{16}}$
30. $\sqrt{\frac{25}{d^8}}$
31. $\sqrt{\frac{a^2b^6}{9}}$
32. $\sqrt{\frac{144x^4}{y^2z^{10}}}$
33. $\sqrt{\frac{25x^4y^8}{4z^{10}}}$
34. $\sqrt{\frac{81a^2d^4}{64x^{12}y^{20}}}$

In exercises 35–42: (a) Simplify each expression. (Leave your answer in radical form.) (b) Estimate the square root, roughly. (Use $\sqrt{2} \approx 1.4$, $\sqrt{3} \approx 1.7$.)

SAMPLE. $\sqrt{12}$	***Solution.*** (a) $\sqrt{12} = \sqrt{4 \cdot 3}$ $= \sqrt{4}\sqrt{3}$ $= 2\sqrt{3}$ (b) $\sqrt{12} \approx 2(1.7)$ $= 3.4$

35. $\sqrt{18}$
36. $\sqrt{32}$
37. $\sqrt{50}$
38. $\sqrt{200}$
39. $\sqrt{27}$
40. $\sqrt{48}$
41. $\sqrt{75}$
42. $\sqrt{300}$

In exercises 43–56, simplify each expression. Leave answers in radical form. Assume $a > 0$, $b > 0$, $c > 0$.

SAMPLE. $\sqrt{20c^5}$	***Solution.*** $\sqrt{20c^5} = \sqrt{20}\sqrt{c^5} = \sqrt{4}\sqrt{5}\sqrt{c^4}\sqrt{c} = 2\sqrt{5}\,c^2\sqrt{c} = 2c^2\sqrt{5c}$

43. $\sqrt{a^3}$
44. $\sqrt{c^7}$
45. $\sqrt{\dfrac{a^5}{b^6}}$
46. $\sqrt{4b^9}$
47. $\sqrt{12a^{12}}$
48. $\sqrt{27b^9}$
49. $\sqrt{a^3b^5}$
50. $\sqrt{\dfrac{c^3}{a^2b^2}}$
51. $\sqrt{\dfrac{9a^9}{b^2}}$
52. $\sqrt{\dfrac{24a^{11}b^7}{c^2}}$
53. $\sqrt{98a^4b^3}$
54. $\sqrt{162a^7b^9}$
55. $\sqrt{\dfrac{50a^3}{49b^8}}$
56. $\sqrt{\dfrac{175c^5}{108a^2b^4}}$

In exercises 57–64, find each square root.

57. $\sqrt{.09}$
58. $\sqrt{.36}$
59. $\sqrt{.0004}$
60. $\sqrt{.0064}$
61. $\sqrt{.000001}$
62. $\sqrt{.0144x^4}$
63. $\sqrt{81x^8y^4}$
64. $\sqrt{.0121x^{12}y^{16}}$

11.4 ADDITION AND SUBTRACTION OF ROOTS

Simplifying Expressions With Radicals

Irrational square roots can be added or subtracted in the same way you combine terms involving variables. Thus

$$2\sqrt{3} + 3\sqrt{3} = 5\sqrt{3}$$

just as

$$2x + 3x = 5x$$

But

$$2\sqrt{3} + 3\sqrt{5}$$

cannot be further simplified, just as

$$2x + 3y$$

cannot be.

EXAMPLE 1
Simplify:

$$\sqrt{7} + 5\sqrt{7} - 3\sqrt{7}$$

Solution. Note that

$$\sqrt{7} = 1\sqrt{7}.$$

Thus

$$\begin{aligned} \sqrt{7} + 5\sqrt{7} - 3\sqrt{7} &= (1 + 5 - 3)\sqrt{7} \\ &= 3\sqrt{7} \end{aligned}$$

EXAMPLE 2
Simplify:

$$2\sqrt{7} + 3\sqrt{5} - (\sqrt{7} - \sqrt{5})$$

Solution.

$$\begin{aligned} 2\sqrt{7} + 3\sqrt{5} - (\sqrt{7} - \sqrt{5}) &= 2\sqrt{7} + 3\sqrt{5} - \sqrt{7} + \sqrt{5} \\ &= (2\sqrt{7} - \sqrt{7}) + (3\sqrt{5} + \sqrt{5}) \\ &= \sqrt{7} + 4\sqrt{5} \end{aligned}$$

EXAMPLE 3
Simplify:

$$\sqrt{8} + \sqrt{18}$$

Solution.

$$\begin{aligned} \sqrt{8} &= \sqrt{4}\sqrt{2} = 2\sqrt{2} \\ \sqrt{18} &= \sqrt{9}\sqrt{2} = 3\sqrt{2} \end{aligned}$$

Therefore

$$\begin{aligned} \sqrt{8} + \sqrt{18} &= 2\sqrt{2} + 3\sqrt{2} \\ &= 5\sqrt{2} \end{aligned}$$

EXAMPLE 4
Let $a > 0$. Simplify:

$$\sqrt{4a^5} + \sqrt{25a^5}$$

Solution.

$$\sqrt{4a^5} = \sqrt{4}\sqrt{a^4}\sqrt{a} = 2a^2\sqrt{a}$$

$$\sqrt{25a^5} = \sqrt{25}\sqrt{a^4}\sqrt{a} = 5a^2\sqrt{a}$$

Thus

$$\sqrt{4a^5} + \sqrt{25a^5} = 2a^2\sqrt{a} + 5a^2\sqrt{a}$$
$$= 7a^2\sqrt{a}$$

EXAMPLE 5

Let $a > 0$. Simplify:

$$\sqrt{50a^3} + \sqrt{98a^3}$$

Solution.

$$\sqrt{50a^3} = \sqrt{25}\sqrt{a^2}\sqrt{2a}$$
$$= 5a\sqrt{2a}$$
$$\sqrt{98a^3} = \sqrt{49}\sqrt{a^2}\sqrt{2a}$$
$$= 7a\sqrt{2a}$$

Thus

$$\sqrt{50a^3} + \sqrt{98a^3} = 5a\sqrt{2a} + 7a\sqrt{2a}$$
$$= 12a\sqrt{2a}$$

Combining Fractions

Fractions can be irrational as well as rational. The rules of Section 6.3 and 6.5 for adding or subtracting fractions apply here.

EXAMPLE 6

Add:

$$\frac{\sqrt{2}}{3} + \frac{2\sqrt{3}}{3}$$

Solution.

$$\frac{\sqrt{2}}{3} + \frac{2\sqrt{3}}{3} = \frac{\sqrt{2} + 2\sqrt{3}}{3}$$

EXAMPLE 7

Subtract:

$$\frac{2}{3} - \frac{\sqrt{2}}{5}$$

Solution. The *lcd* is 15. Thus

$$\frac{2}{3} - \frac{\sqrt{2}}{5} = \frac{5 \cdot 2 - 3\sqrt{2}}{15}$$
$$= \frac{10 - 3\sqrt{2}}{15}$$

EXERCISES

Simplify each expression. Assume $a > 0$, $b > 0$, $x > 0$, $y > 0$.

1. $\sqrt{2} + \sqrt{2}$
2. $2\sqrt{3} + 4\sqrt{3}$
3. $5\sqrt{2} - \sqrt{2}$
4. $\sqrt{7} - 2\sqrt{7}$
5. $3\sqrt{6} + 2\sqrt{6}$
6. $7\sqrt{10} - 4\sqrt{10}$
7. $\sqrt{11} + 4\sqrt{11} + 3\sqrt{11}$
8. $6\sqrt{13} - \sqrt{13} + 4\sqrt{13}$
9. $2\sqrt{5} + 6\sqrt{5} - \sqrt{5}$
10. $\sqrt{7} - 2\sqrt{7} - 5\sqrt{7}$
11. $12\sqrt{2} - (7\sqrt{2} + 2\sqrt{2})$
12. $3\sqrt{7} - (2\sqrt{7} - 5\sqrt{7})$
13. $\sqrt{2} + \sqrt{3} + \sqrt{2} + 2\sqrt{3}$
14. $3\sqrt{5} - (\sqrt{2} + \sqrt{5})$
15. $\sqrt{7} - \sqrt{5} + 4\sqrt{7} + 3\sqrt{5}$
16. $\sqrt{11} - (\sqrt{2} + \sqrt{11} - \sqrt{2})$
17. $\sqrt{2} + \sqrt{8}$
18. $\sqrt{6} + \sqrt{24}$
19. $\sqrt{18} - 3\sqrt{2}$
20. $\sqrt{50} + \sqrt{32}$
21. $\sqrt{7} + \sqrt{28} - 2\sqrt{7}$
22. $\sqrt{18} - \sqrt{2} + \sqrt{50}$
23. $\sqrt{300} + \sqrt{12} - \sqrt{75}$
24. $\sqrt{44} + \sqrt{99} - 5\sqrt{11}$
25. $\sqrt{4a} + \sqrt{9a}$
26. $\sqrt{64x} - \sqrt{49x}$
27. $\sqrt{100y^3} + \sqrt{81y^3}$
28. $\sqrt{a^5} - \sqrt{16a^5}$
29. $\sqrt{144ab^2} + \sqrt{121ab^2}$
30. $\sqrt{36x} + \sqrt{64x} - \sqrt{81x}$
31. $\sqrt{12a} + \sqrt{3a}$
32. $\sqrt{20b} - \sqrt{5b}$
33. $\sqrt{98x^3} + \sqrt{50x^3}$
34. $\sqrt{288yz^4} - \sqrt{72yz^4}$
35. $\sqrt{7ab^3} + \sqrt{28ab^3} + \sqrt{63ab^3}$
36. $\sqrt{125x^5y^3} - \sqrt{20x^5y^3} + \sqrt{45x^5y^3}$
37. $\frac{2}{5} + \frac{\sqrt{5}}{5}$
38. $\frac{1}{8} - \frac{\sqrt{2}}{8}$
39. $\frac{\sqrt{2}}{3} - \frac{\sqrt{5}}{3} + \frac{\sqrt{7}}{3}$
40. $\frac{\sqrt{2}}{2} + \frac{\sqrt{5}}{4}$
41. $\frac{\sqrt{2}}{5} - \frac{\sqrt{3}}{7}$
42. $\frac{6\sqrt{3}}{5} - \frac{12\sqrt{3}}{15}$
43. $\frac{\sqrt{2}}{4} + \frac{\sqrt{3}}{8} - \frac{2\sqrt{2}}{12}$
44. $\frac{\sqrt{2}}{3} + \frac{\sqrt{3}}{6} + \frac{1}{12}$

11.5 MULTIPLICATION OF ROOTS

$\sqrt{a}\sqrt{b}$

Recall that the square root of a product equals the product of the square roots.

$$\sqrt{ab} = \sqrt{a}\sqrt{b}, \quad a > 0, b > 0$$

In Section 11.3 you used this to simplify the square root of a product such as $\sqrt{9x^2}$. Thus if $x > 0$,

$$\sqrt{9x^2} = \sqrt{9}\sqrt{x^2} = 3x$$

Now you will use the above law, in reverse, to simplify the product of square roots.

EXAMPLE 1

Multiply, as indicated. Then simplify:

$$\sqrt{2}\sqrt{3}$$

Solution.

$$\begin{aligned}\sqrt{2}\sqrt{3} &= \sqrt{2 \cdot 3} \\ &= \sqrt{6}\end{aligned}$$

EXAMPLE 2

Multiply, as indicated. Then simplify:

$$\sqrt{5}\sqrt{5}$$

Solution.

$$\begin{aligned}\sqrt{5}\sqrt{5} &= \sqrt{5 \cdot 5} \\ &= \sqrt{25} \\ &= 5\end{aligned}$$

EXAMPLE 3

Multiply, as indicated. Then simplify:

$$\sqrt{a^3}\sqrt{a}$$

Solution.

$$\begin{aligned}\sqrt{a^3}\sqrt{a} &= \sqrt{a^3 \cdot a} \\ &= \sqrt{a^4} \\ &= a^2\end{aligned}$$

EXAMPLE 4

Let $x > 0$. Multiply, as indicated. Then simplify:

$$\sqrt{2x}\sqrt{8x}$$

Solution.

$$\begin{aligned}\sqrt{2x}\sqrt{8x} &= \sqrt{(2x)(8x)} \\ &= \sqrt{16x^2} \\ &= \sqrt{16}\sqrt{x^2} \\ &= 4x\end{aligned}$$

EXAMPLE 5

Let $a > 0$, $b > 0$. Multiply, as indicated. Then simplify:

$$\sqrt{3a}\sqrt{27b}\sqrt{ab}$$

Solution.

$$\begin{aligned}\sqrt{3a}\sqrt{27b}\sqrt{ab} &= \sqrt{(3a)(27b)(ab)} \\ &= \sqrt{81a^2b^2} \\ &= \sqrt{81}\sqrt{a^2}\sqrt{b^2} \\ &= 9ab\end{aligned}$$

Distributive Laws

The Distributive Laws apply to products involving roots. Recall that

$$a(b + c) = ab + ac.$$

EXAMPLE 6

Find:

$$\sqrt{2}(\sqrt{2} + \sqrt{5})$$

Solution.

$$\begin{aligned}\sqrt{2}(\sqrt{2} + \sqrt{5}) &= \sqrt{2}\sqrt{2} + \sqrt{2}\sqrt{5} \\ &= \sqrt{2 \cdot 2} + \sqrt{2 \cdot 5} \\ &= 2 + \sqrt{10}\end{aligned}$$

EXAMPLE 7

Find:

$$(1 + \sqrt{7})(2 + \sqrt{7})$$

Solution.

$$\begin{array}{l} 1 + \sqrt{7} \\ \underline{2 + \sqrt{7}} \\ 2 + 2\sqrt{7} \\ \underline{\quad + \ \sqrt{7} + \sqrt{7}\sqrt{7}} \\ 2 + 3\sqrt{7} + 7 \\ \quad \llcorner\!\!-\!9\!-\!\!\lrcorner \end{array}$$

Thus

$$(1 + \sqrt{7})(2 + \sqrt{7}) = 9 + 3\sqrt{7}$$

EXAMPLE 8

Find:

$$(\sqrt{3} + \sqrt{2})(\sqrt{3} - \sqrt{2})$$

Solution.

$$\begin{array}{rl}
 & \sqrt{3} + \sqrt{2} \\
 & \sqrt{3} - \sqrt{2} \\
\hline
 & 3 + \sqrt{3}\sqrt{2} \\
 & \quad - \sqrt{3}\sqrt{2} - 2 \\
\hline
 & 3 \qquad\qquad - 2 \\
 & \qquad\; 1
\end{array}$$

Thus

$$(\sqrt{3} + \sqrt{2})(\sqrt{3} - \sqrt{2}) = 1$$

EXAMPLE 9

Find:

$$(5\sqrt{7} + \sqrt{2})(2\sqrt{7} - 3\sqrt{2})$$

Solution.

$$\begin{array}{rl}
 & 5\sqrt{7} + \sqrt{2} \\
 & 2\sqrt{7} - 3\sqrt{2} \\
\hline
 & 10 \cdot 7 + 2\sqrt{7}\sqrt{2} \\
 & \qquad - 15\sqrt{7}\sqrt{2} - 3 \cdot 2 \\
\hline
 & 70 \quad - 13\sqrt{14} - 6 \\
 & \qquad 64
\end{array}$$

Thus

$$(5\sqrt{7} + \sqrt{2})(2\sqrt{7} - 3\sqrt{2}) = 64 - 13\sqrt{14}$$

Multiplying Fractions

Recall that to multiply fractions, first divide by factors common to the numerators and denominators. Then multiply the simplified numerators and denominators separately.

EXAMPLE 10

Find:

$$\frac{\sqrt{12}}{5} \cdot \frac{\sqrt{6}}{4}$$

Solution.

$$\frac{\sqrt{12}}{5} \cdot \frac{\sqrt{6}}{4} = \frac{\cancel{2}\sqrt{3}}{5} \cdot \frac{\sqrt{2}\sqrt{3}}{\underset{2}{\cancel{4}}}$$

$$= \frac{3\sqrt{2}}{10}$$

EXERCISES

Multiply, as indicated. Then simplify, if possible. Assume $a > 0$, $b > 0$, $c > 0$, $x > 0$, $y > 0$, $z > 0$.

1. $\sqrt{2}\sqrt{5}$
2. $\sqrt{2}\sqrt{7}$
3. $\sqrt{3}\sqrt{5}$
4. $\sqrt{6}\sqrt{5}$
5. $\sqrt{7}\sqrt{8}$
6. $\sqrt{11}\sqrt{7}$
7. $\sqrt{2}\sqrt{2}$
8. $\sqrt{3}\sqrt{3}$
9. $\sqrt{7}\sqrt{7}$
10. $\sqrt{8}\sqrt{8}$
11. $\sqrt{12}\sqrt{3}$
12. $\sqrt{2}\sqrt{18}$
13. $\sqrt{50}\sqrt{2}$
14. $\sqrt{6}\sqrt{3}$
15. $\sqrt{2}\sqrt{6}\sqrt{5}$
16. $\sqrt{3}\sqrt{21}\sqrt{7}$
17. $\sqrt{a}\sqrt{a}$
18. $\sqrt{x^3}\sqrt{x^3}$
19. $\sqrt{b}\sqrt{b^5}$
20. $\sqrt{ax}\sqrt{ax}$
21. $\sqrt{a}\sqrt{ab}\sqrt{b}$
22. $\sqrt{ab^3}\sqrt{bc^3}\sqrt{ca^3}$
23. $\sqrt{5x}\sqrt{20x}$
24. $\sqrt{18a^3}\sqrt{2a}$
25. $\sqrt{2b^3}\sqrt{12b}$
26. $\sqrt{32ab^3}\sqrt{2ba}$
27. $\sqrt{xyz^3}\sqrt{x^2y}\sqrt{zy^2}$
28. $\sqrt{5abx}\sqrt{2ax}\sqrt{10b}$
29. $\sqrt{3a^2b}\sqrt{2bc}\sqrt{6c^3}$
30. $\sqrt{27xy}\sqrt{3xz}\sqrt{y^3z^3}$
31. $\sqrt{2}(\sqrt{2} + \sqrt{3})$
32. $\sqrt{5}(\sqrt{5} - 1)$
33. $\sqrt{a}(\sqrt{a} + \sqrt{b})$
34. $\sqrt{y}(\sqrt{y} + \sqrt{y^3})$
35. $\sqrt{3}(\sqrt{27} - \sqrt{12})$
36. $\sqrt{2ab}(\sqrt{ab} + \sqrt{2})$
37. $2\sqrt{3}(\sqrt{3} + 12)$
38. $(\sqrt{2} + \sqrt{5})(\sqrt{2} - \sqrt{5})$
39. $(7 + \sqrt{2})(7 - \sqrt{2})$
40. $(\sqrt{3} - \sqrt{11})(\sqrt{3} + \sqrt{11})$
41. $(2\sqrt{5} + 3\sqrt{3})(2\sqrt{5} - 3\sqrt{3})$
42. $(1 + \sqrt{2})(4 + \sqrt{2})$
43. $(2 + 3\sqrt{5})(4 + 2\sqrt{5})$
44. $(\sqrt{6} - \sqrt{2})(2\sqrt{6} - \sqrt{2})$
45. $(4\sqrt{7} - 2\sqrt{5})(2\sqrt{7} + 4\sqrt{5})$
46. $(2\sqrt{3} + 3\sqrt{10})(\sqrt{3} - \sqrt{10})$
47. $(\sqrt{5} - 3\sqrt{11})(2\sqrt{5} + \sqrt{11})$
48. $(6\sqrt{3} + 9\sqrt{5})(2\sqrt{3} + \sqrt{5})$
49. $(1 + \sqrt{2})^2$
50. $(2\sqrt{3} + \sqrt{5})^2$
51. $\dfrac{\sqrt{5}}{2} \cdot \dfrac{\sqrt{5}}{3}$
52. $\dfrac{\sqrt{3}}{4} \cdot \dfrac{2\sqrt{2}}{5}$
53. $\dfrac{\sqrt{8}}{5} \cdot \dfrac{\sqrt{3}}{\sqrt{2}}$
54. $\dfrac{\sqrt{6}}{2} \cdot \dfrac{\sqrt{3}}{4}$
55. $\dfrac{\sqrt{12}}{\sqrt{5}} \cdot \dfrac{\sqrt{6}}{2} \cdot \dfrac{\sqrt{10}}{3}$
56. $\dfrac{\sqrt{3}}{\sqrt{2}} \cdot \dfrac{\sqrt{32}}{3} \cdot \dfrac{\sqrt{27}}{4}$

11.6 DIVISION OF ROOTS

$$\frac{\sqrt{a}}{\sqrt{b}}$$

The square root of a quotient equals the quotient of the square roots.

$$\sqrt{\frac{a}{b}} = \frac{\sqrt{a}}{\sqrt{b}}, \quad a > 0, b > 0$$

In Section 11.3 you simplified *the square root of a quotient*. For example,

$$\sqrt{\frac{4}{9}} = \frac{\sqrt{4}}{\sqrt{9}} = \frac{2}{3}$$

Now you will use the above law, together with the product law

$$\sqrt{ab} = \sqrt{a}\sqrt{b}, \quad a > 0, b > 0$$

to simplify *the quotient of square roots.*

EXAMPLE 1

Simplify:

$$\frac{\sqrt{48}}{\sqrt{3}}$$

Solution.

$$\frac{\sqrt{48}}{\sqrt{3}} = \sqrt{\frac{48}{3}}$$
$$= \sqrt{16}$$
$$= 4$$

EXAMPLE 2

Let $x > 0$. Simplify:

$$\frac{\sqrt{28x}}{\sqrt{7x}}$$

Solution.

$$\frac{\sqrt{28x}}{\sqrt{7x}} = \sqrt{\frac{28x}{7x}}$$
$$= \sqrt{4}$$
$$= 2$$

EXAMPLE 3

Simplify:

$$\frac{\sqrt{3}\sqrt{14}}{\sqrt{6}}$$

Solution.

$$\frac{\sqrt{3}\sqrt{14}}{\sqrt{6}} = \frac{\cancel{\sqrt{3}}\cancel{\sqrt{2}}\sqrt{7}}{\cancel{\sqrt{3}}\cancel{\sqrt{2}}}$$
$$= \sqrt{7}$$

Rationalizing the Denominator

A fraction, such as

$$\frac{1}{\sqrt{5}},$$

is often easier to work with when the denominator is rational. **Rationalize the denominator** of this fraction as follows:

$$\frac{1}{\sqrt{5}} = \frac{1 \cdot \sqrt{5}}{\sqrt{5} \cdot \sqrt{5}} = \frac{\sqrt{5}}{5}$$

Similarly,

$$\frac{4}{\sqrt{2}} = \frac{4\sqrt{2}}{\sqrt{2} \cdot \sqrt{2}} = \frac{4\sqrt{2}}{2} = 2\sqrt{2}$$

EXAMPLE 4

Let $a > 0$. Simplify:

$$\frac{\sqrt{2a}}{\sqrt{3a}}$$

Solution.

$$\begin{aligned} \frac{\sqrt{2a}}{\sqrt{3a}} &= \sqrt{\frac{2\not{a}}{3\not{a}}} \\ &= \sqrt{\frac{2}{3}} \end{aligned}$$

Factoring and Dividing

You can often "isolate the common factor" of an expression containing radicals, just as you isolated the common factor of a polynomial. When this expression is the numerator of a fraction, you may be able to divide by factors common to numerator and denominator.

EXAMPLE 5

(a) Factor: $2 + \sqrt{12}$

(b) Simplify: $\frac{2 + \sqrt{12}}{4}$

Solution.

(a)

$$\begin{aligned} 2 + \sqrt{12} &= 2 + \sqrt{4 \cdot 3} \\ &= 2 + \sqrt{4}\sqrt{3} \\ &= 2 + 2\sqrt{3} \\ &= 2(1 + \sqrt{3}) \end{aligned}$$

(b)

$$\frac{2+\sqrt{12}}{4} = \frac{\overset{1}{\cancel{2}}(1+\sqrt{3})}{\underset{2}{\cancel{4}}}$$

$$= \frac{1+\sqrt{3}}{2}$$

EXAMPLE 6
Let $x > 0$, $y > 0$.
(a) Factor:

$$\sqrt{xy} - \sqrt{x^3}$$

(b) Simplify:

$$\frac{\sqrt{xy} - \sqrt{x^3}}{\sqrt{x^5}}$$

Solution.

(a)
$$\sqrt{xy} - \sqrt{x^3} = \sqrt{x}\sqrt{y} - \sqrt{x}\sqrt{x^2}$$
$$= \sqrt{x}(\sqrt{y} - x)$$

(b)
$$\frac{\sqrt{xy} - \sqrt{x^3}}{\sqrt{x^5}} = \frac{\cancel{\sqrt{x}}(\sqrt{y} - x)}{\cancel{\sqrt{x}}\sqrt{x^4}}$$
$$= \frac{\sqrt{y} - x}{x^2}$$

EXAMPLE 7
Let $a > 0$, $x > 0$, $y > 0$, $x \neq y$.
(a) Factor:

$$\sqrt{a^2x^2} - \sqrt{a^2y^2}$$

(b) Simplify:

$$\frac{\sqrt{a^2x^2} - \sqrt{a^2y^2}}{x^2 - y^2}$$

Solution.

(a)
$$\sqrt{a^2x^2} - \sqrt{a^2y^2} = \sqrt{a^2}\sqrt{x^2} - \sqrt{a^2}\sqrt{y^2}$$
$$= ax - ay$$
$$= a(x - y)$$

(b)
$$\frac{\sqrt{a^2x^2} - \sqrt{a^2y^2}}{x^2 - y^2} = \frac{a(x-y)}{(x+y)(x-y)}$$
$$= \frac{a}{x+y}$$

Dividing Fractions

Recall that

$$\frac{a}{b} \div \frac{c}{d} = \frac{a}{b} \cdot \frac{d}{c}$$

Thus *to divide fractions, invert the second fraction and multiply.*

EXAMPLE 8

Divide:

$$\frac{\sqrt{10}}{3} \div \frac{\sqrt{5}}{9}$$

Solution.

$$\begin{aligned}\frac{\sqrt{10}}{3} \div \frac{\sqrt{5}}{9} &= \frac{\sqrt{10}}{\cancel{3}} \cdot \frac{\overset{3}{\cancel{9}}}{\sqrt{5}} \\ &= \cancel{\sqrt{5}}\sqrt{2} \cdot \frac{3}{\cancel{\sqrt{5}}} \\ &= 3\sqrt{2}\end{aligned}$$

EXERCISES

In exercises 1–20, simplify. If necessary, rationalize the denominator. Assume $x > 0$, $y > 0$, $z > 0$.

1. $\frac{\sqrt{27}}{\sqrt{3}}$
2. $\frac{\sqrt{2}}{\sqrt{8}}$
3. $\frac{-\sqrt{12}}{\sqrt{3}}$
4. $\frac{\sqrt{125}}{\sqrt{5}}$
5. $\frac{\sqrt{24}}{\sqrt{6}}$
6. $\frac{\sqrt{75}}{-\sqrt{50}}$
7. $\frac{6}{\sqrt{3}}$
8. $\frac{\sqrt{12}}{\sqrt{20}}$
9. $\frac{\sqrt{98}}{\sqrt{2}}$
10. $\frac{-10}{\sqrt{5}}$
11. $\frac{\sqrt{18x}}{\sqrt{81x}}$
12. $\frac{\sqrt{12y}}{\sqrt{40y}}$
13. $\frac{\sqrt{9x}}{\sqrt{25z}}$
14. $\frac{\sqrt{16xy}}{\sqrt{8xz}}$
15. $\frac{\sqrt{48xyz}}{\sqrt{27xz^3}}$
16. $\frac{\sqrt{32x^2y}}{\sqrt{18yz^2}}$
17. $\frac{\sqrt{30}}{\sqrt{2}\sqrt{15}}$
18. $\frac{-\sqrt{12}}{\sqrt{15}\sqrt{20}}$
19. $\frac{\sqrt{27}}{\sqrt{6}\sqrt{50}}$
20. $\frac{\sqrt{45}\sqrt{14}}{\sqrt{35}\sqrt{18}}$

In exercises 21–36, factor and simplify, as indicated. Assume $a > 0$, $b > 0$, $c > 0$.

21. (a) Factor: $3 + \sqrt{18}$
    (b) Simplify: $\dfrac{3 + \sqrt{18}}{9}$

22. (a) Factor: $2 - \sqrt{32}$
    (b) Simplify: $\dfrac{2 - \sqrt{32}}{8}$

23. (a) Factor: $5 + \sqrt{50}$
    (b) Simplify: $\dfrac{5 + \sqrt{50}}{-15}$

24. (a) Factor: $4 - \sqrt{48}$
    (b) Simplify: $\dfrac{\sqrt{48} - 4}{12}$

25. (a) Factor: $7 + \sqrt{98}$
    (b) Simplify: $\dfrac{7 + \sqrt{98}}{1 + \sqrt{2}}$

26. (a) Factor: $8 + \sqrt{20}$
    (b) Simplify: $\dfrac{8 + \sqrt{20}}{12 + 3\sqrt{5}}$

27. (a) Factor: $\sqrt{2} + \sqrt{6}$
    (b) Simplify: $\dfrac{\sqrt{2} + \sqrt{6}}{\sqrt{8}}$

28. (a) Factor: $\sqrt{12} + \sqrt{27}$
    (b) Simplify: $\dfrac{\sqrt{12} + \sqrt{27}}{\sqrt{75}}$

29. (a) Factor: $\sqrt{6} - \sqrt{18}$
    (b) Simplify: $\dfrac{\sqrt{6} - \sqrt{18}}{3 - \sqrt{3}}$

30. (a) Factor: $\sqrt{a} - \sqrt{a^3}$
    (b) Simplify: $\dfrac{\sqrt{a} - \sqrt{a^3}}{\sqrt{a^5}}$

31. (a) Factor: $\sqrt{a^3b} + \sqrt{abc}$
    (b) Simplify: $\dfrac{\sqrt{a^3b} + \sqrt{abc}}{\sqrt{a^3b^3}}$

32. (a) Factor: $\sqrt{2ab^2} - \sqrt{8b^2}$
    (b) Simplify: $\dfrac{\sqrt{2ab^2} - \sqrt{8b^2}}{2 - \sqrt{a}}$

33. (a) Factor: $\sqrt{4a^2} - \sqrt{4b^2}$
    (b) Simplify: $\dfrac{\sqrt{4a^2} - \sqrt{4b^2}}{10a - 10b}$

34. (a) Factor: $\sqrt{a^2b^2} + \sqrt{a^2c^2}$
    (b) Simplify: $\dfrac{\sqrt{a^2b^2} + \sqrt{a^2c^2}}{b^2 - c^2}$

35. (a) Factor: $\sqrt{9a^2b^2} - \sqrt{16a^2b^2}$
    (b) Simplify: $\dfrac{15ab - 20ab}{\sqrt{9a^2b^2} - \sqrt{16a^2b^2}}$

36. (a) Factor: $\sqrt{a^2c^4} + \sqrt{25a^2c^2}$
    (b) Simplify: $\dfrac{\sqrt{a^2c^4} + \sqrt{25a^2c^2}}{c^2 + 10c + 25}$

In exercises 37–44, divide, as indicated. Assume $s > 0$, $t > 0$.

37. $\dfrac{\sqrt{2}}{3} \div \dfrac{\sqrt{8}}{6}$

38. $\dfrac{\sqrt{15}}{4} \div \dfrac{\sqrt{45}}{2}$

39. $\dfrac{\sqrt{10}}{7} \div \dfrac{\sqrt{40}}{14}$

40. $\dfrac{\sqrt{50}}{3} \div \dfrac{\sqrt{18}}{5}$

41. $\dfrac{\sqrt{st}}{\sqrt{48}} \div \dfrac{\sqrt{3}}{\sqrt{25s^2t}}$

42. $\dfrac{\sqrt{20s}}{\sqrt{49t}} \div \dfrac{\sqrt{10}}{\sqrt{14t^3}}$

43. $\dfrac{\sqrt{12}}{\sqrt{15}} \div \dfrac{\sqrt{45}}{\sqrt{144}}$

44. $\dfrac{\sqrt{14}}{\sqrt{75}} \div \dfrac{\sqrt{21}}{\sqrt{32}}$

11.7 *n*th ROOTS

Cube Roots

Observe that

$$2^3 = 2 \cdot 2 \cdot 2 = 8,$$

whereas

$$(-2)^3 = (-2)(-2)(-2) = -8.$$

2 is called the "cube root of 8" and -2 the "cube root of -8."

Definition

> *CUBE ROOT. The real number b is called the* ***cube root of*** *a if*
>
> $$b^3 = a.$$

Write

$$b = \sqrt[3]{a}$$

when b is the cube root of a.

EXAMPLE 1

Find:
(a) $\sqrt[3]{1}$, (b) $\sqrt[3]{-1}$, (c) $\sqrt[3]{27}$, (d) $\sqrt[3]{125}$.

Solution.
(a) $\sqrt[3]{1} = 1$ because $1^3 = 1$
(b) $\sqrt[3]{-1} = -1$ because $(-1)^3 = -1$
(c) $\sqrt[3]{27} = 3$ because $3^3 = 27$
(d) $\sqrt[3]{125} = 5$ because $5^3 = 125$

Every real number—positive, negative, or zero—has a cube root.

The cube root of { a positive number is positive,
0 is 0,
a negative number is negative,

because

the cube of { a positive number is positive,
0 is 0,
a negative number is negative.

[Recall that, in contrast, *negative numbers do not have (real) square roots.*]

Except for cube roots of those integers that are cubes [such as 1, $8 (= 2^3)$, $27 (= 3^3)$, . . . or -1, -8 $[= (-2)^3]$, -27 $[= (-3)^3]$, . . .] the cube roots of integers are irrational. Thus

$$\sqrt[3]{2},\ \sqrt[3]{3},\ \sqrt[3]{4},\ \sqrt[3]{9},\ \sqrt[3]{25}$$

are all irrational.

In the following example, you do not have to consider the signs of the variables.

EXAMPLE 2

Find:

(a) $\sqrt[3]{x^3}$ (b) $\sqrt[3]{y^9}$

Solution.

(a) $$\sqrt[3]{x^3} = x$$

because $$x \cdot x \cdot x = x^3$$

(b) $\sqrt[3]{y^9} = y^3$

because $$y^3 \cdot y^3 \cdot y^3 = y^9$$

Let y be any real number and let k be a positive number. Then

$$y^{3k} = y^k \cdot y^k \cdot y^k$$

Therefore

$$\sqrt[3]{y^{3k}} = y^k$$

Higher Roots

Square roots can be regarded as 2nd roots, cube roots as 3rd roots. You can also consider 4th roots, 5th roots, etc. For example, 3 is the 4th root of 81 because (3 is positive, and)

$$3^4 = 3 \cdot 3 \cdot 3 \cdot 3 = 81.$$

Also, 2 is the 5th root of 32 because

$$2^5 = 2 \cdot 2 \cdot 2 \cdot 2 \cdot 2 = 32.$$

Definition

> *n*th *ROOT. Let n be an even positive integer (2, 4, 6, . . .) and let $a \geq 0$. Then b is called the* ***n*th root of** *a if $b^n = a$ and b is nonnegative.*
>
> *Let n be an odd positive integer (1, 3, 5, . . .) and let a be any real number. Then b is called the* ***n*th root of** *a if $b^n = a$.*

Negative numbers do not have (real) n*th roots for* $n = 2, 4, 6, . . .$ *, but every real number—positive, negative, or zero—has an* n*th root for* $n = 1, 3, 5, . . .$

Write

$$\sqrt[n]{a} \qquad \text{[Read this: the } n\text{th root of } a\text{]}$$

for the nth root of a, $n = 3, 4, 5, ...$

EXAMPLE 3

Find:

(a) $\sqrt[4]{16}$ (b) $\sqrt[5]{-32}$ (c) $\sqrt[6]{64}$ (d) $\sqrt[10]{1}$

Solution.

(a)
$$\sqrt[4]{16} = 2$$
because 2 is positive and
$$2^4 = 2 \cdot 2 \cdot 2 \cdot 2 = 16.$$

(b)
$$\sqrt[5]{-32} = -2$$
because
$$(-2)^5 = (-2)(-2)(-2)(-2)(-2) = -32$$

(c)
$$\sqrt[6]{64} = 2$$
because 2 is positive and
$$2^6 = 2 \cdot 2 \cdot 2 \cdot 2 \cdot 2 \cdot 2 = 64.$$

(d)
$$\sqrt[10]{1} = 1$$
because 1 is positive and every power of 1 equals 1.

EXAMPLE 4

Let $a > 0$. Find:

(a) $\sqrt[4]{a^4}$, (b) $\sqrt[9]{b^{18}}$

Solution.

(a)
$$\sqrt[4]{a^4} = a$$
because $a > 0$ and
$$a \cdot a \cdot a \cdot a = a^4$$

(b)
$$\sqrt[9]{b^{18}} = b^2$$
because
$$\underbrace{b^2 \cdot b^2 \cdot b^2 \ldots b^2}_{9 \text{ factors}} = b^{18}$$

In general, for positive integers n and k,
$$a^{nk} = \underbrace{a^k \cdot a^k \cdot a^k \ldots a^k}_{n \text{ factors}} = (a^k)^n$$

Therefore
$$\sqrt[n]{a^{nk}} = \sqrt[n]{(a^k)^n} = a^k,$$
except when a is negative, n is even, and k is odd. (See exercises 45–46.)

Laws for *n*th Roots

The laws for products and quotients of square roots carry over to *n*th roots. Thus (whenever these roots are defined):

The nth *root of a product equals the product of the* nth *roots.*

$$\sqrt[n]{ab} = \sqrt[n]{a}\sqrt[n]{b}$$

The nth *root of a quotient equals the quotient of the* nth *roots.*

$$\sqrt[n]{\frac{a}{b}} = \frac{\sqrt[n]{a}}{\sqrt[n]{b}}, \qquad b \neq 0$$

EXAMPLE 5

Let $z > 0$. Find:
(a) $\sqrt[3]{27x^3}$ (b) $\sqrt[4]{16y^8z^{12}}$

Solution.

(a)
$$\begin{aligned}\sqrt[3]{27x^3} &= \sqrt[3]{27}\sqrt[3]{x^3}\\ &= 3x\end{aligned}$$

(b)
$$\begin{aligned}\sqrt[4]{16y^8z^{12}} &= \sqrt[4]{16}\sqrt[4]{y^8}\sqrt[4]{z^{12}}\\ &= 2y^2z^3\end{aligned}$$

EXAMPLE 6

Suppose $y \neq 0$. Find:
(a) $\sqrt[3]{\frac{27}{8}}$ (b) $\sqrt[5]{\frac{-32x^5}{y^{10}}}$

Solution.

(a)
$$\begin{aligned}\sqrt[3]{\frac{27}{8}} &= \frac{\sqrt[3]{27}}{\sqrt[3]{8}}\\ &= \frac{3}{2}\end{aligned}$$

(b)
$$\begin{aligned}\sqrt[5]{\frac{-32x^5}{y^{10}}} &= \frac{\sqrt[5]{-32}\sqrt[5]{x^5}}{\sqrt[5]{y^{10}}}\\ &= \frac{-2x}{y^2}\end{aligned}$$

EXAMPLE 7

Assume $x > 0$ and $y \neq 0$. Simplify:

$$\frac{\sqrt[4]{32x^5}}{\sqrt[3]{8y^3}}$$

Solution.

$$\begin{aligned}\frac{\sqrt[4]{32x^5}}{\sqrt[3]{8y^3}} &= \frac{\sqrt[4]{16}\sqrt[4]{2}\sqrt[4]{x^4}\sqrt[4]{x}}{\sqrt[3]{8}\sqrt[3]{y^3}}\\ &= \frac{\cancel{2}\sqrt[4]{2}\,x\,\sqrt[4]{x}}{\cancel{2}y}\\ &= \frac{\sqrt[4]{2}\,x\,\sqrt[4]{x}}{y}\end{aligned}$$

EXERCISES

In exercises 1–36, find each root. Assume $a > 0$, $b > 0$, $c > 0$.

1. $\sqrt[3]{64}$
2. $\sqrt[3]{1000}$

3. $\sqrt[4]{1}$

4. $\sqrt[4]{10000}$

5. $\sqrt[5]{-1}$

6. $\sqrt[5]{-32}$

7. $\sqrt[8]{7^8}$

8. $\sqrt[12]{5^{12}}$

9. $\sqrt[7]{(-9)^7}$

10. $\sqrt[9]{(-7)^9}$

11. $\sqrt[3]{a^3}$

12. $\sqrt[4]{b^4}$

13. $\sqrt[3]{a^6}$

14. $\sqrt[4]{b^8}$

15. $\sqrt[3]{c^{12}}$

16. $\sqrt[4]{c^{12}}$

17. $\sqrt[6]{a^{18}}$

18. $\sqrt[8]{b^{24}}$

19. $\sqrt[10]{c^{100}}$

20. $\sqrt[8]{a^{40}}$

21. $\sqrt[3]{8a^3}$

22. $\sqrt[3]{-8a^6}$

23. $\sqrt[4]{16b^4}$

24. $\sqrt[5]{-c^5}$

25. $\sqrt[3]{1000a^3b^6}$

26. $\sqrt[4]{16a^8b^{20}}$

27. $\sqrt[7]{8^7a^7b^{14}}$

28. $\sqrt[6]{1\,000\,000a^{12}c^{18}}$

29. $\sqrt[3]{\frac{-8}{27}}$

30. $\sqrt[3]{\frac{-1}{1000}}$

31. $\sqrt[3]{\frac{125}{64}}$

32. $\sqrt[4]{\frac{16}{81}}$

33. $\sqrt[4]{\frac{81}{10000}}$

34. $\sqrt[5]{\frac{32}{a^5}}$

35. $\sqrt[6]{\frac{b^{12}}{64}}$

36. $\sqrt[4]{\frac{a^{12}b^4}{c^8}}$

In exercises 37–44, simplify.

37. $\frac{\sqrt[4]{32}}{\sqrt[4]{2}}$

38. $\frac{\sqrt[3]{16}}{\sqrt[3]{54}}$

39. $\frac{\sqrt[5]{-32}}{\sqrt[3]{-8}}$

40. $\frac{\sqrt[3]{2}}{2} \cdot \frac{\sqrt[3]{3}}{3}$

41. $\sqrt[4]{8} \cdot \sqrt[4]{2} \cdot \sqrt[4]{81}$

42. $\frac{\sqrt[3]{24}}{5} \div \frac{\sqrt[3]{4}}{10}$

43. $\frac{1}{\sqrt[4]{2}} \div \frac{-1}{\sqrt[4]{24}}$

44. $\frac{\sqrt[5]{64}}{\sqrt[3]{27}} \cdot \frac{\sqrt[5]{16}}{2}$

45. Does $\sqrt[4]{(-1)^{4\cdot 3}} = (-1)^3$?

46. Does $\sqrt[3]{(-1)^{3\cdot 4}} = (-1)^4$?

47. Let $f(x) = \sqrt[3]{x}$. Find:
(a) $f(1)$, (b) $f(-1)$, (c) $f(8)$, (d) $f\left(\frac{1}{8}\right)$

48. Let $g(x) = \sqrt[4]{x}$, for x nonnegative. Find:
(a) $g(0)$, (b) $g(1)$, (c) $g(16)$, (d) $g(81)$

What Have You Learned in Chapter 11?

You have learned that $b = \sqrt{a}$ if b is positive or 0 and if $b^2 = a$.

You have seen that the square roots of integers such as 2, 3, and 5 are irrational. They can be *approximated* by rational numbers.

The square root of a product equals the product of the square roots, and the square root of a quotient equals the quotient of the square roots.

You have learned how to simplify expressions involving radicals.

And you have learned that every real number has an nth root for $n = 1, 3, 5, \ldots$, but that only nonnegative numbers have nth roots for $n = 2, 4, 6, \ldots$.

Let's Review Chapter 11.

11.1 Square Roots

In exercises 1–4 find each square root or its inverse.

1. $\sqrt{25}$
2. $\sqrt{64}$
3. $-\sqrt{900}$
4. $\sqrt{x^4}$
5. Assume $x > 0$ and $y < 0$. Find $\sqrt{(xy)^2}$.
6. Find the length of the hypotenuse of the right triangle in Figure 11.18.

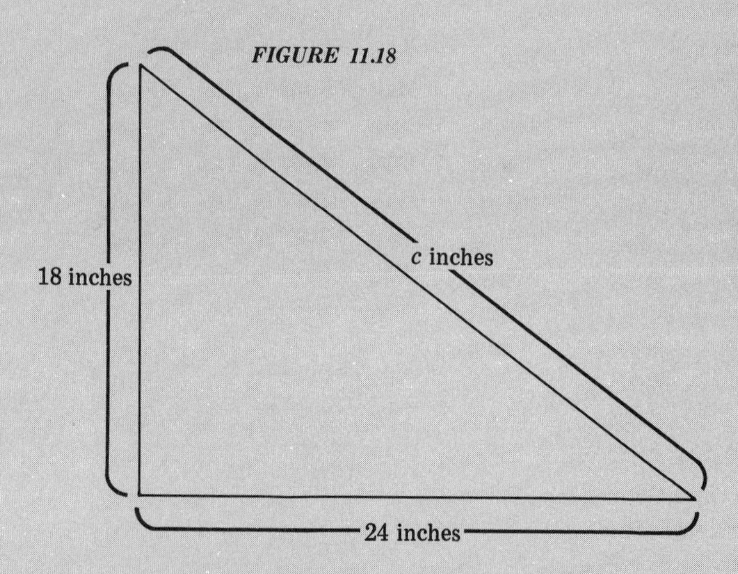

FIGURE 11.18

11.2 *Irrational Square Roots*

7. Find consecutive integers N and $N + 1$ such that

$$N < \sqrt{55} < N + 1.$$

8. Use $\sqrt{2} \approx 1.41$ to approximate $7\sqrt{2}$ to 1 decimal place.
9. Approximate the length of a side of a square whose area is 62 square inches. (Use the next *largest* integer.)
10. Let $f(x) = \sqrt{x}$, for x nonnegative. Find $f(121)$.

11.3 *Roots of Products and Quotients*

In exercises 11–13, find each square root. Assume $a > 0$ and $b > 0$.

11. $\sqrt{25a^2}$
12. $\sqrt{a^4b^6}$
13. $\sqrt{\frac{81}{100}}$
14. Simplify: $\sqrt{\frac{20a^4}{b^8}}$

11.4 *Addition and Subtraction of Roots*

In exercises 15–18, simplify each expression. Assume $a > 0$, $b > 0$, $c > 0$.

15. $5\sqrt{5} - (3\sqrt{5} - \sqrt{5})$
16. $\sqrt{9a} - \sqrt{4a}$
17. $\sqrt{64a^2b^4} + \sqrt{81a^2b^4}$
18. $\frac{4\sqrt{3}}{15} + \frac{2\sqrt{3}}{5}$

11.5 *Multiplication of Roots*

In exercises 19–22, multiply as indicated. Assume $y > 0$.

19. $\sqrt{10}\sqrt{5}\sqrt{2}$
20. $\sqrt{xy}\sqrt{x^3y}$
21. $\sqrt{2}(\sqrt{8} - \sqrt{2})$
22. $\frac{\sqrt{8}}{\sqrt{3}} \cdot \frac{\sqrt{3}}{\sqrt{2}}$

11.6 *Division of Roots*

23. Simplify: $\frac{\sqrt{32}}{\sqrt{8}}$
24. Assume $x > 0$. Simplify: $\frac{\sqrt{12x}}{\sqrt{75x}}$
25. (a) Factor: $7 + \sqrt{98}$

    (b) Simplify: $\frac{7 + \sqrt{98}}{21}$
26. Divide: $\frac{\sqrt{15}}{\sqrt{18}} \div \frac{\sqrt{5}}{\sqrt{32}}$

11.7 nth Roots

In exercises 27–29, find each root. Assume $a > 0$.

27. $\sqrt[3]{125}$

28. $\sqrt[4]{16a^{12}}$

29. $\sqrt[5]{\dfrac{-32a^5}{b^{15}}}$

30. Simplify: $\dfrac{\sqrt[4]{81}}{\sqrt[3]{-27}}$

And these from Chapters 1–10:

31. Multiply: $\dfrac{81}{a^4} \cdot \dfrac{a^3}{27b^2c^3} \cdot \dfrac{bc^2}{2}$

32. Find: $(.02)^2 + (.4)^3$

33. Let $f(x) = x^2$ and $g(x) = \sqrt{x}$. Find: (a) $f(4)$, (b) $g(4)$, (c) $f(4) + g(4)$, (d) $f(4) \cdot g(4)$

34. Fill in "<", "=" or ">".

(a) $2^2 \ \square \ \sqrt{2}$

(b) $1^2 \ \square \ \sqrt{1}$

(c) $\left(\dfrac{1}{4}\right)^2 \ \square \ \sqrt{\dfrac{1}{4}}$

(d) $(.09)^2 \ \square \ \sqrt{.09}$

Try These Exam Questions for Practice.

1. Find each square root:

(a) $\sqrt{144}$ (b) $\sqrt{100x^4}$ (c) $\sqrt{\dfrac{16}{49}}$

2. Simplify: $\sqrt{\dfrac{32a^4b^8}{c^8}}$

3. Simplify: $\dfrac{5\sqrt{2} - 3\sqrt{2}}{8}$

4. Assume $x > 0$, $y > 0$, $z > 0$.
Multiply: $\sqrt{9x^3y}\,\sqrt{3xy^3z}\,\sqrt{12x^2z}$

5. Divide: $\dfrac{\sqrt{20}}{\sqrt{63}} \div \dfrac{\sqrt{7}}{\sqrt{45}}$

6. (a) Factor: $\sqrt{18} - 9$

(b) Simplify: $\dfrac{\sqrt{18} - 9}{6}$

7. Find: $\dfrac{\sqrt[3]{64}}{\sqrt[4]{16}}$

12

Quadratic Equations

12.1 SOLUTIONS BY FACTORING

Form of a Quadratic Equation

A **quadratic,** or **second-degree, equation** (in a single variable, x) is an equation *that can be written in the form*

$$Ax^2 + Bx + C = 0, \qquad (A \neq 0).$$

The quadratic equation

$$x^2 - 3x + 2 = 0$$

is already of this form with $A = 1, B = -3, C = 2$. The quadratic equation

$$3x^2 + 5 = 2x - 3$$

can be transformed into the above form by adding $-2x + 3$ to both sides:

$$3x^2 - 2x + 8 = 0$$

Here $A = 3, B = -2, C = 8$.

Quadratic equations have either 0, 1, or 2 *real* roots.

When is a Product 0?

To solve a quadratic equation such as

$$x^2 - x = 0,$$

factor the left side to obtain

$$x(x-1)=0.$$

A product is 0 if at least one of the factors is 0. Moreover, if neither factor is 0, then a product cannot be 0. For example,

$$\begin{aligned} 0\cdot 6 &= 0\\ -7\cdot 0 &= 0\\ 0\cdot 0 &= 0\\ -7\cdot 6 &\neq 0 \end{aligned}$$

Thus in order that

$$x(x-1)=0.$$

at least one of the factors x or $x-1$ must be 0. Therefore you can solve the two simpler equations

$$x=0,\ x-1=0,$$

separately:

$$x=0 \quad\Big|\quad \begin{aligned} x-1&=0\\ x&=1 \end{aligned}$$

Thus the roots of the quadratic equation are 0 and 1. To check these roots, replace x first by 0 and then by 1 in the given quadratic equation,

$$x^2-x=0:$$

$$\begin{aligned} 0^2-0 &\stackrel{?}{=} 0\\ 0 &\stackrel{\checkmark}{=} 0 \end{aligned} \quad\Big|\quad \begin{aligned} 1^2-1 &\stackrel{?}{=} 0\\ 0 &\stackrel{\checkmark}{=} 0 \end{aligned}$$

EXAMPLE 1

Consider the equation

$$(x-4)(x+2)=0$$

(a) Show that this is a quadratic equation.
(b) Solve this quadratic equation.
(c) Check the roots by using the given equation.
(d) Check the roots by using the form of the equation obtained in Part (a).

Solution.

(a)

$$(x-4)(x+2)=x^2-2x-8$$

Thus the given equation can be written as

$$x^2-2x-8=0.$$

This is the proper form for a quadratic equation: $A=1, B=-2, C=-8$

(b) The original equation,

$$(x-4)(x+2)=0,$$

can be solved by considering

$$x - 4 = 0 \text{ and } x + 2 = 0$$

separately:

$$\begin{array}{r|r} x - 4 = 0 & x + 2 = 0 \\ x = 4 & x = -2 \end{array}$$

The roots of the quadratic equation are 4 and −2.

(c) Check each of these roots by replacing x first by 4 and then by −2 in the given equation,

$$(x - 4)(x + 2) = 0.$$

$$\begin{array}{r|r} \text{CHECK for } 4: & \text{CHECK for } -2: \\ (4 - 4)(4 + 2) \stackrel{?}{=} 0 & (-2 - 4)(-2 + 2) \stackrel{?}{=} 0 \\ 0 \cdot 6 \stackrel{?}{=} 0 & (-6)0 \stackrel{?}{=} 0 \\ 0 \stackrel{\checkmark}{=} 0 & 0 \stackrel{\checkmark}{=} 0 \end{array}$$

(d) Check each of these roots by replacing x first by 4 and then by −2 in the equation

$$x^2 - 2x - 8 = 0.$$

$$\begin{array}{rr} \text{CHECK for } 4: & \text{CHECK for } -2: \\ 4^2 - 2 \cdot 4 - 8 \stackrel{?}{=} 0 & (-2)^2 - 2(-2) - 8 \stackrel{?}{=} 0 \\ 16 - 8 - 8 \stackrel{?}{=} 0 & 4 + 4 - 8 \stackrel{?}{=} 0 \\ 0 \stackrel{\checkmark}{=} 0 & 0 \stackrel{\checkmark}{=} 0 \end{array}$$

Factoring the Left Side

If you can factor the left side of a quadratic equation of the form

$$Ax^2 + Bx + C = 0,$$

you can apply the above method.

EXAMPLE 2

Solve: $x^2 - 3x + 2 = 0$

Solution.

$$x^2 - 3x + 2 = (x - 1)(x - 2)$$

Thus

$$x^2 - 3x + 2 = 0$$

can be written as

$$(x - 1)(x - 2) = 0.$$

$$\begin{array}{r|r} x - 1 = 0 & x - 2 = 0 \\ x = 1 & x = 2 \end{array}$$

The roots are 1 and 2.

EXAMPLE 3

Solve: $t^2 + 4t + 3 = 0$

Solution.

$$t^2 + 4t + 3 = (t + 1)(t + 3)$$

Thus

$$t^2 + 4t + 3 = 0$$

can be written as

$$(t + 1)(t + 3) = 0.$$

$$\begin{array}{r|r} t + 1 = 0 & t + 3 = 0 \\ t = -1 & t = -3 \end{array}$$

The roots are -1 and -3.

EXAMPLE 4

Solve: $2x^2 + 5x + 2 = 0$

Solution.

$$2x^2 + 5x + 2 = (2x + 1)(x + 2)$$

Thus

$$2x^2 + 5x + 2 = 0$$

can be written as

$$(2x + 1)(x + 2) = 0.$$

$$\begin{array}{r|r} 2x + 1 = 0 & x + 2 = 0 \\ 2x = -1 & x = -2 \\ x = \frac{-1}{2} & \end{array}$$

The roots are $\frac{-1}{2}$ and -2.

EXAMPLE 5

Solve: $x^2 - 6x + 9 = 0$

Solution.

$$\begin{aligned} x^2 - 6x + 9 &= (x - 3)(x - 3) \\ &= (x - 3)^2 \end{aligned}$$

Thus

$$x^2 - 6x + 9 = 0$$

can be written as

$$(x - 3)^2 = 0.$$

The *only* root is 3.

EXERCISES

In exercises 1–40, solve each equation. Check the roots, when indicated, by using the given equation.

1. $x(x-4)=0$
2. $x(x+6)=0$
3. $(x-2)(x-3)=0$
4. $(x-1)(x+4)=0$
5. $(y+3)(y-8)=0$ (Check.)
6. $(z+2)(z+4)=0$
7. $\left(x+\frac{1}{2}\right)\left(x-\frac{1}{4}\right)=0$
8. $(x+2)^2=0$
9. $x^2-5x+4=0$
10. $x^2-5x+6=0$
11. $t^2+6t+8=0$ (Check.)
12. $y^2+y-12=0$
13. $x^2-2x-8=0$
14. $y^2+2y-8=0$
15. $x^2+8x+15=0$
16. $x^2-2x-15=0$ (Check.)
17. $z^2+9z+18=0$
18. $t^2-t-30=0$
19. $x^2-10x+25=0$
20. $x^2+8x+16=0$
21. $x^2+4x-21=0$
22. $x^2+7x+10=0$ (Check.)
23. $x^2+8x-9=0$
24. $x^2+4x-12=0$
25. $2x^2+2x-4=0$ (Check.)
26. $3x^2+6x-9=0$
27. $2x^2+3x+1=0$
28. $2x^2+7x+3=0$
29. $2t^2+7t-4=0$
30. $3y^2+7y+2=0$ (Check.)
31. $4x^2+8x+3=0$
32. $6y^2+y-1=0$
33. $4x^2+20x+25=0$
34. $3x^2-10x+3=0$
35. $2y^2+6=7y$
36. $2z^2-13z=7$
37. $x^2-2x=0$
38. $x^2+5x=0$
39. $2x^2-x=0$
40. $3x^2+2x=0$

SAMPLE. (for exercises 41–44): Let $$f(x)=x^2+5x+6.$$ When does $f(x)=0$?	***Solution.*** $f(x)=x^2+5x+6$ Thus to find when $f(x)=0$, set $$x^2+5x+6=0.$$ Solve this equation by factoring the left side. $$(x+2)(x+3)=0$$ $x+2=0$, $x=-2$ | $x+3=0$, $x=-3$ Thus $f(x)=0$ when $x=-2$ and when $x=-3$.

41. Let $f(x) = (x + 5)(x - 3)$. When does $f(x) = 0$?

42. Let $f(x) = x^2 + 7x + 6$. When does $f(x) = 0$?

43. Let $f(x) = x^2 - 10x + 25$. When does $f(x) = 0$?

44. Let $f(x) = 2x^2 + 3x + 1$. When does $f(x) = 0$?

12.2 EQUATIONS OF THE FORM $x^2 = a$

Let n be a positive number, and let

$$\pm n$$

stand for

both n and −n.

For example,

$$\pm 2$$

stands for

both 2 and −2.

If a is *positive,* there are *two* numbers whose square is a. These numbers are

$$\pm\sqrt{a}$$

because

$$\sqrt{a}\sqrt{a} = a \text{ and } (-\sqrt{a})(-\sqrt{a}) = a.$$

Thus if a quadratic equation can be written in the form

$$x^2 = a,$$

its roots are $\pm\sqrt{a}$.

EXAMPLE 1

Solve: $x^2 - 25 = 0$

Solution 1.

$$x^2 - 25 = 0$$
$$x^2 = 25$$
$$x = \pm 5$$

Solution 2.

$$x^2 - 25 = 0$$ Factor the left side.

$$(x + 5)(x - 5) = 0$$

$x + 5 = 0$	$x - 5 = 0$
$x = -5$	$x = 5$

The roots are ± 5.

EXAMPLE 2

Solve: $t^2 = 7$

Solution.

$$t^2 = 7$$
$$t = \pm\sqrt{7}$$

EXAMPLE 3

Solve: $x^2 - 24 = 0$

Solution.

$$x^2 - 24 = 0$$
$$x^2 = 24$$
$$x = \pm\sqrt{24}$$
$$x = \pm\sqrt{4 \cdot 6}$$
$$x = \pm\sqrt{4}\sqrt{6}$$
$$x = \pm 2\sqrt{6}$$

EXAMPLE 4

Solve for u: $(2u + 3)^2 = 25$
Check the roots.

Solution. This second-degree equation is of the form

$$x^2 = 25.$$

Here $$x = 2u + 3$$

Thus $$2u + 3 = \pm 5$$

You obtain two "first-degree equations":

$2u + 3 = 5$	$2u + 3 = -5$
$2u = 2$	$2u = -8$
$u = 1$	$u = -4$

The roots are 1 and -4.

CHECK for 1:	CHECK for -4:
$(2 \cdot 1 + 3)^2 \stackrel{?}{=} 25$	$[2(-4) + 3]^2 \stackrel{?}{=} 25$
$5^2 \stackrel{?}{=} 25$	$(-5)^2 \stackrel{?}{=} 25$
$25 \stackrel{\checkmark}{=} 25$	$25 \stackrel{\checkmark}{=} 25$

EXAMPLE 5

Solve for x:

$$(3x - 1)^2 = 5$$

Check the roots.

Solution.

$$(3x - 1)^2 = 5$$
$$3x - 1 = \pm\sqrt{5}$$

$$\begin{array}{r|r} 3x - 1 = \sqrt{5} & 3x - 1 = -\sqrt{5} \\ 3x = 1 + \sqrt{5} & 3x = 1 - \sqrt{5} \\ x = \frac{1 + \sqrt{5}}{3} & x = \frac{1 - \sqrt{5}}{3} \end{array}$$

The roots differ only in the sign of $\sqrt{5}$. Thus the roots are

$$\frac{1 \pm \sqrt{5}}{3}.$$

$$\begin{array}{r|r} \text{CHECK for } \frac{1+\sqrt{5}}{3}: & \text{CHECK for } \frac{1-\sqrt{5}}{3}: \\ \left[3\left(\frac{1+\sqrt{5}}{3}\right) - 1\right]^2 \stackrel{?}{=} 5 & \left[3\left(\frac{1-\sqrt{5}}{3}\right) - 1\right]^2 \stackrel{?}{=} 5 \\ [(1 + \sqrt{5}) - 1]^2 \stackrel{?}{=} 5 & [(1 - \sqrt{5}) - 1]^2 \stackrel{?}{=} 5 \\ \sqrt{5}^2 \stackrel{?}{=} 5 & (-\sqrt{5})^2 \stackrel{?}{=} 5 \\ 5 \stackrel{\checkmark}{=} 5 & 5 \stackrel{\checkmark}{=} 5 \end{array}$$

EXAMPLE 6

Solve for z:

$$(5z - 2)^2 = 0$$

Solution. The equation

$$x^2 = 0$$

has one root—namely, 0. Thus let $x = 5z - 2$. From

$$(5z - 2)^2 = 0,$$

you obtain

$$5z - 2 = 0,$$
$$5z = 2,$$

and finally,

$$z = \frac{2}{5}.$$

When a is *negative*, the equation

$$x^2 = a$$

does not have any real number as a root. For, the square of a real number is always *at least* 0. In more advanced courses, you will learn about other numbers, known as "complex numbers." If a is negative, the roots of the above equation are complex numbers. For the present, you need only consider real numbers.

To sum up, the equation

$$x^2 = a$$

has

2 (real) roots if $a > 0$,
1 (real) root if $a = 0$,
0 (real) roots if $a < 0$.

EXERCISES

Solve for the indicated variable. Check the roots, when indicated, by using the given equation.

1. $x^2 = 16$
2. $x^2 - 36 = 0$
3. $y^2 - 49 = 0$
4. $t^2 - 100 = 0$
5. $x^2 = 6$
6. $x^2 - 11 = 0$
7. $y^2 - 13 = 0$ (Check.)
8. $z^2 - 15 = 0$
9. $x^2 - 8 = 0$
10. $t^2 - 12 = 0$
11. $u^2 - 18 = 0$
12. $y^2 - 20 = 0$
13. $u^2 - 45 = 0$
14. $x^2 - 40 = 0$
15. $x^2 - \frac{4}{9} = 0$
16. $u^2 - \frac{25}{16} = 0$ (Check.)
17. $x^2 - \frac{2}{3} = 0$
18. $t^2 - \frac{12}{25} = 0$
19. $(2u + 1)^2 = 49$ (Check.)
20. $(2x - 3)^2 = 1$ (Check.)
21. $(5x - 2)^2 = 16$
22. $\left(\frac{y}{2} + 1\right)^2 = 9$
23. $(3z + 4)^2 = 64$
24. $(5x - 2)^2 = 81$
25. $(3x - 2)^2 = 3$ (Check.)
26. $(2x + 1)^2 = 5$
27. $(5t + 4)^2 = 7$
28. $(u - 4)^2 = 11$
29. $(3y + 5)^2 = 6$
30. $(4z - 1)^2 = 13$
31. $(2x - 1)^2 = 8$
32. $(2t - 5)^2 = 12$
33. $(z + 3)^2 = 32$
34. $\left(\frac{t}{2} + 1\right)^2 = 20$ (Check.)
35. $\left(\frac{x - 1}{2}\right)^2 = 98$
36. $\left(\frac{2x + 7}{3}\right)^2 = 80$
37. $(x - 2)^2 = 0$
38. $(2t + 1)^2 = 0$
39. $(5z - 3)^2 = 0$
40. $\left(\frac{4u + 5}{3}\right)^2 = 0$

12.3 THE QUADRATIC FORMULA

Applying the Formula

There is a mechanical procedure for finding the roots of *any quadratic equation*

$$Ax^2 + Bx + C = 0, \qquad (A \neq 0.)$$

This method is particularly useful when the previous methods do not easily apply. First write the equation in the above form. The roots are then given by the **quadratic formula:**

$$x = \frac{-B \pm \sqrt{B^2 - 4AC}}{2A}$$

EXAMPLE 1

Use the quadratic formula to find the roots of the equation

$$x^2 + 5x + 3 = 0.$$

Check the roots.

Solution.

$$x^2 + 5x + 3 = 0$$

is of the form

$$Ax^2 + Bx + C = 0,$$

where $\qquad A = 1, B = 5, C = 3.$

Apply the quadratic formula,

$$x = \frac{-B \pm \sqrt{B^2 - 4AC}}{2A}.$$

$$x = \frac{-5 \pm \sqrt{5^2 - 4 \cdot 1 \cdot 3}}{2 \cdot 1}$$

$$= \frac{-5 \pm \sqrt{25 - 12}}{2}$$

$$= \frac{-5 \pm \sqrt{13}}{2}$$

CHECK for $\frac{-5 + \sqrt{13}}{2}$:

$$\left(\frac{-5 + \sqrt{13}}{2}\right)^2 + 5\left(\frac{-5 + \sqrt{13}}{2}\right) + 3 \stackrel{?}{=} 0$$

$$\frac{25 - 10\sqrt{13} + 13}{4} + \frac{-25 + 5\sqrt{13}}{2} + 3 \stackrel{?}{=} 0$$

$$\frac{25 - 10\sqrt{13} + 13 - 50 + 10\sqrt{13} + 12}{4} \stackrel{?}{=} 0$$

$$\frac{\overbrace{(25 + 13 + 12 - 50)}^{0} + \overbrace{(-10\sqrt{13} + 10\sqrt{13})}^{0}}{4} \stackrel{?}{=} 0$$

$$0 \stackrel{\checkmark}{=} 0.$$

CHECK for $\frac{-5-\sqrt{13}}{2}$:

$$\left(\frac{-5-\sqrt{13}}{2}\right)^2 + 5\left(\frac{-5-\sqrt{13}}{2}\right) + 3 \stackrel{?}{=} 0$$
$$\frac{25+10\sqrt{13}+13}{4} + \frac{-25-5\sqrt{13}}{2} + 3 \stackrel{?}{=} 0$$
$$\frac{25+10\sqrt{13}+13-50-10\sqrt{13}+12}{4} \stackrel{?}{=} 0$$
$$\frac{\overbrace{(25+13+12-50)}^{0}+\overbrace{(10\sqrt{13}-10\sqrt{13})}^{0}}{4} \stackrel{?}{=} 0$$
$$0 \stackrel{\checkmark}{=} 0.$$

EXAMPLE 2

Solve the quadratic equation

$$x^2 - 7x + 10 = 0$$

(a) by factoring the left side, and
(b) by applying the quadratic formula.

Solution.

(a)
$$x^2 - 7x + 10 = 0$$
$$(x-2)(x-5) = 0$$

$$x - 2 = 0 \quad\Big|\quad x - 5 = 0$$
$$x = 2 \quad\Big|\quad x = 5$$

The roots are 2 and 5.

(b)
$$x^2 - 7x + 10 = 0$$

Here $A = 1, B = -7, C = 10$.
Apply the quadratic formula:

$$x = \frac{-B \pm \sqrt{B^2 - 4AC}}{2A}$$
$$x = \frac{-(-7) \pm \sqrt{(-7)^2 - 4 \cdot 1 \cdot 10}}{2 \cdot 1}$$
$$= \frac{7 \pm \sqrt{49 - 40}}{2}$$
$$= \frac{7 \pm \sqrt{9}}{2}$$
$$= \frac{7 \pm 3}{2}$$

$$x = \frac{7+3}{2} \quad\Big|\quad x = \frac{7-3}{2}$$
$$x = 5 \quad\Big|\quad x = 2$$

As you see in Example 2, if you can factor the left side of an equation

$$Ax^2 + Bx + C = 0,$$

you obtain the roots more easily than by applying the quadratic formula.

The Discriminant

In the quadratic formula,

$$x = \frac{-B \pm \sqrt{B^2 - 4AC}}{2A},$$

the expression within the radical sign

$$B^2 - 4AC$$

is known as the **discriminant.** For the equation

$$2x^2 - 6x + 1 = 0$$

the discriminant is given by

$$\begin{aligned} B^2 - 4AC &= (-6)^2 - 4 \cdot 2 \cdot 1 \\ &= 36 - 8 \\ &= 28 \end{aligned}$$

When you evaluate the discriminant, you can tell the number of (real) roots of the given equation.

$B^2 - 4AC$ (discriminant)	Number of Roots
positive	2 (real) roots
0	1 (real) root
negative	no (real) roots

For, consider the quadratic equation,

$$x = \frac{-B \pm \sqrt{B^2 - 4AC}}{2A}$$

If the discriminant, $B^2 - 4AC$, is positive, you add and subtract its square root in the numerator and obtain 2 different roots.

If the discriminant is 0, then because $\sqrt{0} = 0$, the only root is $\frac{-B}{2A}$.

If the discriminant is negative, the equation has no *real* square root. (The roots obtained by applying the formula are not real numbers.)

EXAMPLE 3

Consider the equation

$$x^2 + 7x + 1 = 0.$$

(a) Evaluate the discriminant.
(b) How many (real) roots are there?
(c) Apply the quadratic formula to solve the equation.

Solution.

(a) $A = 1, B = 7, C = 1$. The discriminant is given by

$$\begin{aligned} B^2 - 4AC &= 49 - 4 \\ &= 45 \end{aligned}$$

(b) There are 2 roots because the discriminant is positive.

(c)
$$\begin{aligned} x &= \frac{-B \pm \sqrt{B^2 - 4AC}}{2A} \\ &= \frac{-7 \pm \sqrt{45}}{2} \\ &= \frac{-7 \pm 3\sqrt{5}}{2} \end{aligned}$$

EXAMPLE 4

Consider the equation

$$9y^2 + 12y + 4 = 0.$$

(a) Evaluate the discriminant.
(b) How many (real) roots are there?
(c) Apply the quadratic formula to solve the equation.

Solution.

(a) $A = 9, B = 12, C = 4$. The discriminant is given by

$$\begin{aligned} B^2 - 4AC &= 12^2 - 4 \cdot 9 \cdot 4 \\ &= 144 - 144 \\ &= 0 \end{aligned}$$

(b) There is 1 root because the discriminant is 0.

(c)
$$\begin{aligned} x &= \frac{-B \pm \sqrt{B^2 - 4AC}}{2A} \\ &= \frac{-12 \pm \sqrt{0}}{2 \cdot 9} \\ &= \frac{-2}{3} \end{aligned}$$

EXAMPLE 5

Consider the equation

$$2x^2 - 3x + 2 = 0$$

(a) Evaluate the discriminant.
(b) How many (real) roots are there?

Solution.

(a) $A = 2, B = -3, C = 2$

$$\begin{aligned} B^2 - 4AC &= (-3)^2 - 4 \cdot 2 \cdot 2 \\ &= 9 - 16 \\ &= -7 \end{aligned}$$

(b) There are no (real) roots because the discriminant is negative.

In later courses you will learn to apply the quadratic formula in the case where the discriminant is negative. You will also verify that the formula is correct in all cases.

EXERCISES

In exercises 1–16: (a) Evaluate the discriminant. (b) How many (real) roots are there? (c) If there are (real) roots, use the quadratic formula to solve the equation.

1. $x^2 + 4x + 2 = 0$
2. $x^2 + 6x + 9 = 0$
3. $2x^2 + 3 = 5x$
4. $5x^2 + 10x + 8 = 0$
5. $t^2 - 3t = 1$
6. $t^2 - 8t + 15 = 0$
7. $2y^2 - 6y + 3 = 0$
8. $z^2 + 20 = 9z$
9. $4x^2 - 4x + 1 = 0$
10. $2u^2 + 7u + 9 = 0$
11. $4z^2 - 20z + 5 = 0$
12. $u^2 + u + 1 = 0$
13. $y^2 + y - 1 = 0$
14. $x^2 - 2x + 1 = 0$
15. $3z^2 + z + 3 = 0$
16. $y^2 - 5y + 1 = 0$

In exercises 17–28, apply the quadratic formula to solve the equation. In exercises 17–20, check the roots.

17. $x^2 + 5x - 2 = 0$
18. $y^2 + 9y + 1 = 0$
19. $3t^2 - 5t + 1 = 0$
20. $u^2 + 12u + 36 = 0$
21. $x^2 + 10x + 5 = 0$
22. $2z^2 - 9z + 6 = 0$
23. $u^2 + 3u - 1 = 0$
24. $u^2 + 3u + 1 = 0$
25. $2t^2 + 5t + 3 = 0$
26. $3x^2 - 7x + 4 = 0$
27. $3y^2 = 10y - 3$
28. $(y - 2)^2 = y + 3$

In exercises 29–36: (a) Apply the quadratic formula to solve the equation. (b) Solve the equation by some other method.

20\. $x^2 - 3x + 2 = 0$

30\. $x^2 + 8x + 16 = 0$

31\. $y^2 + y - 20 = 0$

32\. $z^2 - 81 = 0$

33\. $(t + 1)^2 = 5$

34\. $2x^2 + 3x + 1 = 0$

35\. $6y^2 - y = 1$

36\. $10u^2 + 11u + 2 = 0$

37\. Let $f(x) = x^2 + 6x + 6$. When does $f(x) = 0$?

38\. Let $f(x) = x^2 + 3x + 3$. Show that $f(x)$ is never 0.

12.4 WORD PROBLEMS

Stated problems can often be translated into quadratic equations. Your first task is to translate the problem into algebraic symbols. You then apply the techniques of this chapter to solve the corresponding quadratic equation.

Number Problems

EXAMPLE 1

The sum of two numbers is 16 and the product is 48. Find these numbers.

Solution. Let x be one of these numbers. Then the other is

$$16 - x$$

because

$$\underbrace{\text{the sum of (the) two numbers}}_{x + (16 - x)} \ \underset{=}{\text{is}} \ \underset{16}{16.}$$

The quadratic equation is obtained from the second piece of information:

$$\underbrace{\text{The product (of the two numbers)}}_{x(16 - x)} \ \underset{=}{\text{is}} \ \underset{48}{48.}$$

Multiply, as indicated, to obtain the quadratic equation

$$16x - x^2 = 48$$

or

$$x^2 - 16x + 48 = 0. \qquad \text{Factor the left side.}$$

$$(x - 4)(x - 12) = 0$$

$x - 4 = 0$	$x - 12 = 0$
$x = 4$	$x = 12$
$16 - x = 12$	$16 - x = 4$

In either case, the two numbers are 4 and 12.

Geometry Problems

EXAMPLE 2

The perimeter (or boundary) of a rectangle is 36 inches. Its area is 80 square inches. Find the dimensions.

Solution. Let l and w be the dimensions of the rectangle. Then

$$2l + 2w = 36$$

and

$$lw = 80$$

(See Figure 12.1.)

FIGURE 12.1

From the first equation,

$$l + w = 18,$$
$$w = 18 - l.$$

The second equation,

$$lw = 80,$$

becomes

$$l(18 - l) = 80,$$
$$18l - l^2 = 80,$$

or

$$l^2 - 18l + 80 = 0 \qquad \text{Factor the left side.}$$

$$(l - 10)(l - 8) = 0$$

$l - 10 = 0$	$l - 8 = 0$
$l = 10$	$l = 8$
$w = 18 - l$	$w = 18 - l$
$w = 8$	$w = 10$

The rectangle is 8 inches by 10 inches.

Motion Problems

EXAMPLE 3

An object falls from a tower 320 feet high. As it is falling, its height is given by

$$h = 320 - 16t^2 \text{ (feet)},$$

where t, the time falling, is measured in seconds. How long does it take for the object to hit the ground?

Solution.

$$h = 320 - 16t^2$$

Because h represents height, the object hits the ground when $h = 0$ (Fig. 12.2). Thus replace h by 0 and solve for t.

$$\begin{aligned} 0 &= 320 - 16t^2 \\ 16t^2 &= 320 \\ t^2 &= 20 \\ t &= \pm\sqrt{20} \\ &= \pm\sqrt{4 \cdot 5} \\ &= \pm\sqrt{4}\sqrt{5} \\ &= \pm 2\sqrt{5} \end{aligned}$$

Because of the physical nature of the problem, t *must be positive.*

$$\begin{aligned} \sqrt{5} &\approx 2.23 \\ 2\sqrt{5} &\approx 4.46 \end{aligned}$$

It takes slightly less than $4\frac{1}{2}$ seconds for the object to hit the ground.

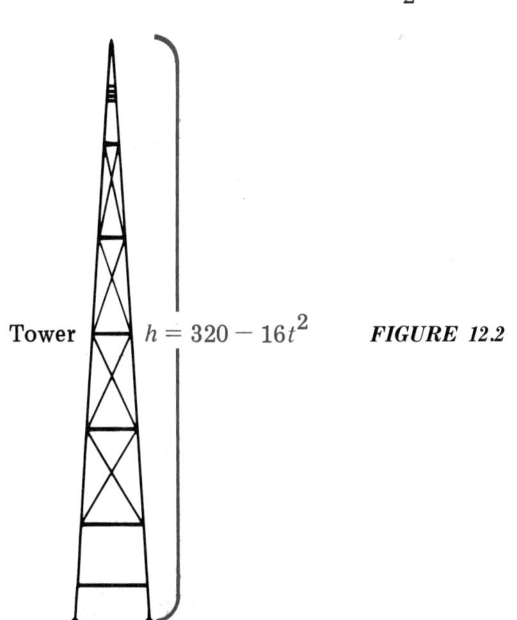

FIGURE 12.2

Profit Problems

The profit derived from manufacturing certain items can be described by means of a quadratic equation. Because of limited plant facilities, the profit will rise to a maximum and then taper off. This profit is often described by an inverted parabola (Figure 12.3). The maximum profit is the y-coordinate of the vertex.

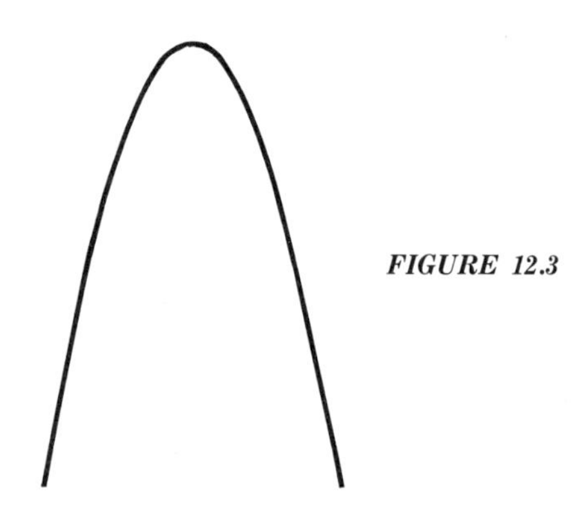

FIGURE 12.3

EXAMPLE 4

The annual profit P (in thousands of dollars) derived from manufacturing industrial machines is given by

$$P = 100 - x^2 + 20x.$$

How many machines must be made in order to derive a profit of \$200 000?

Solution. Let $P = 200$ in the equation

$$P = 100 - x^2 + 20x.$$

(Note that P is measured in thousands of dollars.) Thus

$$200 = 100 - x^2 + 20x$$

$$\begin{aligned} x^2 - 20x + 100 &= 0 \\ (x - 10)^2 &= 0 \\ x - 10 &= 0 \\ x &= 10 \end{aligned}$$

Thus 10 machines must be produced.

EXERCISES

1. The sum of two numbers is 9 and the product is 20. Find these numbers.
2. The sum of two numbers is 20 and the product is 96. Find these numbers.

3. The sum of two numbers is 3 and the product is -70. Find these numbers.

4. The sum of two numbers is -14 and the product is 48. Find these numbers.

5. The sum of two numbers is 0 and the product is -36. Find these numbers.

6. The product of two consecutive *negative* integers is 72. Find these integers.

7. Four times a certain *positive* number is five less than its square. Find this number.

8. The square of a certain *negative* number is three more than twice the number. Find this number.

9. The perimeter of a rectangle is 64 feet and the area is 240 square feet. Find the dimensions of the rectangle.

10. The perimeter of a rectangle is 70 inches and the area is 300 square inches. Find the dimensions of the rectangle.

11. The area of a garden is 1200 square feet. If the length of the garden were increased by 5 feet and the width decreased by 5 feet, the area would be decreased by 75 square feet. Find the dimensions of the garden.

12. The area of a Chinese rug is 360 square feet. A Persian rug, which is 4 feet longer but 3 feet narrower, has the same area. Find the dimensions of the Chinese rug.

13. An object falls from a window 576 feet high. As it is falling, its height is given by

$$h = 576 - 16t^2 \text{ (feet)},$$

where t is the time falling (in seconds). How long does it take to hit the ground?

14. A flower pot falls from a window 192 feet high. As it falls, its height is given by

$$h = 192 - 16t^2 \text{ (feet)},$$

where t is the time falling (in seconds). How long does it take to hit the ground?

15. A baseball is thrown straight up from ground level. Its height, h feet, after t seconds is given by

$$h = 128t - 16t^2.$$

(a) How long does it take to rise 112 feet?
(b) At what time, t, is it 112 feet above the ground on its way down?

16. The monthly profit P (in *thousands* of dollars) derived from manufacturing prefabricated houses is given by

$$P = -x^2 + 16x - 14,$$

where x is the number of houses constructed. How many houses must be constructed in order to make a monthly profit of \$50 000?

17. In exercise 16, how many houses must be constructed in order to make a monthly profit of $46 000?

18. In exercise 16, because of limited plant facilities, it is unprofitable to build too many houses per month. *To the nearest integer,* what is the largest number of houses the firm can construct per month before it loses money? (Hint: Replace P by 0 in the equation of exercise 16 and apply the quadratic formula.)

19. The cost C (in dollars) of building a bookcase is given by

$$C = \frac{l^2}{10} - 3l,$$

where l is the total length of the shelves (in yards). Suppose there are to be 5 equally long shelves. How long will each shelf be if the cost is to be $100?

What Have You Learned in Chapter 12?

You have learned various methods of solving quadratic equations. You know that when the determinant,

$$B^2 - 4AC,$$

of a quadratic equation

$$Ax^2 + Bx + C = 0, \qquad A \neq 0,$$

is positive, there are 2 (real) roots; when the discriminant is 0, there is 1 (real) root, and when the discriminant is negative, there are no (real) roots. In the first two cases, you can apply the quadratic formula to obtain the roots:

$$x = \frac{-B \pm \sqrt{B^2 - 4AC}}{2A}$$

And you can apply these methods of solving quadratic equations to solve stated problems.

Let's Review Chapter 12.

12.1 *Solutions by Factoring*

Solve the equations in exercises 1–4 by factoring the left side. Check the roots in exercise 1.

1. $x^2 + 7x + 12 = 0$
2. $x^2 - 6x + 9 = 0$
3. $2x^2 + 7x + 3 = 0$
4. $3x^2 + 5x - 2 = 0$

12.2 *Equations of the Form $x^2 = a$*

Solve the equations in exercises 5–8. Use the method:

If $x^2 = a$, $a > 0$, then $x = \pm\sqrt{a}$

Check exercise 7.

5. $x^2 - 17 = 0$
6. $x^2 - 80 = 0$
7. $(2u + 1)^2 = 9$
8. $(3y - 5)^2 = 2$

12.3 *The Quadratic Formula*

9. (a) Evaluate the discriminant of the equation

$$x^2 + 3x + 5 = 0.$$

(b) Without solving, indicate how many (real) roots there are.

In exercises 10–12, apply the quadratic formula to solve the equation. Check the roots in equation 10.

10. $x^2 + 5x + 5 = 0$

11. $4x^2 + 4x + 1 = 0$

12. $y^2 + 6y - 1 = 0$

12.4 Word Problems

13. Find two integers whose sum is 4 and whose product is −45.
14. The perimeter of a rectangle is 30 inches. The area is 36 square inches. Find the dimensions.

And these from Chapters 1–11:

15. Let $f(x) = x^2 + 3x + 2$. Find: (a) $f(1)$, (b) $f(2)$, (c) $f(3)$. (d) For which values of x is $f(x) = 0$?
16. Let $f(x) = (x+2)(x-5)$. Find: (a) $f(0)$, (b) $f(-1)$, (c) $f\left(\frac{1}{2}\right)$. (d) For which values of x is $f(x) = 0$?
17. Let $f(x) = x^2 + 6x + 1$. Find: (a) $f(0)$, (b) $f\left(\frac{1}{2}\right)$, (c) $f\left(\frac{1}{6}\right)$. (d) For which values of x is $f(x) = 0$?
18. Factor $5x^2 - 4x - 1$.
19. Factor $10x^2 + 9x + 2$.
20. Find three consecutive integers whose sum is 66.

Try These Sample Questions for Practice.

1. Solve by factoring: $x^2 - 5x - 14 = 0$
2. Solve by using the method:

$$\text{If } x^2 = a,\ a > 0, \text{ then } x = \pm\sqrt{a}$$

$$(2y - 1)^2 = 5$$

3. Solve by applying the quadratic formula: $x^2 + 9x + 3 = 0$
4. Solve by any method: $2y^2 - 3y - 5 = 0$
5. The sum of two numbers is 17 and the product is 60. Find these numbers.

Answers to Odd-Numbered Problems

CHAPTER 1

1.1, p. 2

1–8.

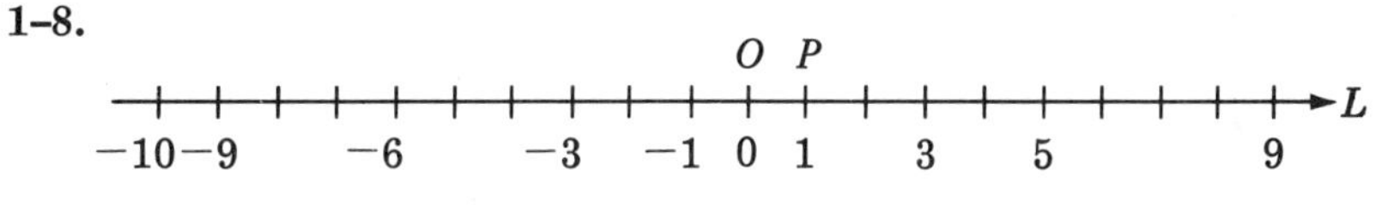

FIGURE 1A

9. Q: 4, R: 7, S: −2, T: −5, U: −6, V: −7
11. 19 **13.** 37

1.2, p. 7

1–8.

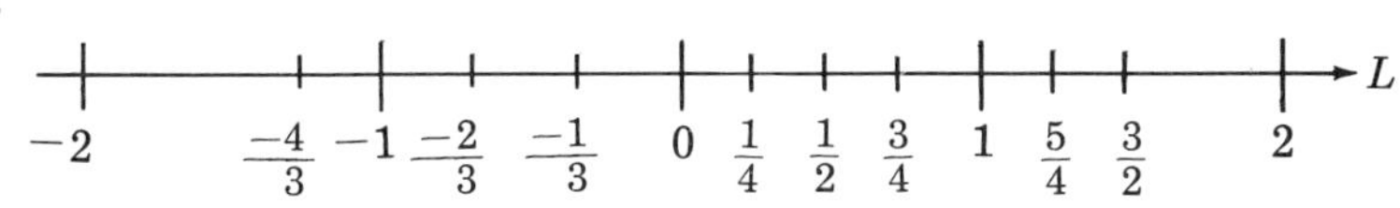

FIGURE 1B

9. Q: $\frac{7}{5}$, R: $\frac{-2}{5}$, S: $\frac{-1}{5}$, T: $\frac{1}{5}$, U: $\frac{-7}{5}$, V: $1\frac{1}{5}$, W: $\frac{4}{5}$

11. $\frac{5}{4}$ **13.** $\frac{7}{5}$ **15.** $\frac{-1}{1}$ **17.** $\frac{13}{4}$ **19.** $\frac{-4}{1}$

21. $\frac{1}{10}$ **23.** $\frac{31}{100}$ **25.** $\frac{-83}{100}$ **27.** $\frac{481}{100}$ **29.** $\frac{-599}{1000}$

1.3, p. 10

1. $<$ **3.** $<$ **5.** $<$ **7.** $<$ **9.** $<$ **11.** $<$
13. $>$ **15.** $>$ **17.** $<$ **19.** $<$ **21.** left
23. right **25.** left **27.** left **29.** left
31. $2 < 3 < 4 < 10 < 12 < 18$ **33.** $-12 < -9 < -7 < 0 < 7 < 10$
35. $\frac{1}{5} < \frac{2}{3} < 1 < \frac{4}{3} < \frac{9}{5} < 2$ **37.** $10 > 8 > 7 > 6 > 5 > 2$
39. $1 > \frac{1}{2} > \frac{1}{5} > \frac{1}{7} > \frac{1}{10} > \frac{1}{12}$ **41.** $1 > .3 > .2 > -2 > -4 > -9$

1.4, p. 14

1. -2 **3.** 0 **5.** $-\pi$
7. 3 **9.** 200 **11.** -5
13. 8 **15.** 0 **17.** $\frac{1}{4}$
19. 10 000 **21.** $=$ **23.** $=$
25. $=$ **27.** $>$ **29.** $>$
31. 8 and -8 **33.** 11 **35.** -44

Let's Review Chapter 1, p. 15

1. Q: 3, R: 8, S: 10, T: -1, U: -6, V: -10
2. (a) -12 (b) 12
3–6.

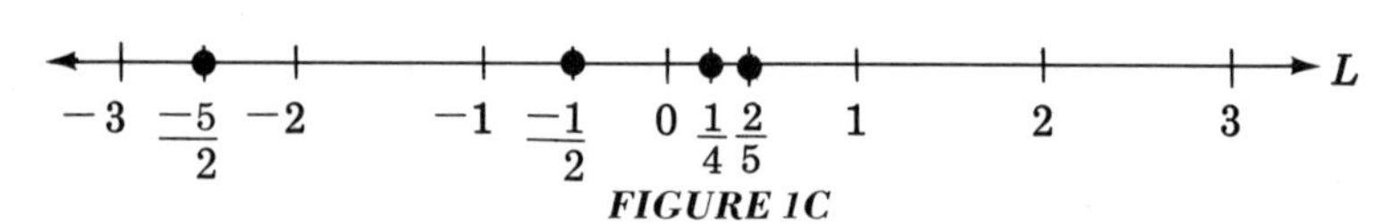

FIGURE 1C

7. $\frac{5}{3}$ **8.** $\frac{37}{100}$
9. (a) $<$ (b) $>$ (c) $<$
10. (a) left (b) right
11. $-5 < -1 < -\frac{1}{5} < \frac{1}{5} < 1 < 5$
12. $1 > \frac{1}{2} > \frac{1}{4} > 0 > -7 > -14$
13. (a) $\frac{-1}{4}$ (b) 2 (c) 0
14. (a) 3 (b) $\frac{2}{5}$ (c) 0
15. (a) $=$ (b) $<$
16. 7 and -7

Try These Exam Questions for Practice, p. 16

1. -6 **2.** $\frac{5}{2}$ **3.** $\frac{-9}{10}$

4. $<$ **5.** right

6. $-12 < -2 < \frac{-1}{2} < \frac{1}{12} < \frac{1}{2} < 12$

7. 19 **8.** 19

CHAPTER 2

2.1, p. 22

1. 22 **3.** 22 **5.** -36 **7.** 3
9. -8 **11.** 648 **13.** -291 **15.** 207
17. -297 **19.** 1392 **21.** 7 **23.** -13
25. 19 **27.** -16 **29.** -7 **31.** -321
33. 4536 **35.** -7892 **37.** 5 **39.** -13
41. 12 **43.** -35 **45.** -2 **47.** $>$
49. 1 **51.** 31 **53.** 10 miles south **55.** $21
57. $6

2.2, p. 26

1. 22 **3.** -17 **5.** 39 **7.** 130
9. -47 **11.** -927 **13.** 11 **15.** 15
17. 45 **19.** -21 **21.** -115 **23.** 1014
25. 9 **27.** 1 **29.** 4 **31.** 9
33. 10 **35.** 4 **37.** 18 **39.** 2
41. 17 **43.** 17 **45.** 17 **47.** 4
49. -14 **51.** 6 **53.** 0 **55.** 12 miles west

2.3, p. 30

1. 72 **3.** 434 **5.** 2077 **7.** 30 415
9. 533 112 **11.** -54 **13.** 70 **15.** -546
17. 8184 **19.** 114 366 **21.** 120 **23.** 6
25. -112 **27.** 40 **29.** 0 **31.** 2160
33. $<$ **35.** $>$ **37.** -45 **39.** 1500
41. $66

2.4, p. 35 (The checks follow no. 71.)

1. 2 **3.** 6 **5.** 4 **7.** -4
9. 2 **11.** -4 **13.** -7 **15.** -16

17. 12 **19.** 77 **21.** 36 **23.** 202
25. 303 **27.** 9 **29.** 27 **31.** 38
33. 121 **35.** 842
37. The quotient is 20 and the remainder is 2.
39. The quotient is 10 and the remainder is 10.
41. The quotient is 48 and the remainder is 8.
43. The quotient is 349 and the remainder is 11.
45. The quotient is 9 and the remainder is 728.
47. The quotient is 59 and the remainder is 135.
49. The quotient is 11 and the remainder is 6.
51. The quotient is 9 and the remainder is 5.
53. The quotient is 27 and the remainder is 0.
55. The quotient is 52 and the remainder is 30.
57. The quotient is 186 and the remainder is 297.
59. 0 **61.** not defined **63.** 2
65. −2 **67.** −9 **69.** $12 009 **71.** 108

Checks:

27.
```
  12
   9
 ---
 108
```

29.
```
  25
  27
 ---
 175
 50
 ---
 675
```

31.
```
   73
   38
  ---
  584
 219
 ----
 2774
```

33.
```
   117
   121
  ----
   117
  234
 117
 -----
 14157
```

35.
```
    272
    842
    ---
    544
  1088
 2176
 ------
 229024
```

49.
```
   11
   11
  ---
  121
 +  6
  ---
  127
```

51.
```
   43
    9
  ---
  387
 +  5
  ---
  392
```

53.
```
   38
   27
  ---
  266
  76
 ----
 1026
```

55.
```
   132
    52
   ---
   264
  660
  ----
  6864
 +  30
  ----
  6894
```

57.
```
    389
    186
   ----
   2334
  3112
  389
  -----
  72354
 +  297
  -----
  72651
```

2.5, p. 39

1. 9 **3.** 64 **5.** 9
7. 100 **9.** 144 **11.** 625
13. 8 **15.** 125 **17.** −1

19. -125
21. -8000
23. 81
25. 729
27. -243
29. 64
31. -125
33. 1
35. 1
37. 16
39. 10 000 000
41. the second power
43. the second power

2.6, p. 44

1. 23
3. 16
5. 3
7. -16
9. 0
11. 4
13. 48
15. 5
17. 14
19. 1
21. -27
23. -16
25. 35
27. 0
29. 27
31. -21
33. -12
35. 10
37. -24
39. 21
41. 45
43. 80
45. (a) 3 (b) -1
47. 20
49. 4

Let's Review Chapter 2, p. 46

1. -8
2. -1412
3. 115
4. -14
5. -17
6. -383
7. -1
8. 21
9. 34844
10. -108
11. 72
12. -210
13. (a) -12 (b) 12
14. 7, Check: $\begin{array}{r} 61 \\ \underline{7} \\ 427 \end{array}$
15. The quotient is 5 and the remainder is 12. Check: $\begin{array}{r} 27 \\ \underline{5} \\ 135 \\ +\ \underline{12} \\ 147 \end{array}$
16. -15
17. 81
18. -27
19. -1
20. the fourth power
21. 11
22. 21
23. 8
24. -8
25. 40
26. 95

And these from Chapter 1, p. 47

27. $\frac{-9}{10}$
28. (a) $>$ (b) $>$
29. 3
30. 20 and -20

Try These Exam Questions for Practice, p. 47

1. 5455
2. -120

3. The quotient is 29 and the remainder is 3. Check:

$$\begin{array}{r} 34 \\ \underline{29} \\ 306 \\ \underline{68} \\ 986 \\ + \underline{3} \\ 989 \end{array}$$

4. -4 **5.** 25 **6.** -1
7. the third power **8.** -21

CHAPTER 3

3.1, p. 52

1. $t \cdot t$ **3.** $5 \cdot x \cdot y \cdot y$ **5.** $\frac{1}{2} \cdot a \cdot a \cdot x \cdot x$
7. $\frac{3}{5} \cdot u \cdot u \cdot v \cdot v \cdot v$ **9.** $\frac{-1}{2} \cdot u \cdot u \cdot v \cdot v \cdot w$ **11.** 7
13. 5 **15.** $\frac{1}{2}$ **17.** $\frac{3}{4}$
19. 0 **21.** 2 **23.** π
25. like **27.** like **29.** unlike
31. like **33.** unlike **35.** like
37. unlike

3.2, p. 55

1. $3x$ **3.** x **5.** $4xy$
7. $5abc$ **9.** $6xy$ **11.** $5x$
13. 0 **15.** $8w$ **17.** $6xyz$
19. $2x + 2y$ **21.** $6u - 3v$ **23.** $5x$
25. $6a + 12b$ **27.** $7ab - 4cd$ **29.** $6a - 3c$
31. $-2m + 3n$ **33.** $13xyz - 8abc$ **35.** $3y^2 - 2z^2$
37. 14 **39.** $4n - 10p$ **41.** $14x$
43. $5a$

3.3, p. 58

1. (*i*) **3.** (*i*) **5.** (*iv*)
7. (*iii*) **9.** (*i*) **11.** (*i*)
13. $3x + 5y$ **15.** $16m + 16n$ **17.** $a + 3$
19. $3a + 2b + 3c$ **21.** $5x$ **23.** $7a - 2c$
25. $5r + 2s + 2u$ **27.** $22a + 5b + 5c + 11d$
29. $7a - b + 2c + 6d + 2e$

31.	$2a + b$	**33.**	$5c - 11d$	**35.**	$16t$
37.	$5a - 3b + 3c$	**39.**	$3a - 3b + 2c - 2d$		
41.	$-2r + 2t - 4u + 2v$				
43.	$a - 4b$	**45.**	$3m + n$	**47.**	a
49.	$-4r + s + 2t$	**51.**	$6x - y$	**53.**	$7 + 3a - 6b + c$
55.	$10a - b$	**57.**	$4x + y$	**59.**	$7a + 2b$

3.4, p. 63

1.	12	**3.**	7	**5.**	9
7.	36	**9.**	14	**11.**	13
13.	1	**15.**	3	**17.**	-12
19.	10	**21.**	25	**23.**	32
25.	10	**27.**	8	**29.**	1 009 702
31.	(a) 12		(b) 8		(c) 0
33.	(a) 9		(b) 19		(c) 59
35.	(a) 0		(b) 24		(c) 8
37.	(a) 21		(b) 45		(c) 55
39.	(a) 1		(b) 13		(c) 73
41.	(a) 121 sq. in.		(b) 225 sq. in.		(c) 900 sq. in.
43.	880 square feet				
45.	(a) 100π sq. in.		(b) 144π sq. in.		(c) 169π sq. in.
47.	(a) 480 cubic inches		(b) 2400 cubic inches		

Let's Review Chapter 3, p. 66

1. (a) $5 \cdot x \cdot y \cdot y \cdot y$ (b) $\frac{1}{2} \cdot u \cdot v \cdot v \cdot w \cdot w \cdot w$

2.	-12	**3.**	yes	**4.**	no
5.	$10x$	**6.**	$4uv$	**7.**	$2m + 15n$
8.	$5x + 2y - 3z$	**9.**	$8a + 7b$	**10.**	$4w + x + 4y - z$
11.	$a + b + c$	**12.**	$3x + 3z$	**13.**	$13a$
14.	$3m - 7n$	**15.**	50	**16.**	45
17.	(a) 4		(b) -12	**18.**	400 square inches

And these from Chapters 1 and 2, p. 67

19.	(a) -2		(b) 2		
20.	1	**21.**	0	**22.**	24

Try These Exam Questions for Practice, p. 67

1.	$\frac{1}{2}$	**2.**	$10m$	**3.**	$7a$
4.	$12a + 11b - 13c + 11d$				
5.	$3x + 3y - z$	**6.**	$-2m - n - 3p$	**7.**	45
8.	325				

CHAPTER 4

4.1, p. 72

1. a^2
3. c^3
5. y^7
7. x^4
9. b^5
11. m^{18}
13. $4a^6$
15. $4m^6$
17. $6x^3$
19. a^4
21. c^9
23. a^7
25. x^{11}
27. $-2a^9$
29. x^2y^2
31. x^3y^3
33. a^5b^3
35. $30x^2y^2$
37. $2x^3y^3$
39. $-6a^3b^2$
41. $-3a^2x^4$
43. $30x^5y^2$
45. $1000a^2b^2$
47. $30am^3n^3$
49. $-24a^4b^3c^4d^2$
51. $18x^3y^2$
53. 32
55. $4x$
57. $-4x^2$
59. y^4
61. xy
63. m^2n^4
65. x^3yz
67. $2x$
69. $3a^4$
71. $4a^3b$
73. $3xy^2$
75. $4a^9b^2c$

4.2, p. 75

1. $4x + 4y$
3. $-10a - 10b$
5. $2a - 2b$
7. $-3m + 3n$
9. $5x + 5y$
11. $2ax + 6y$
13. $-10a^2 + 20b$
15. $a^2 + 4a$
17. $a^3 + a^2b$
19. $2a^3 + 2a^4$
21. $5x^3yz + 5xy^3z$
23. $6a^3b - 10ab^2$
25. $-2m^3n + 3mn^3$
27. $10u^3v - 6u^3v^3$
29. $50x^6y^2z^6 + 80x^3y^3z^5$
31. $8x + 8y + 8z$
33. $5a - 5b - 5c$
35. $10x^2 + 20x + 30$
37. $3x^2 + 3xy + 3xz$
39. $-2a^3b - 10a^2b^2 + 2ab^3$
41. $-x^3yz^2 - 2x^3y^2z^2 + 3x^3y^2z^3$
43. 45
45. $3x - 2y - 10z$
47. $3xy - 6xz - 4x$
49. $a^2 - 2ab + b^2$
51. $4a + 27b$
53. $x + y$
55. $x - 2z$
57. $a + 1$
59. $m^4 + m$
61. $xy + 1$
63. $x - a$
65. $b - 3$
67. $4a + 6 + 3b$

4.3, p. 81

1. composite
3. prime
5. prime
7. prime
9. prime
11. $2 \cdot 5$
13. $2 \cdot 3^2$
15. $-2^3 \cdot 3$
17. $2^2 \cdot 7$
19. $2^2 \cdot 3^2$
21. $2 \cdot 3 \cdot 7$
23. $5 \cdot 13$
25. $2 \cdot 5 \cdot 7$
27. $2^2 \cdot 3^3$
29. 13^2
31. 2
33. 3
35. 5
37. 12
39. 24
41. 2
43. 5
45. 8
47. 9
49. 6
51. 3
53. 7
55. 1
57. 6
59. 6
61. 6
63. 2, 3, 5, 7, 11, 13, 17, 19, 23, 29, 31, 37, 41, 43, 47

4.4, p. 85

1. $2(a+1)$ **3.** $6(x+1)$ **5.** $5(x^2+2)$
7. $10(2x^2+5)$ **9.** $16a+25$ **11.** $t(t+1)$
13. $x(2x+3)$ **15.** $y^2(y-1)$ **17.** $a^7(a^3-1)$
19. $10x^2(x-2)$ **21.** $4a^3(8a+3)$ **23.** $xy(x-y)$
25. $xyz^2(x-1)$ **27.** $2ab(ac+2b)$
29. $4m^2n(4p+5m^2n^3)$ **31.** $ab^3c^2(a^2b+a+1)$
33. $2(x^2+2x+1)$ **35.** $10x(x^2-2x+5)$
37. $5a^2(a^2-2a+3)$ **39.** $2x^3y(10x^3-12x+15)$
41. $18mn^3(m+3n-4)$ **43.** $2xy(x^4+6x^3y+4x^2y^2-3xy^3+2y^4)$

4.5, p. 89

1. x^2+3x+2 **3.** x^2+7x+6 **5.** y^2+5y+6
7. $z^2+7z+12$ **9.** a^2+4a-5 **11.** $b^2-4b-21$
13. $x^2-6x-27$ **15.** $m^2+8m+12$ **17.** $c^2+3c-10$
19. $x^2-7x-18$ **21.** y^2+9y+8 **23.** $-2a^2+8a+90$
25. $-c^2-3c+18$ **27.** x^2+2x+1 **29.** a^2-4a+4
31. $c^2+12c+36$ **33.** $x^2+20x+100$ **35.** x^2-64
37. v^2-1 **39.** x^2-144 **41.** x^4+8x^2+16
43. a^8-1 **45.** x^6+2x^3+1 **47.** x^4+5x^2+6
49. c^6-5c^3+4 **51.** $x^2+2xy+y^2$ **53.** $m^2+3mn+2n^2$
55. $m^2-2mn-8n^2$ **57.** $-10a^2+36ab-18b^2$
59. $20m^2-43mn+6n^2$ **61.** $2x^2+10x+10$
63. $2y^2+15y+11$ **65.** $2x^2+7x+5$
67. -676

4.6, p. 92

1. $(x+3)(x-3)$ **3.** $(x+7)(x-7)$ **5.** $(m+10)(m-10)$
7. $(y+9)(y-9)$ **9.** $(1+x)(1-x)$ **11.** $(5+b)(5-b)$
13. $(x+y)(x-y)$ **15.** $(a+x)(a-x)$ **17.** $(2x+1)(2x-1)$
19. $(6x+7)(6x-7)$ **21.** $(1+3u)(1-3u)$
23. $(2x+y)(2x-y)$ **25.** $(3m+7n)(3m-7n)$
27. $(12y+5x)(12y-5x)$ **29.** $(20t+u)(20t-u)$
31. $2(x+1)(x-1)$ **33.** $3(s+5)(s-5)$
35. $3(10+x)(10-x)$ **37.** $2(x+4y)(x-4y)$
39. $8(a+5b)(a-5b)$ **41.** $6(m+3n)(m-3n)$
43. $6(2x+7y)(2x-7y)$ **45.** $a(a+1)(a-1)$
47. $b(b+7)(b-7)$ **49.** $a^2(a+1)(a-1)$
51. $x(x+y)(x-y)$ **53.** $m^2(m+7n)(m-7n)$
55. $2x(x+1)(x-1)$ **57.** $3x(x+y)(x-y)$
59. $2x(2x+5y)(2x-5y)$ **61.** (c)
63. (d)

4.7, p. 97

1. $(x + 3)(x + 1)$
3. $(z + 6)(z + 1)$
5. $(b + 1)^2$
7. $(x + 3)^2$
9. $(a + 2)(a - 1)$
11. $(b + 1)(b - 2)$
13. $(s + 3)(s - 4)$
15. $(y + 6)(y + 2)$
17. $(a - 6)^2$
19. $(x - 7)(x - 2)$
21. $(a + 7)(a - 2)$
23. $(c + 5)(c + 4)$
25. $(x + 8)(x + 5)$
27. $(y + 3)(y - 10)$
29. $(x + 12)(x - 3)$
31. $2(x + 2)(x + 1)$
33. $3(x + 4)(x - 2)$
35. $-(y - 5)(y - 4)$
37. $a(a + 5)(a - 2)$
39. $4(m + 5)^2$
41. $3(6 - x)(4 + x)$
43. $v^2(u + 9)^2$
45. $2t(t - 7)(t - 5)$
47. $(x + a)^2$
49. $(x + 4a)(x + a)$
51. $(m + 2n)^2$
53. $(x + 3y)^2$
55. $2(x + 6y)^2$
57. $a(x + 4y)(x + 2y)$
59. (d)

4.8, p. 102

1. $(3x + 1)(x + 1)$
3. $(7a + 1)(a + 1)$
5. $(3x - 1)(x + 1)$
7. $(5y - 1)(y - 1)$
9. $(2x + 1)(x + 3)$
11. $(2y + 3)(y + 1)$
13. $(2a + 5)(a + 1)$
15. $(2m - 1)(m - 2)$
17. $(2n - 1)(n - 3)$
19. $(2a - 1)(a + 4)$
21. $(2a + 5)(a + 5)$
23. $(3y - 1)^2$
25. $(2a + 1)^2$
27. $(3y - 7)(y + 1)$
29. $(2m + 3)(2m + 1)$
31. $(3x - 2)(2x + 3)$
33. $(7a + 2)(a + 1)$
35. $(5x + 4)(2x - 1)$
37. $(3x - 13)(3x + 1)$
39. $(5a - 3)(a + 4)$
41. $2(2x + 1)(x + 1)$
43. $(2b - 1)(b + 12)$
45. $3(2b + 1)(b + 1)$
47. $2(2a - 1)(a - 1)$
49. $2x(2x + 3)(x + 1)$
51. $(2x + y)^2$
53. $(2y + z)(y + z)$
55. $(2x + y)(x + 2y)$
57. $(3a + 2b)(3a - 10b)$
59. $2(3x + y)(x + 2y)$

Let's Review Chapter 4, p. 103

1. $6x^3y^4$
2. $8m^4n^4$
3. $2x^2y$
4. $8a^3$
5. $x^3 - 5x^2y$
6. $8a^4b^2c + 12a^3b^2c^2$
7. $2a + 2b$
8. $4x - 3y$
9. 2, 3, 5, 7, 11, 13, 17, 19
10. (a) $2^2 \cdot 3^2$ (b) $2^2 \cdot 3 \cdot 11$
11. 12
12. 4
13. $2(2x + 3)$
14. $x^2(x + 1)$
15. $3a^4(3a^2 - 5)$
16. $x^2y^3z(x^3yz^2 + xz^2 + 1)$

17. $x^2 + 6x + 5)$

18. $a^2 - 5a - 14$

19. $z^4 - 3z^2 - 4$

20. $m^2 - mn - 6n^2$

21. $(a + 5)(a - 5)$

22. $(2y + 7z)(2y - 7z)$

23. $a(a + 1)(a - 1)$

24. $3x(y + 2)(y - 2)$

25. $(c + 4)(c + 3)$

26. $(m + 4)(m - 5)$

27. $t(t + 3)^2$

28. $(x - 3)(x - 2)$

29. $(4a + 1)(a + 1)$

30. $(2x + 5)(x + 1)$

31. $(2y - 1)(y - 3)$

32. $(4x - 3y)(x + y)$

33. $5(m - 2n)$

34. $(6 + a)(6 - a)$

35. $(x + 7)(x + 2)$

36. $7(a + 2b)(a - 2b)$

37. $x(x + 2)(x - 2)$

38. $x(3x + 1)(x - 2)$

39. $(m + 5n)(m + 3n)$

40. $4(m + 2n^2)(m - 2n^2)$

41. $s(s + 7)(s - 1)$

42. $t^2(t + 4)(t - 4)$

43. $(a - 8)(a - 2)$

44. $(5x + 2)(x - 2)$

And these from Chapters 1–3, p. 104

45. $2 < 7 < 17 < 23 < 29 < 41$

46. 15 **47.** 39 **48.** 78

Try These Exam Questions for Practice, p. 105

1. $-12x^3y^8$

2. $8x^2 + 14x - 15$

3. $3x - 5y^2$

4. $2^2 \cdot 3^3$

5. $2x(3x - 4)$

6. $(3a + 4b)(3a - 4b)$

7. $(x + 6)(x + 3)$

8. $(2y + 1)(y - 1)$

CHAPTER 5

5.1, p. 110

1. $4 \cdot 4 \overset{?}{=} 16 \cdot 1$
$16 \overset{\checkmark}{=} 16$
Thus $\frac{4}{16} = \frac{1}{4}$

3. $-4 \cdot 3 \overset{?}{=} 6(-2)$
$-12 \overset{\checkmark}{=} -12$
Thus $\frac{-4}{6} = \frac{-2}{3}$

5. $14 \cdot 1 \overset{?}{=} 2 \cdot 7$
$14 \overset{\checkmark}{=} 14$
Thus $\frac{14}{2} = \frac{7}{1}$

7. $(-8)(-7) \overset{?}{=} 14 \cdot 4$
$56 \overset{\checkmark}{=} 56$
Thus $\frac{-8}{14} = \frac{4}{-7}$

9. $13 \cdot 3 \overset{?}{=} 39 \cdot 1$
$39 \overset{\checkmark}{=} 39$
Thus $\frac{13}{39} = \frac{1}{3}$

11. $\frac{1}{2}$

13. $\frac{-1}{3}$

15. $\frac{1}{4}$

17. $\frac{-4}{3}$

19. $\frac{-4}{5}$

21. $\frac{-2}{3}$

23. $\frac{1}{2}$

25. $\frac{1}{6}$

27. $\frac{3}{4}$

29. $\frac{3}{2}$

31. $\frac{3}{8}$

33. $\frac{1}{2}$

35. $\frac{-1}{3}$

37. $\frac{6}{7}$

39. (a), (c)

5.2, p. 113

1. (a) x (b) $x + 1$

3. (a) $y + 3$ (b) $y - 4$

5. (a) $2x^2 - 1$ (b) 7

7. (a) 8 (b) 9

9. (a) $x^2 + 2xy + 3y$ (b) $2x + 5y$

11. $\frac{20}{11}$

13. $\frac{5}{2}$

15. -1

17. $\frac{17}{15}$

19. 2

21. 2

23. $\frac{1}{2}$

25. $\frac{2}{3}$

27. 3

29. 3

31. $\frac{-3}{4}$

33. 30

35. $\frac{15}{4}$

37. $\frac{1}{2}$

39. $\frac{10}{9}$

41. (a) 3, (b) 5, (c) 7

43. (a) $\frac{7}{4}$ (b) 4 (c) $\frac{5}{2}$

45. (a) 5 (b) 7 (c) $\frac{1}{3}$

47. (a) 0 (b) $\frac{4}{5}$ (c) $\frac{117}{50}$

5.3, p. 119

1. 5

3. $\frac{49}{11}$

5. 1
7. x^3
9. z^6
11. $\frac{1}{a}$
13. $\frac{m^7}{n}$
15. abc
17. $\frac{x^2y}{z^{15}}$
19. $\frac{b}{2}$
21. $\frac{ab}{2c}$
23. $\frac{3}{5a^2c}$
25. $\frac{2}{m^2n}$
27. $\frac{3m^9p}{2n}$
29. $\frac{2}{3}$
31. $\frac{11}{24}$
33. $\frac{-16}{21}$
35. $\frac{15}{8}$
37. $\frac{11a}{4}$
39. $\frac{8}{3y^6}$
41. $\frac{2a^3}{3}$
43. $x + 1$
45. $(a + b)^2$
47. 1
49. $\frac{1}{(a - b)^3}$
51. 2
53. $-6(a + b)^3$
55. $x + y$
57. $\frac{(x + y)^2}{(x - y)^2}$
59. $\frac{(a + b)\,(x + y)^6}{4(x - y)}$
61. $\frac{1}{25x(a + 2)}$

5.4, p. 122

1. $x + y$
3. $m + 2n$
5. $x + y$
7. $x + y$
9. $1 + y$
11. $\frac{a + b}{2}$
13. $\frac{2a - b}{4}$
15. $x + y$
17. $\frac{b + c^2}{a}$
19. $y(1 + y)$
21. $\frac{2x + 3y}{2}$
23. $3(a + 2b)$
25. $\frac{3 + 2b}{2}$
27. $\frac{3b - 4ac}{a}$
29. $\frac{-(4r + 5t)}{3}$
31. $x + y + z$
33. $\frac{a - 2b + c}{5}$
35. $\frac{s + 2t - u}{6}$

37. $\dfrac{a - 2x + 3ax}{4ax}$

39. $\dfrac{4bc + 5a^2 - 6c}{5b}$

41. $\dfrac{3b^5c^2 - 4ac + 5a^2b^2}{10b^2}$

43. $\dfrac{1}{2 + 3x}$

45. $\dfrac{yz}{2y^2 - 3z^2}$

47. $\dfrac{20xy}{2y - 5x + 10xy}$

5.5, p. 125

1. $\dfrac{a+b}{x+y}$

3. $\dfrac{x+y}{x-y}$

5. $\dfrac{x+y}{x-y}$

7. $\dfrac{x+y}{x-2y}$

9. $\dfrac{6}{5}$

11. $\dfrac{a}{b}$

13. $\dfrac{x-y}{3}$

15. $\dfrac{3}{x+y}$

17. $\dfrac{a-3}{4}$

19. $\dfrac{3}{5(m-n)}$

21. $\dfrac{a+2b}{2}$

23. $\dfrac{2x-3y}{6}$

25. $\dfrac{x+1}{x+3}$

27. $\dfrac{y-3}{y+1}$

29. $\dfrac{u-6}{u+1}$

31. $\dfrac{m+4}{m+2}$

33. $\dfrac{x-2}{x-4}$

35. $\dfrac{a-\mathrm{b}}{2(a+2b)}$

37. $\dfrac{x+4}{2(x-2)}$

39. $\dfrac{2x+5}{x+3}$

5.6, p. 130

1. (a) $x^2 + 2x + 1$ (b) 2
3. (a) $-t^5 + t^3 + t$ (b) 5
5. (a) $-t + 1$ (b) 1
7. (a) $z^{10} + z^7 - z^4 + z$ (b) 10

(The checks follow no. 35.)

9. $x + 4$
11. $x + 5$
13. $t - 9$
15. $t + 5$
17. $x^2 + x + 1$
19. $3t^2 - 1$
21. $x + 9$
23. $z^2 + 1$
25. $x - 4$
27. $x + 2$
29. $x^2 + x + 1$
31. $2a^2 - a + 2$
33. $x + 2$
35. $x^2 + 2x + 1$

Checks:

9.
$$\begin{array}{r} x + 4 \\ \underline{x + 1} \\ x^2 + 4x \\ \underline{+\ x + 4} \\ x^2 + 5x + 4 \end{array}$$

11.
$$\begin{array}{r} x + 5 \\ \underline{x + 3} \\ x^2 + 5x \\ \underline{+ 3x + 15} \\ x^2 + 8x + 15 \end{array}$$

13.
$$\begin{array}{r} t - 9 \\ \underline{t - 1} \\ t^2 - 9t \\ \underline{-\ t + 9} \\ t^2 - 10t + 9 \end{array}$$

15.
$$\begin{array}{r} t + 5 \\ \underline{t + 9} \\ t^2 + 5t \\ \underline{+ 9t + 45} \\ t^2 + 14t - 45 \end{array}$$

17.
$$\begin{array}{r} x^2 + x + 1 \\ \underline{x + 4} \\ x^3 + x^2 + x \\ \underline{+ 4x^2 + 4x + 4} \\ x^3 + 5x^2 + 5x + 4 \end{array}$$

19.
$$\begin{array}{r} 3t^2 - 1 \\ \underline{t + 4} \\ 3t^3 \quad - t \\ \underline{+ 12t^2 \quad - 4} \\ 3t^3 + 12t^2 - t - 4 \end{array}$$

5.7, p. 133

1. (a) 4 (b) 3 (c) $\underbrace{\text{degree } (3)}_{0} < \underbrace{\text{degree } (2x)}_{1}$

3. (a) 3 (b) 1 (c) $\underbrace{\text{degree } (1)}_{0} < \underbrace{\text{degree } (3x + 1)}_{1}$

5. (a) $2x$ (b) 1 (c) $\underbrace{\text{degree } (1)}_{0} < \underbrace{\text{degree } (x)}_{1}$

7. (a) $x + 1$ (b) 5 (c) $\underbrace{\text{degree } (5)}_{0} < \underbrace{\text{degree } (x + 3)}_{1}$

9. (a) $2x$ (b) 5 (c) $\underbrace{\text{degree } (5)}_{0} < \underbrace{\text{degree } (x^2 - 2)}_{2}$

(The checks follow no. 35.)

11. (a) 3 (b) -1 **13.** (a) 5 (b) -19

15. (a) 10 (b) -9 **17.** (a) $x + 3$ (b) -2

19. (a) $x^2 + 2x + 1$ (b) 4

21. (a) $3x + 5$ (b) $10x + 2$

23. (a) $x + 5$ (b) 5 (c) $x + 5 + \dfrac{5}{x + 3}$

25. (a) $2x + 2$ (b) 3 (c) $2x + 2 + \dfrac{3}{2x + 1}$

27. (a) $x + 5$ (b) -41 (c) $x + 5 + \dfrac{-41}{x + 10}$

29. (a) $x^2 - 4x + 6$ (b) -5 (c) $x^2 - 4x + 6 + \dfrac{-5}{x+1}$

31. (a) $2x^2 + 2x - 3$ (b) -3 (c) $2x^2 + 2x - 3 + \dfrac{-3}{x-1}$

33. (a) $4x - 3$ (b) $-3x + 13$ (c) $4x - 3 + \dfrac{-3x+13}{x^2+x+1}$

35. (a) $x^2 - 1$ (b) 0 (c) $x^2 - 1 + \dfrac{0}{x^2+1}$

Checks:

11.
$$\begin{array}{rr} 3(x+1) = & 3x+3 \\ & -\ 1 \\ \hline & 3x+2 \end{array}$$

13.
$$\begin{array}{rr} 5(x+5) = & 5x+25 \\ & -\ 19 \\ \hline & 5x+\ \ 6 \end{array}$$

15.
$$\begin{array}{rr} 10(1+x) = & 10+10x \\ & -9\qquad \\ \hline & 1+10x \end{array}$$

17.
$$\begin{array}{r} x+3 \\ x+4 \\ \hline x^2+3x\qquad \\ +4x+12 \\ \hline x^2+7x+12 \\ -2 \\ \hline x^2+7x+10 \end{array}$$

19.
$$\begin{array}{r} x^2+2x+1 \\ x+1 \\ \hline x^3+2x^2+x\qquad \\ +x^2+2x+1 \\ \hline x^3+3x^2+3x+1 \\ +4 \\ \hline x^3+3x^2+3x+5 \end{array}$$

21.
$$\begin{array}{r} 3x+5 \\ x^2-x \\ \hline 3x^3+5x^2\qquad\qquad \\ -3x^2-5x\qquad \\ \hline 3x^3+2x^2-5x\qquad \\ +10x+2 \\ \hline 3x^3+2x^2+5x+2 \end{array}$$

Let's Review Chapter 5, p. 135

1. $3(-15) \stackrel{?}{=} 5(-9)$

$-45 \stackrel{\checkmark}{=} -45$

Thus

$\dfrac{3}{5} = \dfrac{-9}{-45}$

2. $\dfrac{2}{3}$

3. $\dfrac{-5}{6}$ **4.** $\dfrac{4}{5}$ **5.** (a) $2y - 1$ (b) $y^2 + 3y - 5$

6. 32 **7.** $\dfrac{3}{4}$ **8.** (a) -2 (b) -2 (c) $\dfrac{5}{2}$

9. $\dfrac{a^2}{b^4}$ **10.** $\dfrac{50z^3}{x^3}$ **11.** $\dfrac{4a^3}{3c^2}$ **12.** $\dfrac{(x+y)^2}{27(x-y)}$

13. $2(x-y)$ **14.** $\dfrac{a+b}{x}$ **15.** $\dfrac{a+2c}{3}$ **16.** $\dfrac{2}{1-5y}$

17. $\dfrac{a+b}{5(a+2b)}$ **18.** $\dfrac{m-3}{m+3}$ **19.** $\dfrac{1}{s}$ **20.** $\dfrac{2(x+y)}{x-y}$

21. (a) $7x^6 + 5x^5 - x^4 + 2x^2$ (b) 6

22. $y + 5$ Check:

$$\begin{array}{r} y + 5 \\ \underline{y + 2} \\ y^2 + 5y \\ \underline{+ 2y + 10} \\ y^2 + 7y + 10 \end{array}$$

23. $x^2 - x + 1$ **24.** $a^2 - 1$

25. (a) 5 (b) 2 (c) $\underbrace{\text{degree } (2)}_{0} < \underbrace{\text{degree } (x + 1)}_{1}$

26. (a) $4x - 15$ (b) 39 (c)

$$\begin{array}{r} 4x - 15 \\ \underline{x + 2} \\ 4x^2 - 15x \\ \underline{+ 8x - 30} \\ 4x^2 - 7x - 30 \\ \underline{+ 39} \\ 4x^2 - 7x + 9 \end{array}$$

27. (a) $a - 3$ (b) -11 (c) $a - 3 + \dfrac{-11}{a + 7}$

28. (a) $x^2 - 5x + 20$ (b) -79 (c) $x^2 - 5x + 20 + \dfrac{-79}{x + 4}$

And these from Chapters 1–4, p. 136

29. (a) $2x + 4$ (b) $\dfrac{x + 2}{4}$

30. (a) $a^2 - 1$ (b) $\dfrac{a - 1}{6}$

31. (a) $y^2 + 4y + 4$ (b) $y + 2$

32. (a) -4 (b) $\dfrac{-1}{2}$

(c) The evaluation is larger for 3 because $\dfrac{-1}{2} > -4$.

Try These Exam Questions for Practice, p. 137

1. $\dfrac{6}{7}$ **2.** $\dfrac{x + 1}{3}$

3. $\dfrac{x^2 + 2}{x}$ **4.** $\dfrac{x - 3}{x + 2}$

5. $\dfrac{4}{3}$ **6.** $x + 6$ Check:

$$\begin{array}{r} x + 6 \\ \underline{x + 3} \\ x^2 + 6x \\ \underline{3x + 18} \\ x^2 + 9x + 18 \end{array}$$

7. (a) $x + 5$ (b) -8 (c)

$$\begin{array}{r} x + 5 \\ x + 2 \\ \hline x^2 + 5x \\ + 2x + 10 \\ \hline x^2 + 7x + 10 \\ - 8 \\ \hline x^2 + 7x + 2 \end{array}$$

CHAPTER 6

6.1, p. 141

1. $\frac{1}{6}$ **3.** $\frac{2}{15}$ **5.** $\frac{3}{2}$ **7.** -2

9. $\frac{1}{4}$ **11.** $\frac{2}{3}$ **13.** $\frac{-3}{2}$ **15.** $\frac{1}{24}$

17. $\frac{-8}{5}$ **19.** $\frac{1}{30}$ **21.** $\frac{4}{9}$ **23.** $\frac{2}{3}$

25. $\frac{ax}{y}$ **27.** $\frac{a^2b}{c^2}$ **29.** $\frac{3x^2}{2b}$ **31.** $\frac{20bx}{ay}$

33. $\frac{amx}{bny}$ **35.** $\frac{48a^2}{x^2}$ **37.** $\frac{x+2}{x+4}$ **39.** $\frac{3}{2y}$

41. $x - 1$ **43.** $\frac{2(a-2)}{3(a+2)}$ **45.** a **47.** $\frac{3}{a+3}$

49. 24 **51.** $\frac{1}{3}$ **53.** $\frac{x-1}{x^2}$

6.2, p. 145

1. 2 **3.** $\frac{-1}{4}$ **5.** $\frac{3}{10}$ **7.** $\frac{2}{3}$

9. 4 **11.** $\frac{1}{15}$ **13.** -6 **15.** 6

17. $\frac{1}{60}$ **19.** $\frac{22}{25}$ **21.** $\frac{b}{a}$ **23.** $\frac{1}{a}$

25. $\frac{a^2}{b^2}$ **27.** $\frac{b^2}{ac^2}$ **29.** $\frac{c^5}{b^2}$ **31.** $\frac{2}{x-a}$

33. $\frac{a-1}{a+1}$ **35.** $y(x-1)$ **37.** $\frac{2}{a-b}$ **39.** $\frac{x+2}{a^2b}$

41. $\frac{2}{x+3}$ **43.** a^2 **45.** $2a^2$ **47.** $\frac{5}{2}$

49. $\frac{3}{x}$

6.3, p. 149

1. $\frac{2}{5}$ **3.** 1 **5.** 1 **7.** $\frac{1}{2}$

9. $\frac{2\pi}{3}$ **11.** $\frac{2}{\pi}$ **13.** $\frac{3}{4}$ **15.** 1

17. $\frac{8}{9}$ **19.** $\frac{3}{4}$ **21.** $\frac{3}{2}$ **23.** 0

25. $\frac{2}{x}$ **27.** $\frac{1}{y}$ **29.** $\frac{x+1}{2}$ **31.** $\frac{11}{x+1}$

33. $\frac{8a}{x+2}$ **35.** $\frac{9}{a}$ **37.** $\frac{6}{x}$ **39.** 1

41. 3 **43.** -1 **45.** $\frac{1}{x+2}$ **47.** $\frac{b^2+1}{b+1}$

49. $y+1$

6.4, p. 153

1. 6 **3.** 4 **5.** 6
7. 25 **9.** 12 **11.** 30
13. 60 **15.** 108 **17.** 30
19. 12 **21.** 300 **23.** ax
25. $5x$ **27.** x^2 **29.** $(x+1)^2$
31. a^2x^2 **33.** $5x^2yz$ **35.** $12a^2bxy$
37. $(x+a)(x-a)$ or x^2-a^2 **39.** $(x+2)(x-2)$ or x^2-4
41. $(a+5)^2(a-5)$ **43.** $8(x-1)^2$
45. $a^2b^2(x-y)$ **47.** $a^2b^7x^4y^3z$ **49.** $25(a+2)(a-2)$

6.5, p. 159

1. (a) 3 (b) $\frac{1}{3}=\frac{1}{3}, 1=\frac{3}{3}$

3. (a) 6 (b) $\frac{1}{2}=\frac{3}{6}, \frac{5}{6}=\frac{5}{6}$

5. (a) 20 (b) $\frac{5}{4}=\frac{25}{20}, \frac{7}{10}=\frac{14}{20}$

7. (a) 140 (b) $\frac{7}{20}=\frac{49}{140}, \frac{3}{28}=\frac{15}{140}$

9. (a) 108 (b) $\frac{5}{36}=\frac{15}{108}, \frac{1}{27}=\frac{4}{108}$

11. (a) xy (b) $\frac{1}{x}=\frac{y}{xy}, \frac{1}{y}=\frac{x}{xy}$

13. (a) $x+a$
(b) $\frac{2}{x+a}=\frac{2}{x+a}, -1=\frac{-(x+a)}{x+a}$

15. (a) $x^2y^4z^2$

(b) $\dfrac{x-a}{x^2y^3z^2} = \dfrac{y(x-a)}{x^2y^4z^2}, \dfrac{a}{xy^4z} = \dfrac{axz}{x^2y^4z^2}$

17. (a) $(x+1)(x-1)^2$

(b) $\dfrac{1}{x^2-1} = \dfrac{x-1}{(x+1)(x-1)^2}, \dfrac{x}{(x-1)^2} = \dfrac{x(x+1)}{(x+1)(x-1)^2}$

19. (a) $a(x+2)(x-2)$

(b) $\dfrac{1}{ax-2a} = \dfrac{x+2}{a(x+2)(x-2)}, \dfrac{a}{x^2-4} = \dfrac{a^2}{a(x+2)(x-2)},$

$\dfrac{-1}{ax+2a} = \dfrac{2-x}{a(x+2)(x-2)}$

21. $\dfrac{5}{6}$ **23.** $\dfrac{3}{4}$ **25.** $\dfrac{4}{3}$

27. $\dfrac{11}{9}$ **29.** $\dfrac{23}{24}$ **31.** $\dfrac{53}{120}$

33. $\dfrac{163}{288}$ **35.** $\dfrac{25}{18}$ **37.** $\dfrac{71}{50}$

39. $\dfrac{x+y}{xy}$ **41.** $\dfrac{a(x+1)}{x^2}$ **43.** $\dfrac{2+x^2y^2z^2}{xyz}$

45. $\dfrac{x+a+2}{(x+a)^2}$ **47.** $\dfrac{(1+a)(1-a)}{a(t+1)}$ **49.** $\dfrac{x+1}{x(x+2)}$

51. $\dfrac{10a+9}{a(a+9)(a-9)}$ **53.** $\dfrac{3x^2-12x+11}{(x-1)(x-2)(x-3)}$

55. $\dfrac{ax+4x+a}{a^3(a+4)(a-4)}$ **57.** $\dfrac{1}{x-1}$

6.6, p. 163

1. $\dfrac{1}{4}$ **3.** $\dfrac{-1}{4}$ **5.** 3

7. $\dfrac{-1}{6}$ **9.** $\dfrac{3}{4}$ **11.** 3

13. 6 **15.** 5 **17.** $\dfrac{20}{23}$

19. $\dfrac{25}{3}$ **21.** $\dfrac{a}{xy}$ **23.** $\dfrac{-2}{xy}$

25. $\dfrac{a^3}{b^3}$ **27.** $-b$ **29.** $\dfrac{x^2}{2y^3}$

31. $\dfrac{4ab^2}{3d^2}$ **33.** $\dfrac{x-1}{x-2}$ **35.** $\dfrac{1+a}{a^3}$

37. -1 **39.** $\dfrac{-(1+a-b)}{a}$

6.7, p. 171

1. .3	**3.** $-.3$	**5.** .003
7. .21	**9.** 71.3	**11.** .713
13. .25	**15.** $-.6$	**17.** .05
19. $-.075$	**21.** .666 666 . . .	**23.** .090 909 . . .
25. .833 333 3 . . .	**27.** .083 333 3 . . .	**29.** $\frac{9}{10}$
31. $\frac{119}{1000}$	**33.** $\frac{1}{50}$	**35.** $\frac{1}{500}$
37. $\frac{5}{4}$	**39.** 1.1	**41.** 1.1
43. .5	**45.** .6	**47.** .888
49. 6.35	**51.** 635	**53.** .065
55. $-.012$	**57.** .003	**59.** 2
61. 2000	**63.** .9	**65.** .0002
67. 500	**69.** $.9x$	**71.** $.91ab$
73. $.15xy$	**75.** $.0005a^3b^4$	**77.** $2x$
79. 4		

Let's Review Chapter 6, p. 173

1. $\frac{3}{8}$	**2.** $\frac{4}{45}$	**3.** ab^2
4. $\frac{2y(x+3)}{z(x-3)}$	**5.** -3	**6.** 8
7. axy	**8.** $(x+1)(x-y)$	**9.** $\frac{1}{3}$
10. $\frac{1}{4}$	**11.** 2	**12.** $\frac{1}{x+3}$
13. 30	**14.** 12	**15.** x^2y
16. $(x-1)(x+2)(x-2)$		
17. (a) 15	(b) $\frac{1}{3}=\frac{5}{15}, \frac{2}{5}=\frac{6}{15}$	
18. $\frac{5}{4}$	**19.** $\frac{3x-1}{x^2}$	**20.** $\frac{x^2-ax+a}{(x+a)(x-a)}$
21. $\frac{3}{2}$	**22.** $\frac{4}{5}$	**23.** $\frac{ab}{c}$
24. $\frac{1}{(x-1)(x+a)}$	**25.** .8	**26.** .97
27. .418	**28.** .003	

And these from Chapter 1–5, p. 174

29. $\frac{-4}{5}$	**30.** $\frac{x^2}{x-1}$	**31.** .64

32. $\frac{1}{4} < \frac{5}{16} < \frac{3}{8} < \frac{7}{16} < \frac{1}{2} < \frac{5}{8}$
33. $.07 < .69 < .7 < .71 < .77 < .9$
34. $\frac{3}{5} < .65 < \frac{2}{3} < .7 < \frac{3}{4} < .8$

Try These Exam Questions for Practice, p. 175

1. $\frac{2}{3}$	**2.** $x(x - 1)$	**3.** $\frac{11}{8}$
4. 2	**5.** $\frac{2a - 17}{20a^2x^2}$	**6.** $\frac{b^3}{a}$
7. .15	**8.** .02	

CHAPTER 7

7.1, p. 178

1. root	**3.** root	**5.** root
7. root	**9.** not a root	**11.** root
13. not a root	**15.** root	**17.** not a root
19. not a root	**21.** not a root	**23.** not a root
25. not a root	**27.** not a root	**29.** root
31. not a root		

7.2, p. 182

1. 7	**3.** 3	**5.** 14
7. 2	**9.** 3	**11.** -64
13. -1	**15.** -7	**17.** 4
19. 9	**21.** 2	**23.** $\frac{2}{3}$
25. $\frac{2}{3}$		

(The checks follow.)

27. 4	**29.** -12	**31.** 1
33. $\frac{1}{4}$	**35.** 5	**37.** 6

Checks:

27. $3 \cdot 4 \overset{?}{=} 12$
$12 \overset{\checkmark}{=} 12$

29. $\dfrac{-12}{6} \overset{?}{=} -2$
$-2 \overset{\checkmark}{=} -2$

31. $2(1) + 1 \overset{?}{=} 3$
$3 \overset{\checkmark}{=} 3$

33. $1 + 4\left(\dfrac{1}{4}\right) \overset{?}{=} 2$
$1 + 1 \overset{?}{=} 2$
$2 \overset{\checkmark}{=} 2$

35. $6(5) - 3 \overset{?}{=} 5(5) + 2$
$27 \overset{\checkmark}{=} 27$

37. $9 - 3(6) \overset{?}{=} -2(6) + 3$
$-9 \overset{\checkmark}{=} -9$

7.3, p. 184

1. 4 **3.** 6 **5.** 1

7. -2 **9.** 1 **11.** $\dfrac{5}{3}$

13. $\dfrac{9}{2}$ **15.** $\dfrac{-9}{2}$ **17.** 1

19. $\dfrac{7}{3}$ **21.** 8 **23.** 1

25. -2 **27.** 1 **29.** $\dfrac{-4}{3}$

(The checks follow.)

31. 4 **33.** 1 **35.** $\dfrac{-1}{4}$

37. -2 **39.** 2

Checks:

31. $4 - [12 - 2(4)] \overset{?}{=} 0$
$4 - 4 \overset{?}{=} 0$
$0 \overset{\checkmark}{=} 0$

33. $-3(1 + 2) \overset{?}{=} 1 - 10$
$-3\,(3) \overset{?}{=} -9$
$-9 \overset{\checkmark}{=} -9$

35. $1 - \left(\dfrac{-1}{4} - \left[2 - 3\left(\dfrac{-1}{4}\right)\right]\right) \overset{?}{=} 4$
$1 + \dfrac{1}{4} + 2 + \dfrac{3}{4} \overset{?}{=} 4$
$\dfrac{16}{4} \overset{?}{=} 4$
$4 \overset{\checkmark}{=} 4$

37. $(-2 + 4)^2 \overset{?}{=} (-2)^2$
$2^2 \overset{?}{=} 4$
$4 \overset{\checkmark}{=} 4$

39. $(2 - 2)^2 + 5 \overset{?}{=} 2^2 + 1$
$0 + 5 \overset{?}{=} 4 + 1$
$5 \overset{\checkmark}{=} 5$

7.4, p. 190

1. 6 **3.** 2 **5.** 15
7. -6 **9.** \$8.10 **11.** 18
13. the 8-ounce box **15.** 4 **17.** 20
19. 7 **21.** -10 **23.** 10
25. 10 **27.** 5 **29.** 9
31. 2 **33.** 20 **35.** 2
37. 0 **39.** -3 **41.** 7
43. -3 **45.** $\frac{6}{5}$ **47.** 1
49. -16

(The checks follow.)

51. 5 **53.** 18 **55.** $\frac{-8}{7}$
57. 20 **59.** 20 **61.** 200
63. -200

Checks:

51.
$$\frac{2(5)}{5} \stackrel{?}{=} \frac{5+1}{3}$$
$$2 \stackrel{?}{=} \frac{6}{3}$$
$$2 \stackrel{\checkmark}{=} 2$$

53.
$$\frac{18}{18-9} \stackrel{?}{=} 2$$
$$\frac{18}{9} \stackrel{?}{=} 2$$
$$2 \stackrel{\checkmark}{=} 2$$

55.
$$\frac{2}{\frac{-8}{7}+2} \stackrel{?}{=} \frac{5}{1-\left(\frac{-8}{7}\right)}$$
[Cross-multiply.]
$$2\left(1+\frac{8}{7}\right) \stackrel{?}{=} 5\left(\frac{-8}{7}+2\right)$$
$$\frac{2(7+8)}{7} \stackrel{?}{=} \frac{5(-8+14)}{7}$$
$$2(15) \stackrel{?}{=} 5(6)$$
$$30 \stackrel{\checkmark}{=} 30$$

7.5, p. 195

1. $\frac{y}{5}$ **3.** $y-2$ **5.** $-y-4$
7. $\frac{-7t}{5}$ **9.** $1-y-z$ **11.** $\frac{-2z}{3}$
13. $\frac{5-2x}{3}$ **15.** $\frac{2u-6}{3}$ **17.** $\frac{12-b+3c}{2}$
19. $\frac{3}{x-1}$ **21.** (a) $\frac{y+20}{5}$ (b) $5x-20$
23. (a) $\frac{4c-1}{10}$ (b) $\frac{10b+1}{4}$

25. (a) $\dfrac{11z - 1}{3}$ (b) $\dfrac{3y + 1}{11}$

27. (a) $\dfrac{1 - 4y}{y}$ (b) $\dfrac{1}{x + 4}$

29. (a) $\dfrac{10 + y - 2z}{5}$ (b) $5x + 2z - 10$

31. (a) b (b) a

33. $\dfrac{y + 10}{2}$ **35.** $\dfrac{6}{u}$ **37.** $x + 2z - 8$

Checks:

33.
$$2\left(\frac{y + 10}{2}\right) - y \stackrel{?}{=} 10$$
$$y + 10 - y \stackrel{?}{=} 10$$
$$10 \stackrel{\checkmark}{=} 10$$

35.
$$\frac{3}{u} \stackrel{?}{=} \frac{\frac{6}{u}}{2}$$
$$\frac{3}{u} \stackrel{?}{=} \frac{6}{u} \cdot \frac{1}{2}$$
$$\frac{3}{u} \stackrel{\checkmark}{=} \frac{3}{u}$$

37.
$$x - (x + 2z - 8) + 2z \stackrel{?}{=} 8$$
$$x - x - 2z + 8 + 2z \stackrel{?}{=} 8$$
$$8 \stackrel{\checkmark}{=} 8$$

39. $\dfrac{A}{w}$ **41.** (a) $\dfrac{2A}{b}$ (b) 50 inches

43. (a) $\dfrac{d}{t}$ (b) 60 miles per hour

Let's Review Chapter 7, p. 198

(The checks follow.)

1.	root	**2.**	not a root	**3.**	not a root
4.	root	**5.**	-3	**6.**	12
7.	$\dfrac{1}{2}$	**8.**	2	**9.**	12
10.	6	**11.**	-3	**12.**	5
13.	9	**14.**	14	**15.**	4
16.	300	**17.**	$\dfrac{y + 3}{2}$	**18.**	$\dfrac{6 + 2x}{7}$
19.	(a) $\dfrac{-2b}{3}$		(b) $\dfrac{-3a}{2}$	**20.**	$\dfrac{7u}{3}$

Checks:

8.
$$10 - 4(2) \stackrel{?}{=} 3(2) - 4$$
$$2 \stackrel{\checkmark}{=} 2$$

12.
$$2 \cdot 5 - [7 - (5 - 2)] \stackrel{?}{=} 6$$
$$10 - [7 - 3] \stackrel{?}{=} 6$$
$$10 - 4 \stackrel{?}{=} 6$$
$$6 \stackrel{\checkmark}{=} 6$$

20. (b) $7u - 3\left(\frac{7u}{3}\right) \stackrel{?}{=} 0$

$7u - 7u \stackrel{?}{=} 0$

$0 \stackrel{\checkmark}{=} 0$

And these from Chapters 1–6, p. 199

21. 22 **22.** $5x^2 - 12x + 7$

23. -8 **24.** $\frac{1}{2(x + 2)}$ **25.** $(t - 5)(t - 2)$

26. $.89 < .899 < .9 < .909 < .91 < .98$

Try These Exam Questions for Practice, p. 199

1. root **2.** 1 **3.** 2 Check: $2(2) + 5 \stackrel{?}{=} 5(2) - 1$

$9 \stackrel{\checkmark}{=} 9$

4. $\frac{1}{2}$ **5.** 12 **6.** .4

7. $\frac{33 - x}{2}$

CHAPTER 8

8.1, p. 204

1. (a) $x + 2$ (b) $x - 6$ (c) $3x$ (d) $\frac{x}{3}$

3. (a) $x + 9$ (b) $x - 2$ (c) $\frac{x}{2}$ (d) $x + 1$

5. $x - 19$ **7.** $2x - 5$ **9.** 13

11. 32 **13.** 15 and 16 **15.** 20 and 22

17. 7 Check: $2(7) + 1 \stackrel{?}{=} 15$

$15 \stackrel{\checkmark}{=} 15$

19. 11 Check: $4(11) + 7 \stackrel{?}{=} 51$

$51 \stackrel{\checkmark}{=} 51$

21. 22 and 28 Check: $22 + 28 \stackrel{?}{=} 50$ and $28 \stackrel{?}{=} 22 + 6$

$50 \stackrel{\checkmark}{=} 50$ $28 \stackrel{\checkmark}{=} 28$

23. $25

8.2, p. 208

1. $x + 6$ **3.** (a) $x + 5$ (b) $\frac{x}{2}$

5. (a) $y + 3$ (b) $y + 10$ **7.** 13

9. 25 **11.** 28 Check: $28 - 3 \stackrel{?}{=} 5(8 - 3)$

$25 \stackrel{?}{=} 5 \cdot 5$

$25 \stackrel{\checkmark}{=} 25$

13. 25 Check: $25 - 15 \overset{?}{=} 2[(25 - 5) - 15]$
$10 \overset{?}{=} 2[5]$
$10 \overset{\checkmark}{=} 10$

15. 17 Check: $31 + 17 \overset{?}{=} 2(7 + 17)$
$48 \overset{?}{=} 2(24)$
$48 \overset{\checkmark}{=} 48$

8.3, p. 215

1. 90 miles **3.** 180 miles **5.** 300 miles per hour

7. (a) 25 miles (b) 75 miles

9. 690 miles. Check: $(50 + 65)6 \overset{?}{=} 690$
$115 \cdot 6 \overset{?}{=} 690$
$690 \overset{\checkmark}{=} 690$

11. A mile and a half. Check: $(8 - 6)\left(\frac{3}{4}\right) \overset{?}{=} \frac{3}{2}$
$2\left(\frac{3}{4}\right) \overset{?}{=} \frac{3}{2}$
$\frac{3}{2} \overset{\checkmark}{=} \frac{3}{2}$

13. 18 miles. Check: $\frac{18}{6} + \frac{18}{9} \overset{?}{=} 5$
$3 + 2 \overset{?}{=} 5$
$5 \overset{\checkmark}{=} 5$

8.4, p. 219

1. .17 **3.** .42 **5.** .105
7. 2.5 **9.** 600 **11.** 960
13. 325 **15.** $42

(The checks follow.)

17. 6% **19.** $1200 **21.** $1000 at $7\frac{1}{2}$%
23. $25.63 **25.** $3500 **27.** $300 at 7% and $200 at 9%

Checks:

17. $.06(750) \overset{?}{=} 45$
$6(750) \overset{?}{=} 4500$
$4500 \overset{\checkmark}{=} 4500$

19. $.09(1200) \overset{?}{=} 108$
$9(1200) \overset{?}{=} 10\,800$
$10\,800 \overset{\checkmark}{=} 10\,800$

25. $.10(3500) + .07(4500 - 3500) \overset{?}{=} 420$
$10(3500) + 7(1000) \overset{?}{=} 42\,000$
$42\,000 \overset{\checkmark}{=} 42\,000$

8.5, p. 224

1. \$198 **3.** 8% Check: $.08(5000) \stackrel{?}{=} 400$

$400 \stackrel{\checkmark}{=} 400$

5. \$2500 **7.** (a) \$1680 (b) \$40320

9. \$440 **11.** (a) \$12 (b) \$14.40

13. \$125 **15.** (a) \$80 (b) \$320

17. \$15.40 Check: $\frac{20 - 14}{20} \stackrel{?}{=} \frac{22 - 15.40}{22}$

$\frac{6}{20} \stackrel{?}{=} \frac{6.60}{22}$

$\frac{3}{10} \stackrel{?}{=} \frac{660}{2200}$

$6600 \stackrel{\checkmark}{=} 6600$

19. (a) \$2.50 (b) \$12.50

Let's Review Chapter 8, p. 227

1. 7 **2.** 28, 30, 32 Check: $28 + 30 + 32 \stackrel{?}{=} 90$

$90 \stackrel{\checkmark}{=} 90$

3. 22 **4.** 50 **5.** $3\frac{1}{2}$ hours

6. 2 hours **7.** \$58.50 **8.** \$1500 at 8% and \$1000 at 9%

9. \$2000 **10.** (a) \$40 (b) \$160

Now try these unclassified problems, p. 228

11. 8 feet and 4 feet **12.** 20

13. $22\frac{1}{2}$ minutes **14.** 1215

15. 14 inches by 11 inches **16.** \$8.56

And these from Chapters 1–7, p. 228

17. \$2 **18.** $3x^3 - 4x^2 - 6x + 8$ **19.** .0051

20. .006 **21.** 0 **22.** $\frac{2x + 3z - 1}{8}$

Try These Exam Questions For Practice, p. 228

1. 5 **2.** 42 **3.** 15 miles **4.** \$82

CHAPTER 9

9.1, p. 234

1.

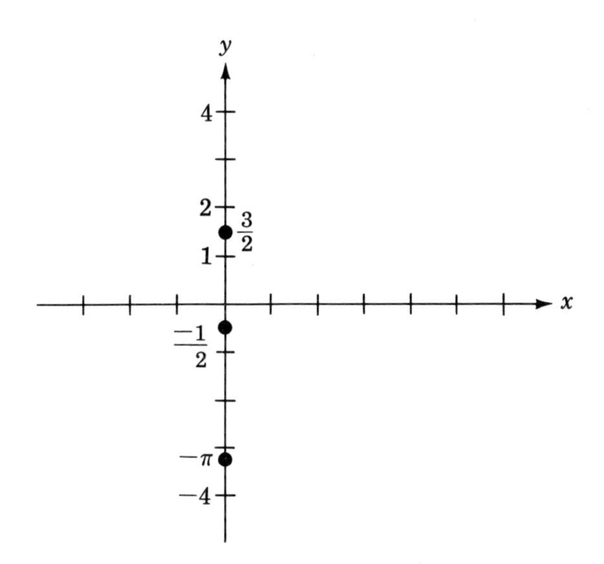

FIGURE 9A

3.				
	P:	(a) 5	(b) 1	(c) $P = (5, 1)$
	Q:	(a) 2	(b) 8	(c) $Q = (2, 8)$
	R:	(a) -3	(b) 2	(c) $R = (-3, 2)$
	S:	(a) -3	(b) -2	(c) $S = (-3, -2)$
	T:	(a) 3	(b) -2	(c) $T = (3, -2)$

5–27.

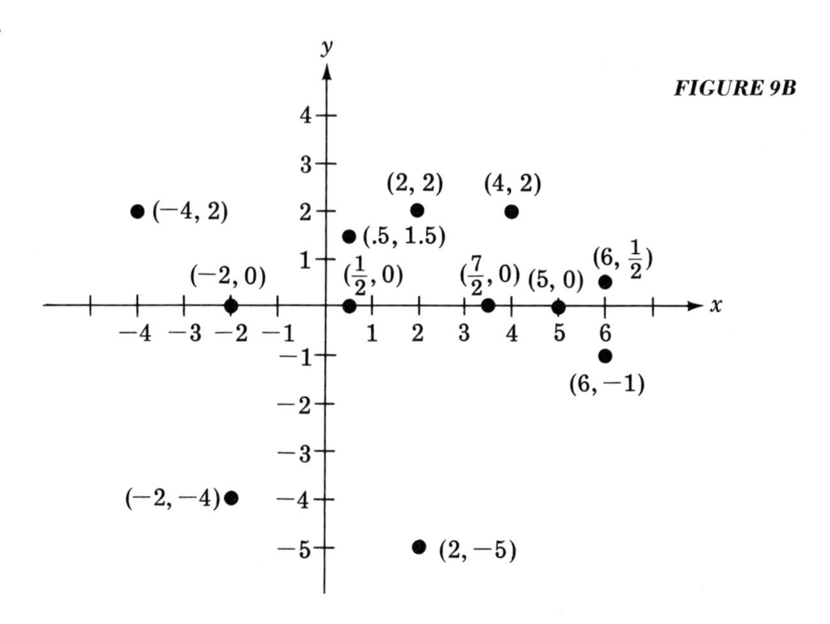

FIGURE 9B

9.2, p. 243

	(a)	(b)	(c)	(d)
1.	6	10	4	0
3.	3	15	-15	0
5.	1	5	-2	$\frac{1}{2}$
7.	0	-100	-1000	20 000
9.	3	-1	11	-9
11.	1	$\frac{-3}{2}$	-2	-4
13.	0	0	0	0
15.	1	1	-1	-1
17.	1	25	25	100
19.	1	10	5	101
21.	1	$\frac{1}{3}$	$\frac{1}{10}$	$\frac{-1}{2}$
23.	-1	1	$\frac{-1}{2}$	$\frac{1}{4}$
25.	2	4	$\frac{4}{3}$	-2

27. (a) and (c)

	(a)	(b)	(c)	(d)
29.	2	0	2	3
31.	$\frac{1}{2}$	$\frac{-1}{2}$	$\frac{3}{2}$	$\frac{7}{2}$

33. (a) $f(t) = 45t$ (b) 135 miles $[f(3) = 135]$
(c) 195 miles $\left[f\left(\frac{13}{3}\right) = 195\right]$

35. (a) $F(r) = 2\pi r$ (b) $F(8) = 16\pi$

37. (a) $g(x) = x + 24$ (b) $g(24) = 48$
(c) Let $g(x) = 60$.
Then $x + 24 = 60$
$x = 36$

9.3, p. 256

1–31. See Figures 9C–9R, pages A-31–A-33.

Let's Review Chapter 9, p. 258

1.

	(a)	(b)	(c)
P:	1	4	$P = (1, 4)$
Q:	-2	3	$Q = (-2, 3)$
R:	-4	-2	$R = (-4, -2)$
S:	3	-3	$S = (3, -3)$
T:	4	0	$T = (4, 0)$

1.

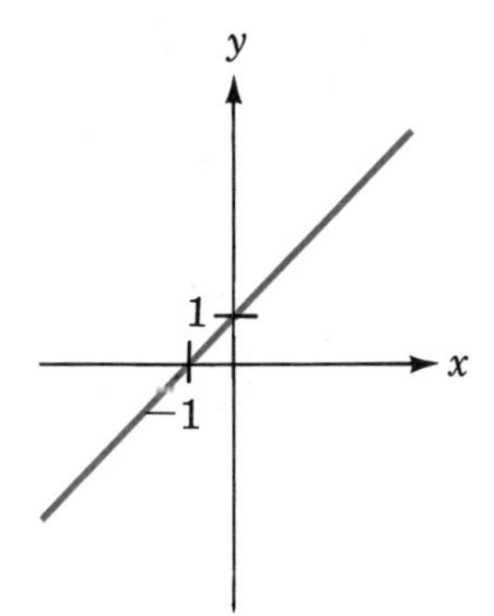

FIGURE 9C. The graph of $y = x + 1$

3.

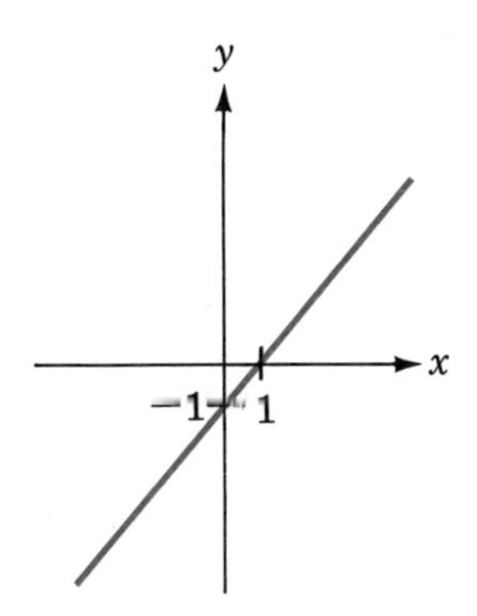

FIGURE 9D. The graph of $y = x - 1$

5.

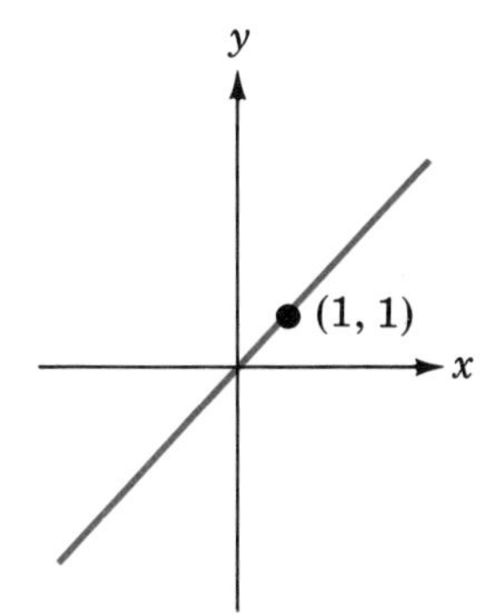

FIGURE 9E. The graph of $y = x$

7.

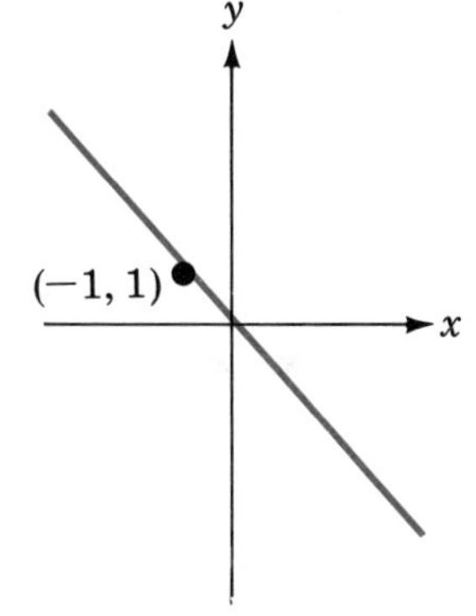

FIGURE 9F. The graph of $y = -x$

9.

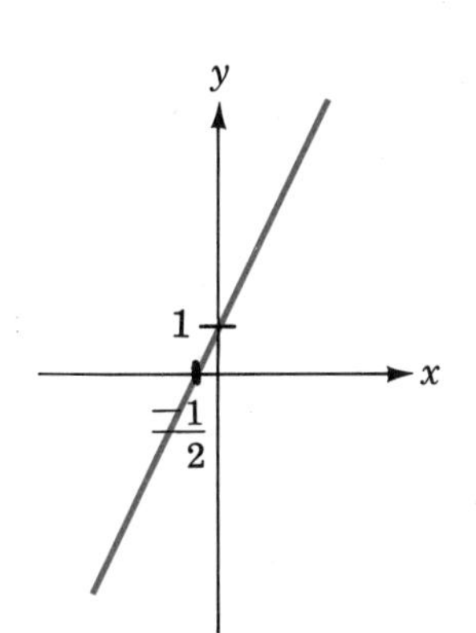

FIGURE 9G. The graph of $y = 2x + 1$

11.

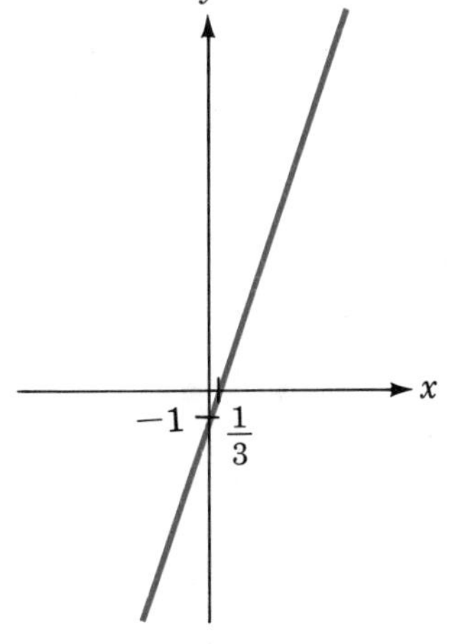

FIGURE 9H. The graph of $y = 3x - 1$

13.

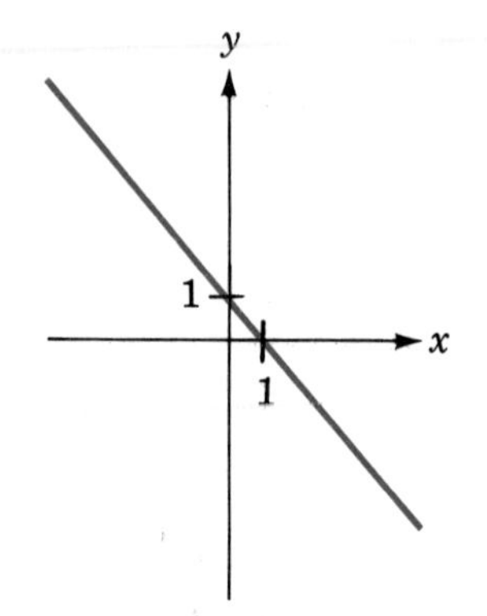

FIGURE 9I. The graph of $y = 1 - x$

15.

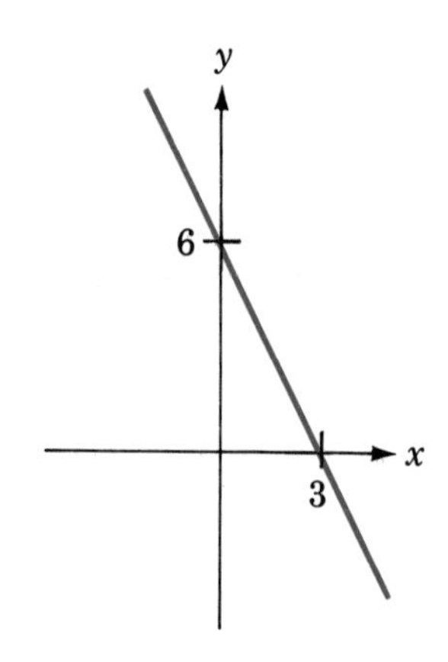

FIGURE 9J. The graph of $y = 6 - 2x$

17.

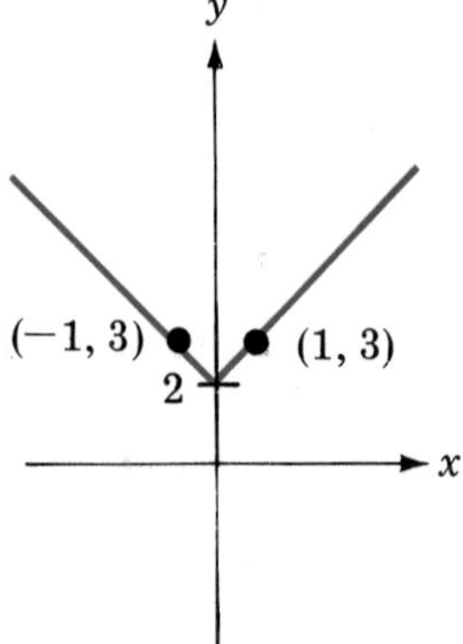

FIGURE 9K. The graph of $f(x) = |x| + 2$

19.

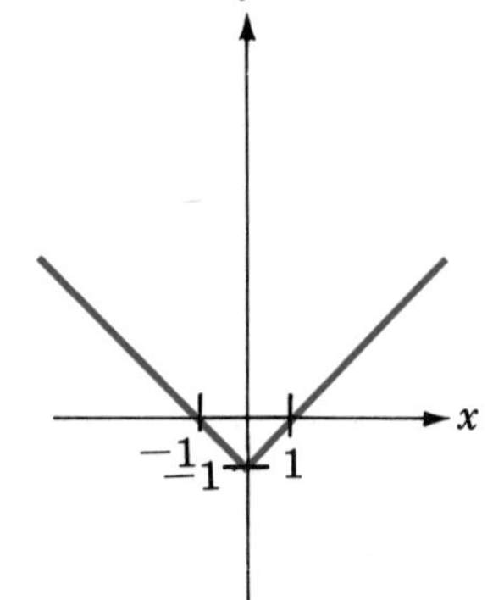

FIGURE 9L. The graph of $F(x) = |x| - 1$

21.

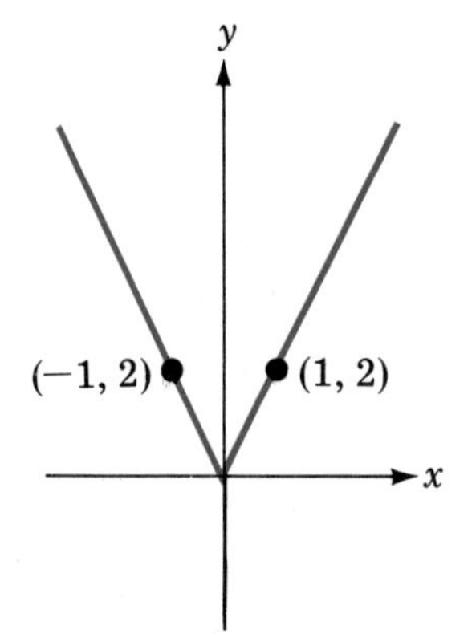

FIGURE 9M. The graph of $f(x) = 2|x|$

23.

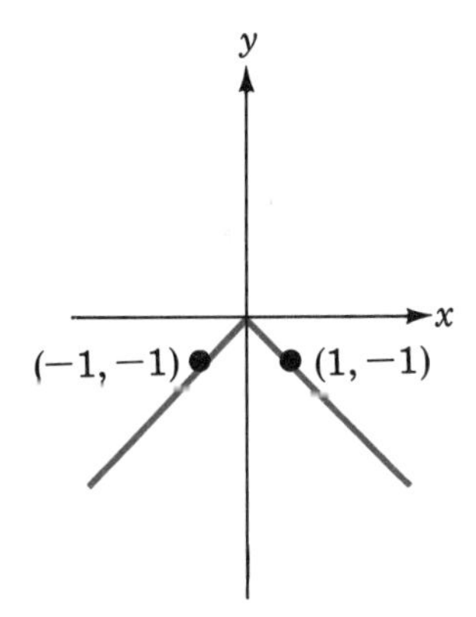

FIGURE 9N. The graph of $F(x) = -|x|$

25.

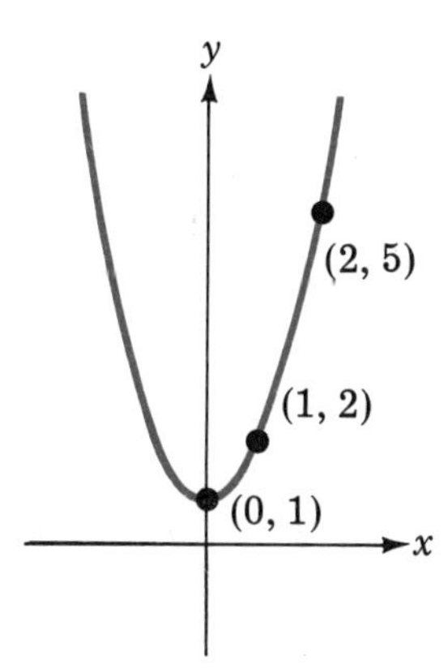

FIGURE 9O. The graph of $f(x) = x^2 + 1$

27.

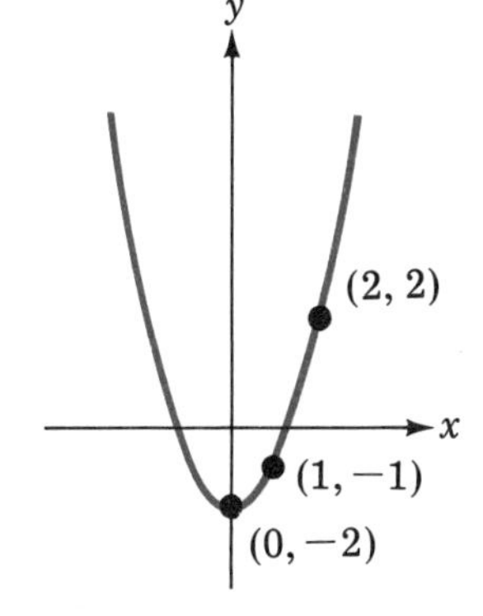

FIGURE 9P. The graph of $F(x) = x^2 - 2$

29.

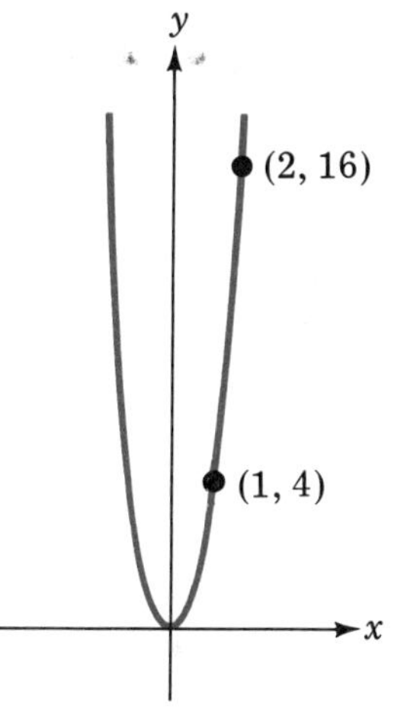

FIGURE 9Q. The graph of $f(x) = 4x^2$

31.

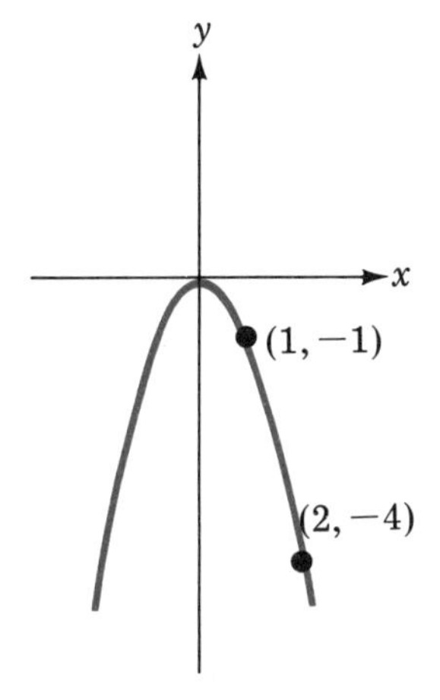

FIGURE 9R. The graph of $F(x) = -x^2$

2–6.

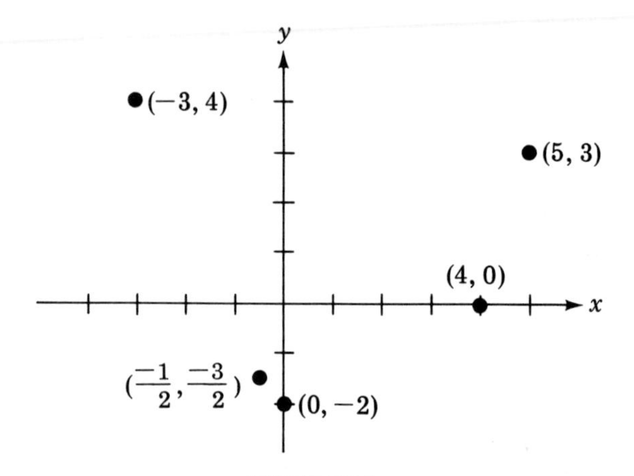

FIGURE 9S

7.	(a) 3	(b) 6	(c) $\frac{7}{2}$	(d) 0
8.	(a) -1	(b) 0	(c) 1	(d) $\frac{1}{2}$
9.	(a) 6	(b) -9	(c) -2	(d) -5
10.	(a) 6	(b) 6	(c) 6	(d) 6
11.	(a) 0	(b) 2	(c) -2	

12. (a) $f(t) = 60t$ (b) 45 miles $\left[f\left(\frac{3}{4}\right) = 60 \cdot \frac{3}{4} = 45\right]$

(c) 90 miles $\left[f\left(\frac{3}{2}\right) = 60 \cdot \frac{3}{2} = 90\right]$

13.

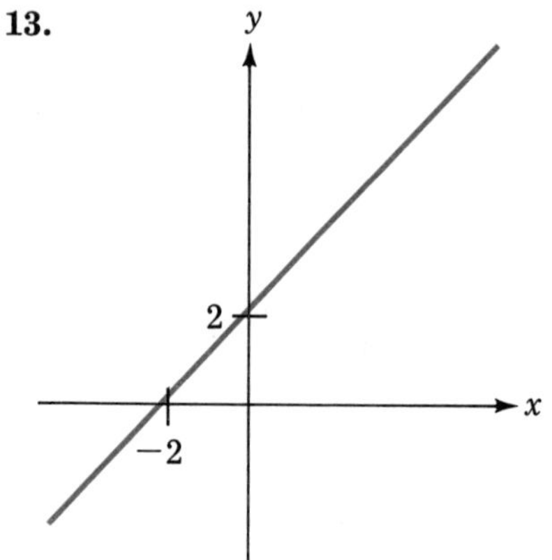

FIGURE 9T. The graph of $y = x + 2$

14.

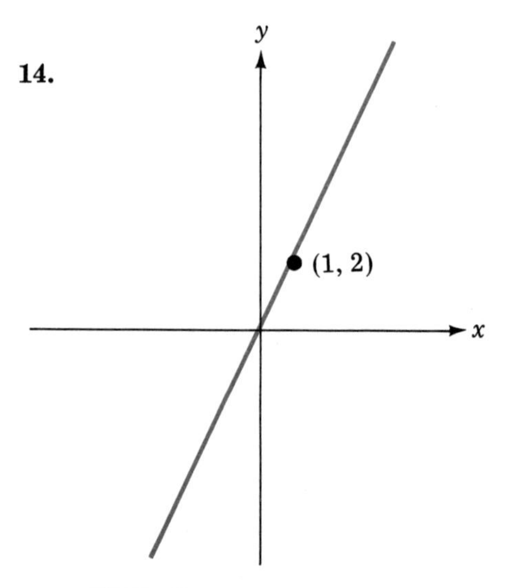

FIGURE 9U. The graph of $y = 2x$

15.

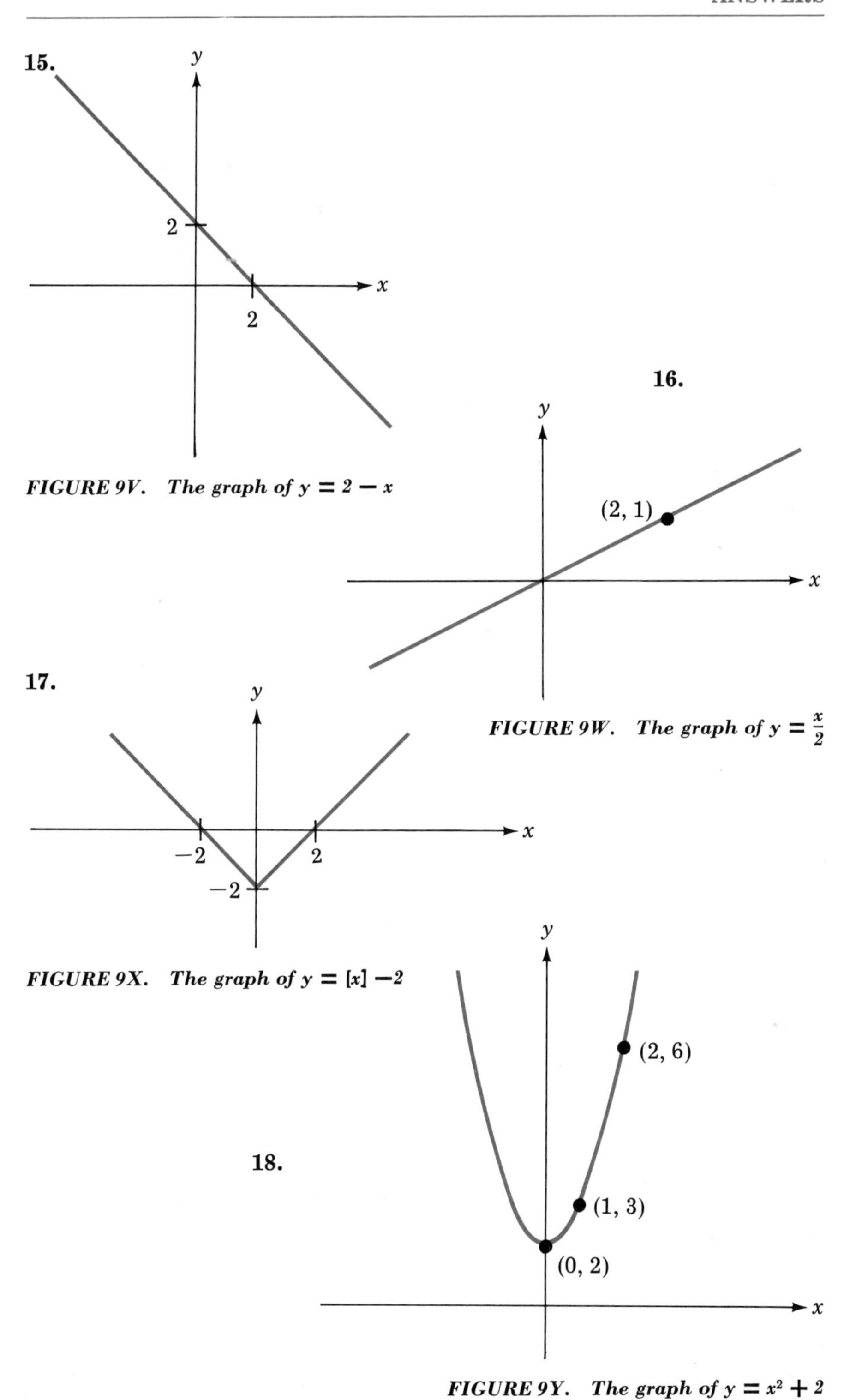

FIGURE 9V. ***The graph of*** $y = 2 - x$

FIGURE 9W. ***The graph of*** $y = \frac{x}{2}$

FIGURE 9X. ***The graph of*** $y = [x] - 2$

FIGURE 9Y. ***The graph of*** $y = x^2 + 2$

And these from Chapters 1–8, p. 259

19. (a) 8 (b) -8 (c) -8

20. (a) 125 (b) 200 (c) 500

21. -24

22. Let x represent the wife's age and $f(x)$ the husband's age. Then $f(x) = x + 2$

23. (a) $<$ (b) $=$ (c) $=$ (d) $=$

24. (a) $<$ (b) $<$ (c) $=$ (d) $>$

Try These Exam Questions for Practice, p. 260

1. P: (a) 3 (b) 2 (c) $P = (3, 2)$
Q: (a) 3 (b) -2 (c) $Q = (3, -2)$
R: (a) -1 (b) -4 (c) $R = (-1, -4)$
S: (a) 0 (b) 5 (c) $R = (0, 5)$
T: (a) -5 (b) 1 (c) $R = (-5, 1)$

2. (a) 5 (b) 1 (c) -7 (d) $\frac{1}{2}$

3. (a) $\frac{1}{4}$ (b) $\frac{1}{2}$ (c) -3

4. Let n be the number of visits, where n is a positive integer. Let $f(n)$ be the patient's fee for n visits. Then $f(n) = 25 + 20(n - 1)$

5.

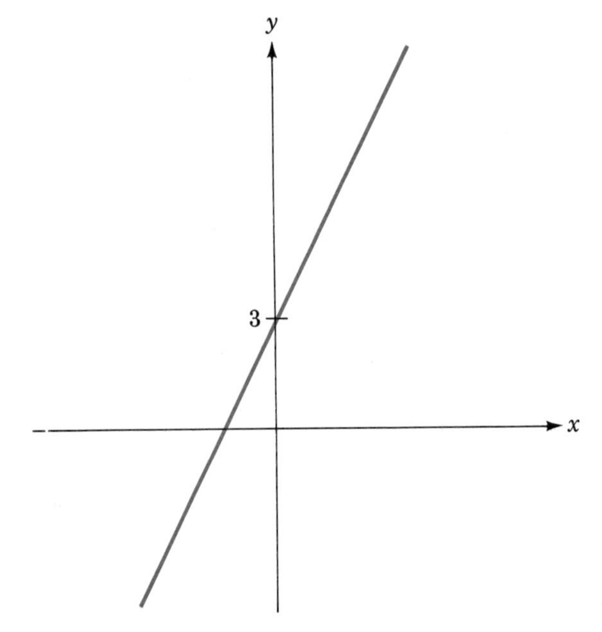

FIGURE 9Z. The graph of $y = 2x + 3$

CHAPTER 10

10.1, p. 272

1. (a) 1 (b) 1 (c) 1

3. (a) 2 (b) 1 (c) 2

5. (a) 1 (b) 2 (c) $\frac{1}{2}$

7. (a) -2 (b) 1 (c) -2

9. (a) -1 (b) -3 (c) $\frac{1}{3}$

11. (a) 2 (b) -4 (c) $\frac{-1}{2}$

13. (a) 1 (b) 4 (c) $\frac{1}{4}$

15. (a) $\frac{3}{2}$ (b) $\frac{1}{2}$ (c) 3

17. (a) 0 (b) 2 (c) 0

19. (a) -2 (b) 14 (c) $\frac{-1}{7}$

21.

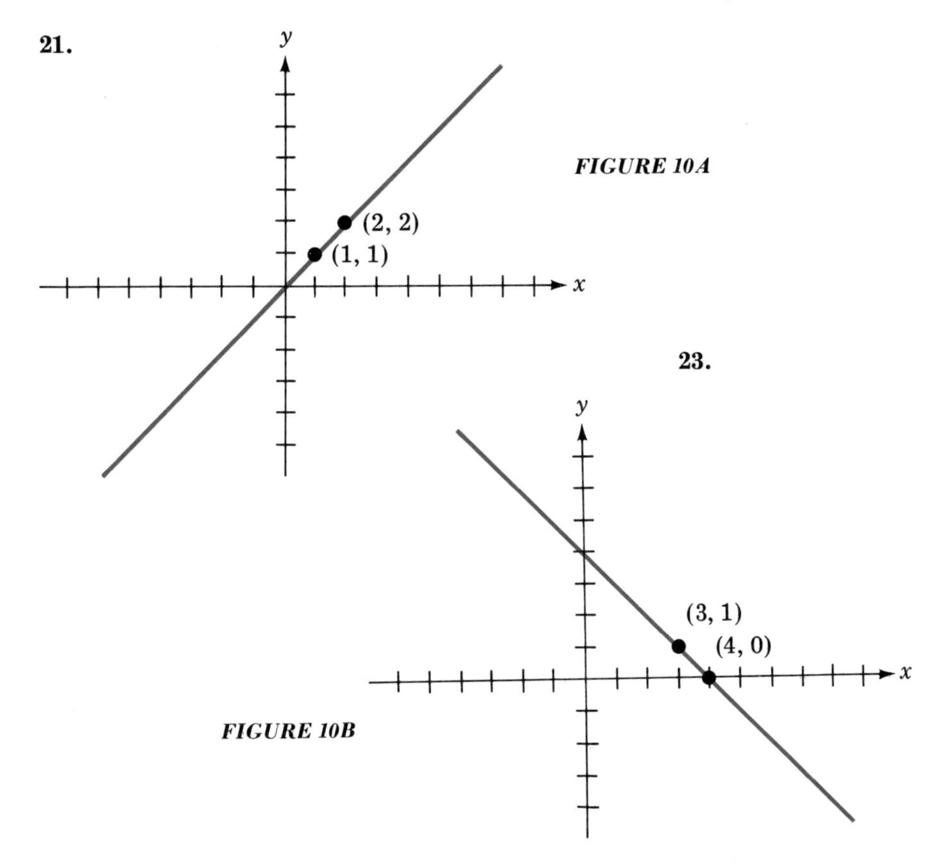

FIGURE 10A

FIGURE 10B

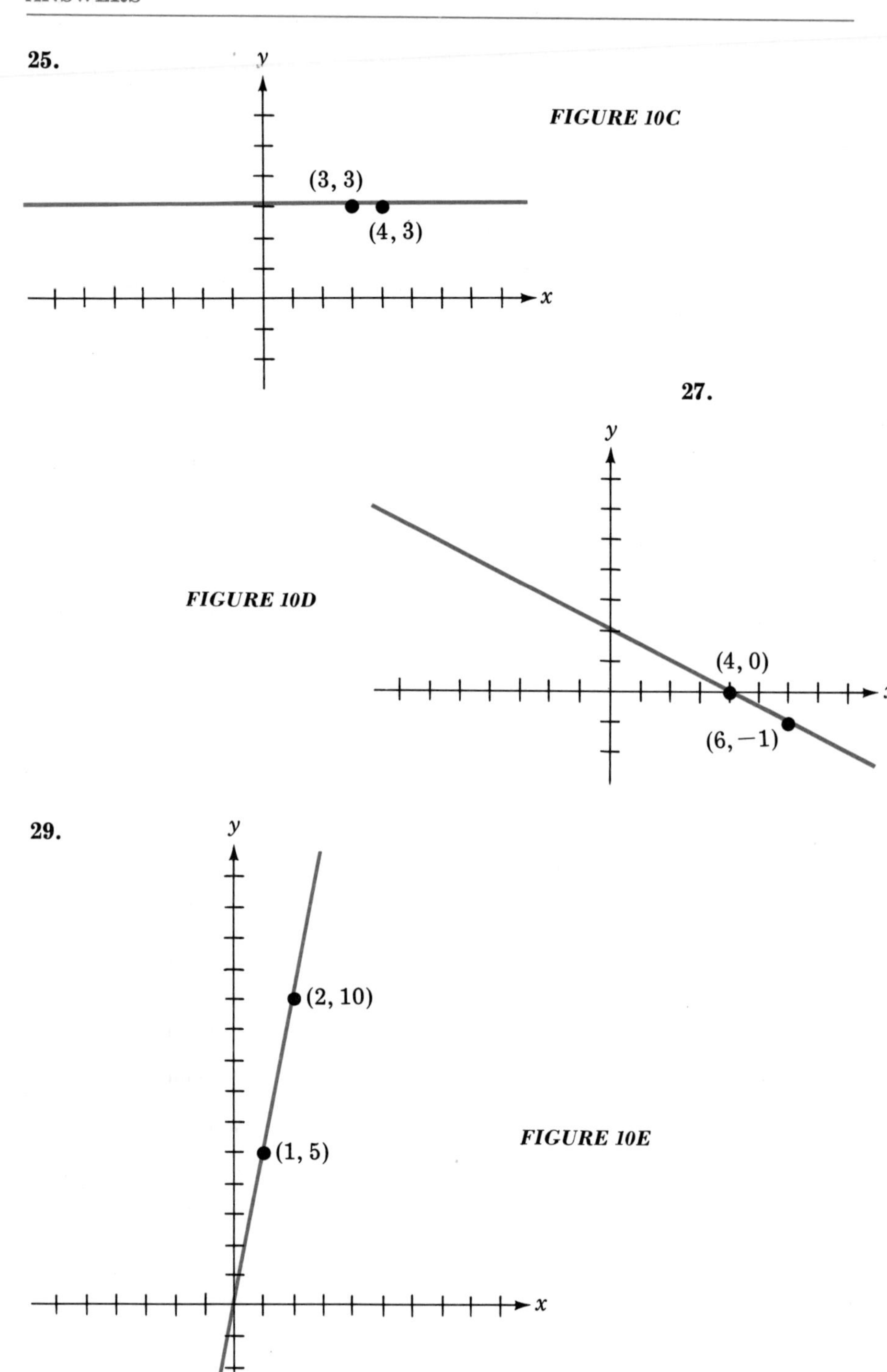

FIGURE 10C

FIGURE 10D

FIGURE 10E

31. 9 **33.** $\frac{3}{2}$ **35.** -2 **37.** 0
39. 12 **41.** 0 **43.** (*iii*) **45.** (*i*)
47. (*i*) **49.** (*iv*)

10.2, p. 282

1. (a) 2 (b) 4
3. (a) $\frac{1}{2}$ (b) 5
5. (a) 1 (b) 2
7. (a) $\frac{1}{2}$ (b) 1
9. (a) 3 (b) 3
11. (a) 3 (b) -2
13. (a) $\frac{1}{2}$ (b) 4
15. (a) 2 (b) 10
17. $y = 4x + 1$
19. $y = -x + 5$
21. $y = \frac{1}{2}x + \frac{1}{4}$
23. $y = -2x + \frac{1}{2}$
25. (a) 2 (b) 1 (c) (See Figure 10F.)
27. (a) -2 (b) 1 (c) (See Figure 10G.)
29. (a) $\frac{1}{2}$ (b) $\frac{1}{2}$ (c) (See Figure 10H.)
31. $y = -x + 3$
33. $y = \frac{-1}{2}x + 1$
35. $y = 3$
37. $y = 2$
39. $x = 2$
41. $y = 2x$
43. (a) $y = \frac{-3}{2}x + 3$ (b) (See Figure 10I.)
45. (a) $y = -3$ (b) (See Figure 10J.)
47. (a) $x = 4$ (b) (See Figure 10K.)

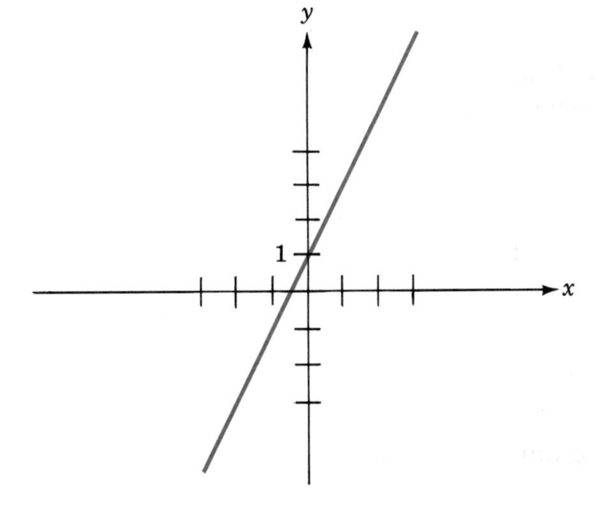

FIGURE 10F

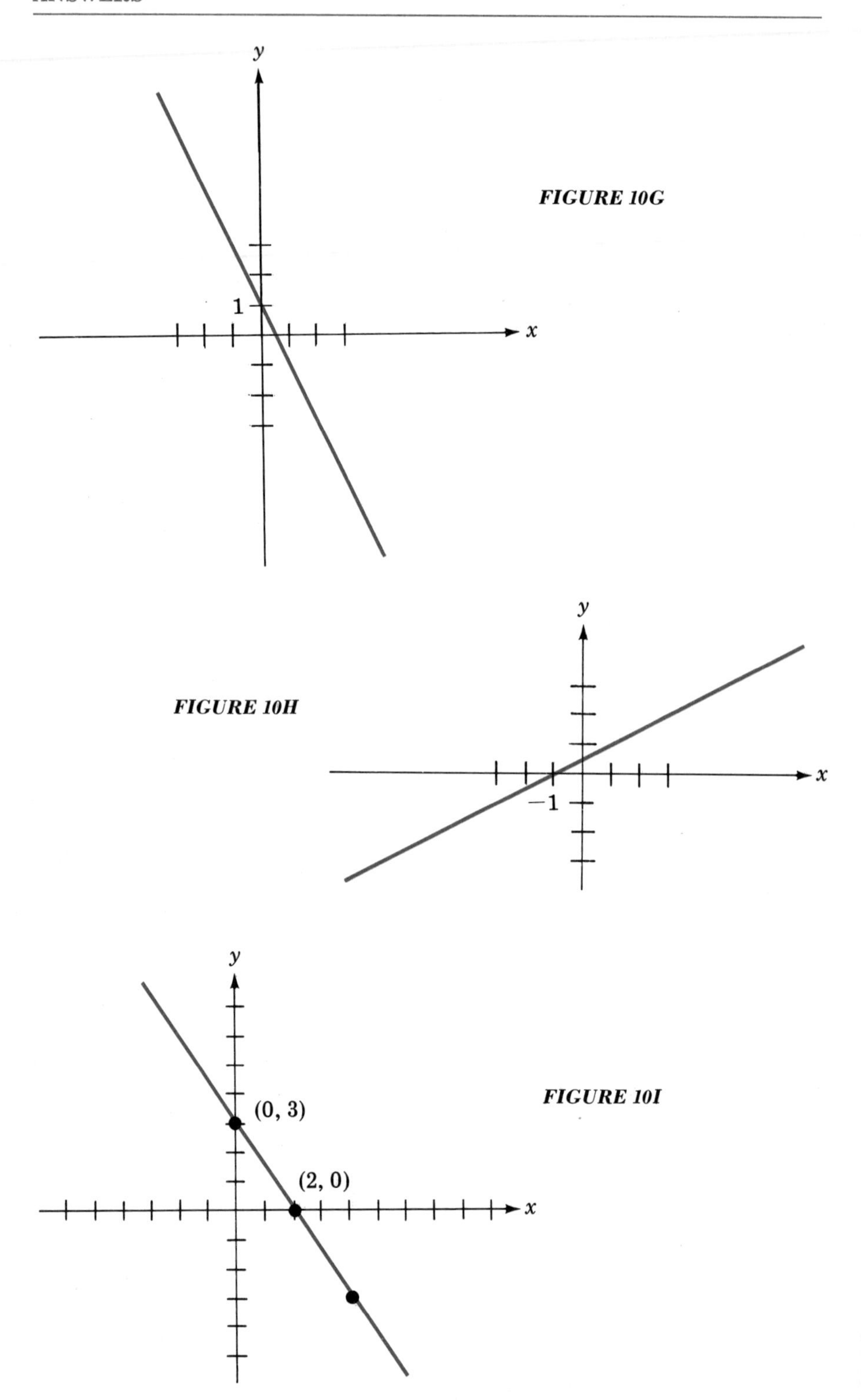

FIGURE 10G

FIGURE 10H

FIGURE 10I

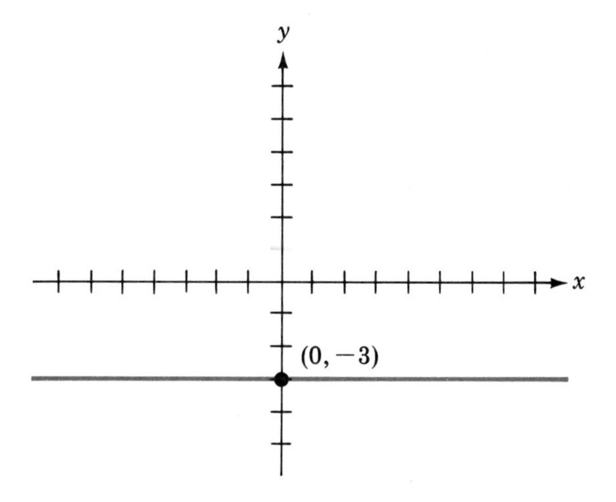

FIGURE 10J

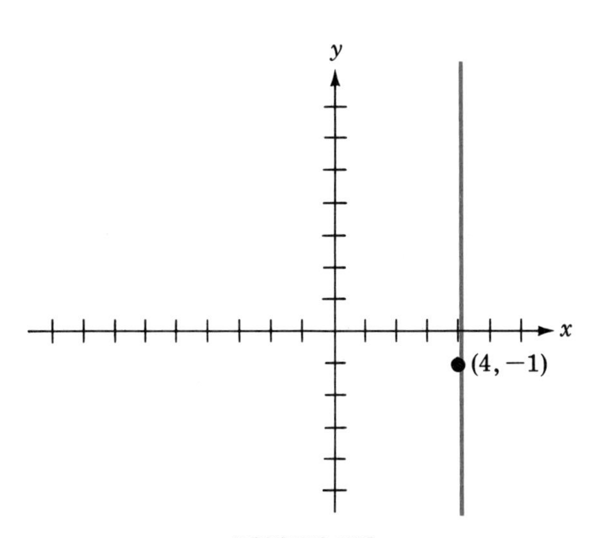

FIGURE 10K

10.3, p. 289

1. (See Figure 10L.) Check: L_1: $2 \overset{\checkmark}{=} 2$ L_2: $2 \overset{?}{=} 4 - 2$, $2 \overset{\checkmark}{=} 2$

3. (See Figure 10M.) Check: L_1: $4 - 4 \overset{?}{=} 0$, $0 \overset{\checkmark}{=} 0$ L_2: $4 - 2 \overset{?}{=} 2(0 + 1)$, $2 \overset{\checkmark}{=} 2$

5–27. See Figures 10N–10Y, pages A-42–A-46.

29. L_1 and L_4

31. L_1 and L_3

1.

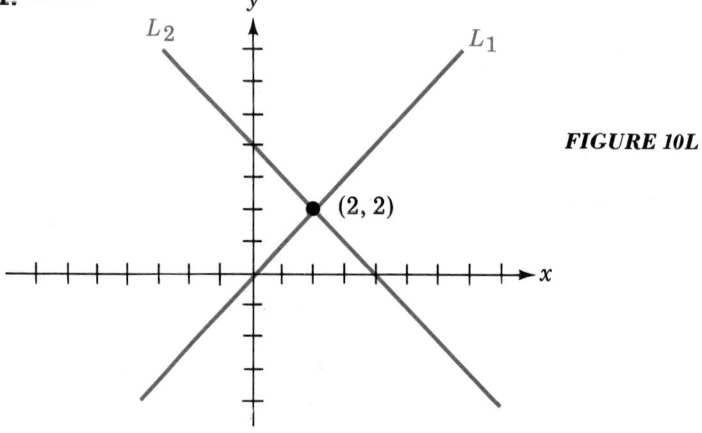

FIGURE 10L

3.

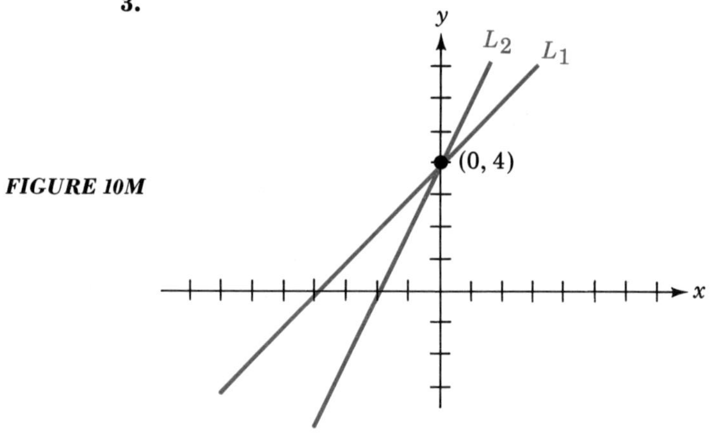

FIGURE 10M

5.

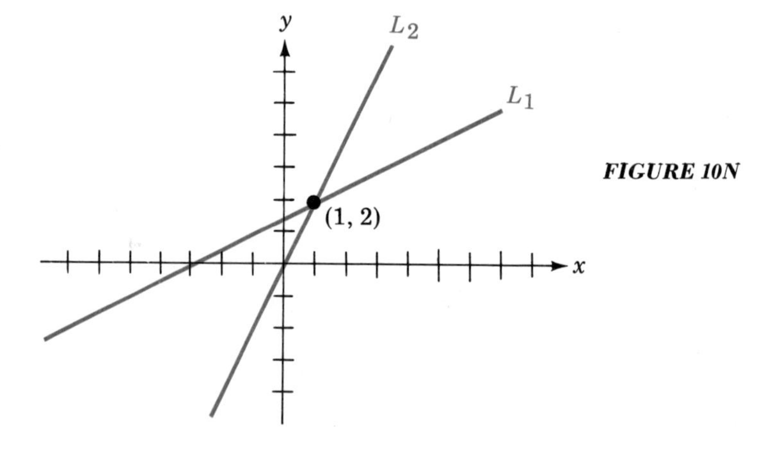

FIGURE 10N

7.

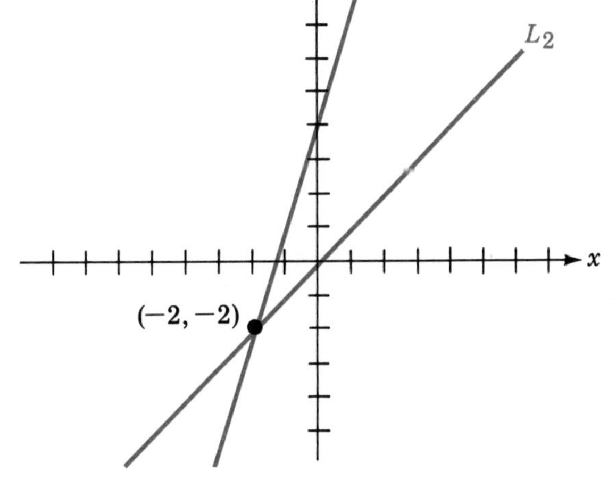

FIGURE 10O

9.

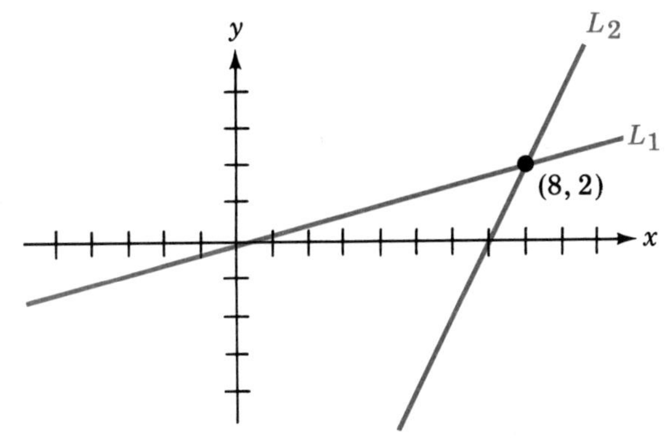

FIGURE 10P

11.

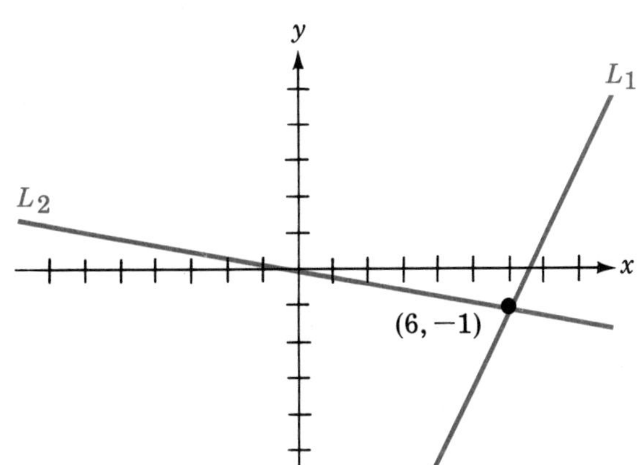

FIGURE 10Q

13.

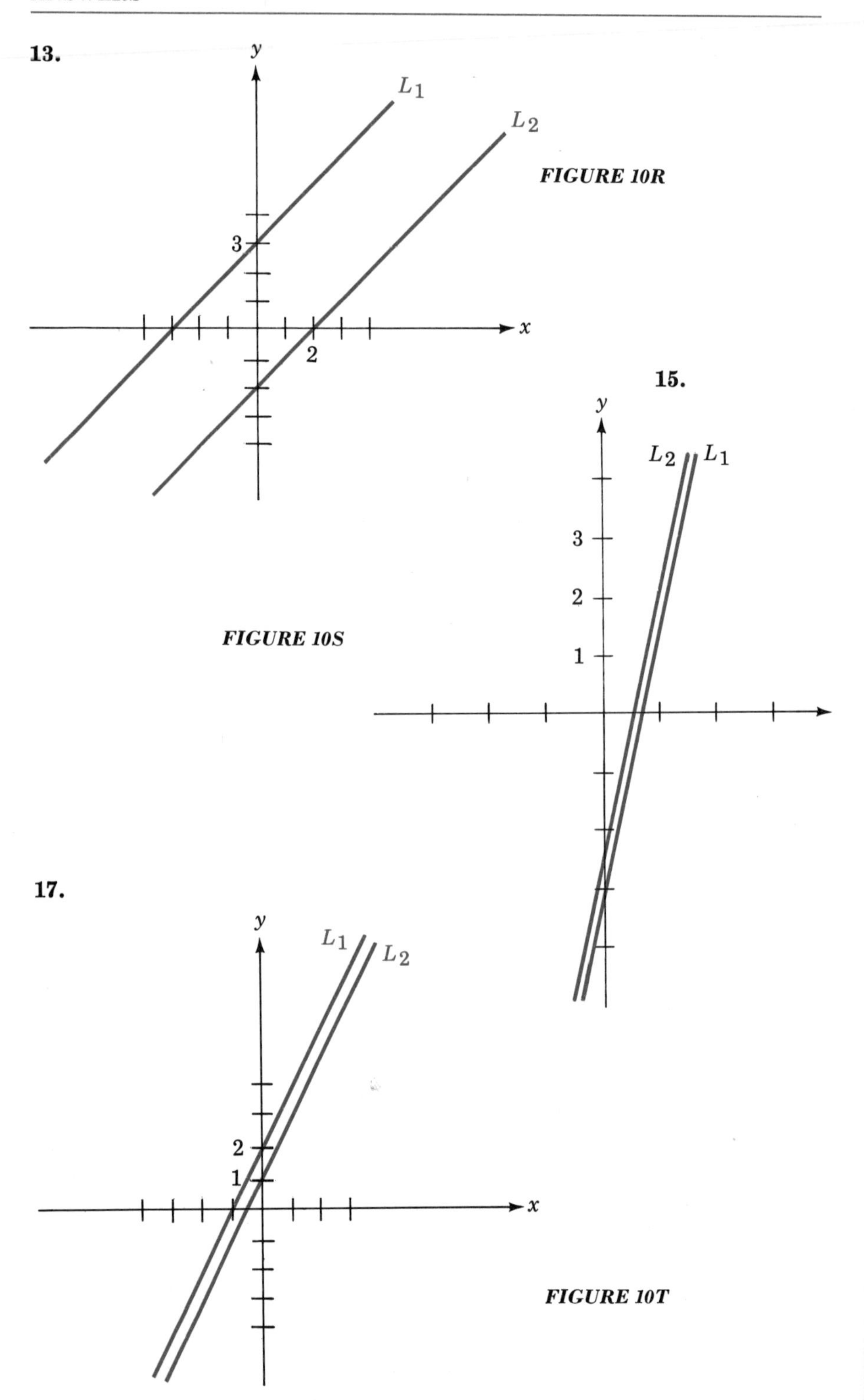

FIGURE 10R

15.

FIGURE 10S

17.

FIGURE 10T

19.

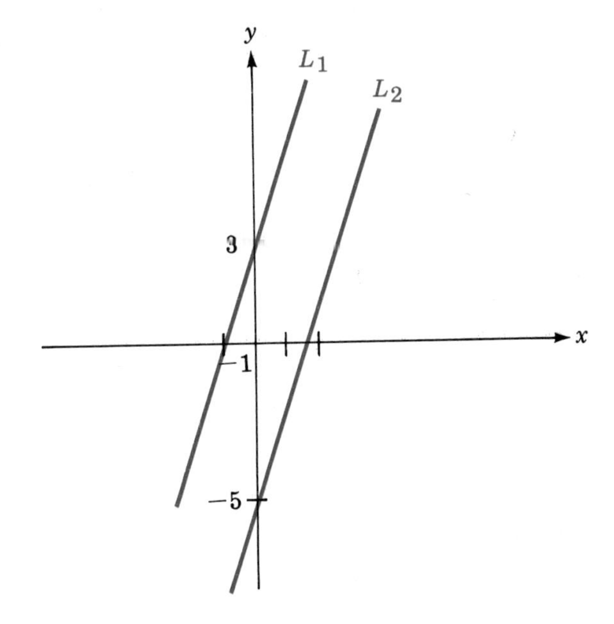

FIGURE 10U

21.

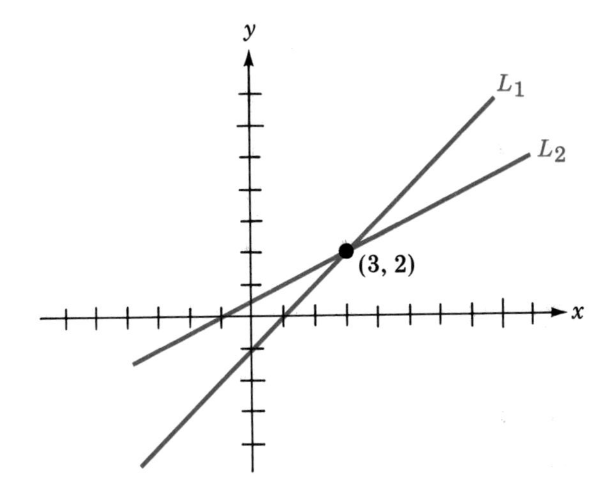

FIGURE 10V

23.

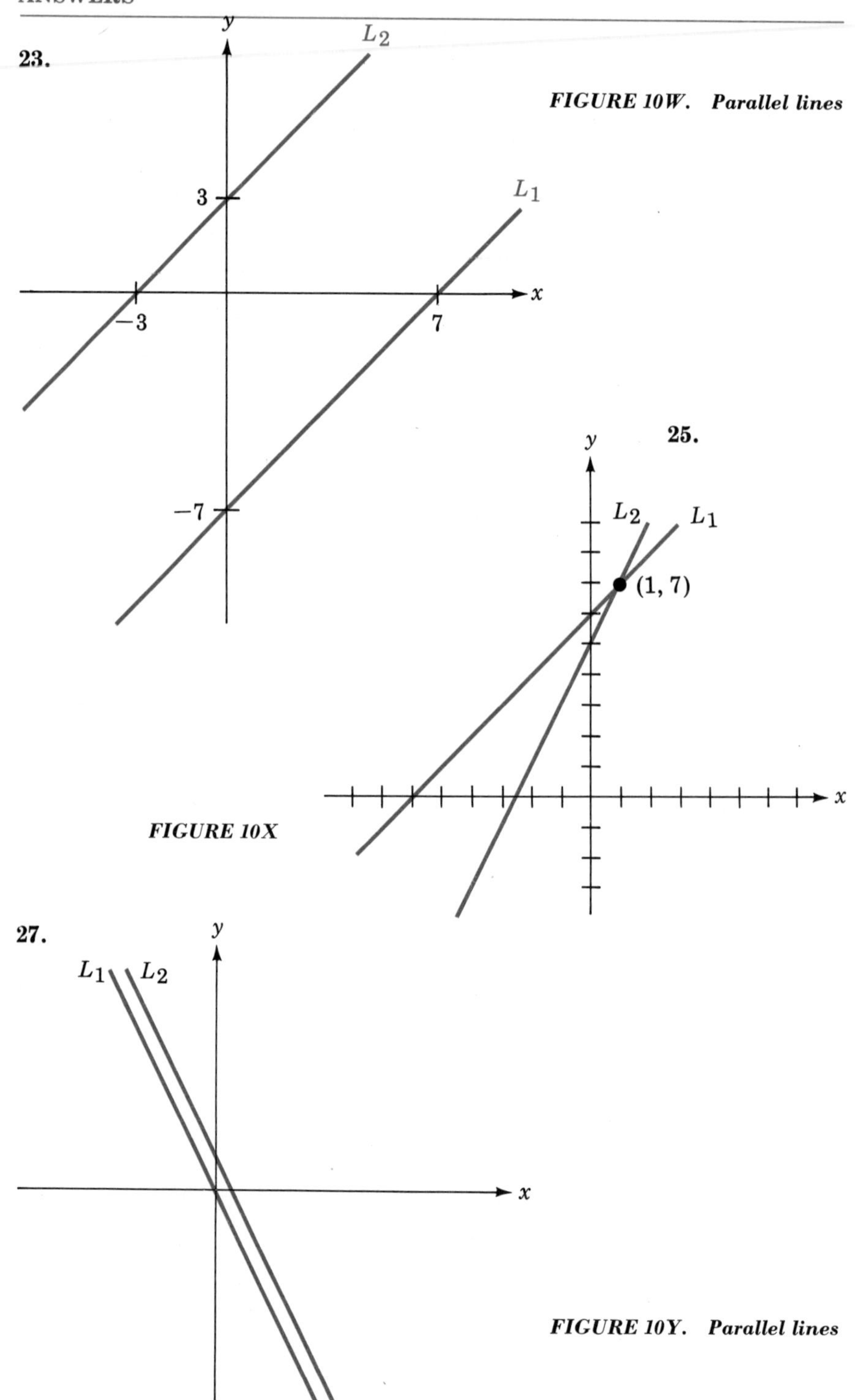

FIGURE 10W. Parallel lines

25.

FIGURE 10X

27.

FIGURE 10Y. Parallel lines

10.4, p. 300 (The checks follow.)

1. (3, 4)
3. (4, 5)
5. (1, 1)
7. (5, 6)
9. (5, −10)
11. (3, 1)
13. $\left(1, \frac{-3}{2}\right)$
15. $\left(1, \frac{1}{5}\right)$
17. (2, 3)
19. (2, 1)
21. (2, 2)
23. (18, 24)
25. (3, 1)
27. (−3, −2)
29. (5, 1)

Checks:

1. $4 \stackrel{?}{=} 3 + 1$
$4 \stackrel{\checkmark}{=} 4$

$4 \stackrel{?}{=} 3(3) - 5$
$4 \stackrel{?}{=} 9 - 5$
$4 \stackrel{\checkmark}{=} 4$

3. $5 \stackrel{?}{=} 2(4) - 3$
$5 \stackrel{?}{=} 8 - 3$
$5 \stackrel{\checkmark}{=} 5$

$5 \stackrel{?}{=} 3(4) - 7$
$5 \stackrel{?}{=} 12 - 7$
$5 \stackrel{\checkmark}{=} 5$

11. $3 + 1 \stackrel{?}{=} 4$
$4 \stackrel{\checkmark}{=} 4$

$3 - 1 \stackrel{?}{=} 2$
$2 \stackrel{\checkmark}{=} 2$

13. $1 + 2\left(\frac{-3}{2}\right) \stackrel{?}{=} -2$
$1 - 3 \stackrel{?}{=} -2$
$-2 \stackrel{\checkmark}{=} -2$

$3(1) - 2\left(\frac{-3}{2}\right) \stackrel{?}{=} 6$
$3 + 3 \stackrel{?}{=} 6$
$6 \stackrel{\checkmark}{=} 6$

10.5, p. 308

1. 2
3. 3
5. −2
7. 2
9. 10
11. 10
13. 40
15. −144
17. 8
19. 2
21. 2
23. 32
25. 1
27. 2
29. −12
31. 1
33. directly
35. inversely
37. $2.10
39. 1000

Let's Review Chapter 10, p. 310

1. (a) 5, (b) 2, (c) $\frac{5}{2}$

2. See Figure 10Z, page A-48.

3. 8

4. (a) (iv), (b) (i)

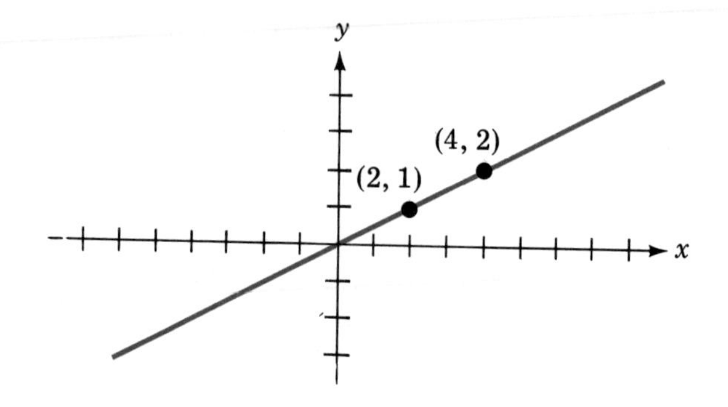

FIGURE 10Z

5. (a) $\frac{-1}{2}$ (b) 7

6. $y = 4x$

7. $y = \frac{2}{3}x - 2$

8. $y = 5$

9.

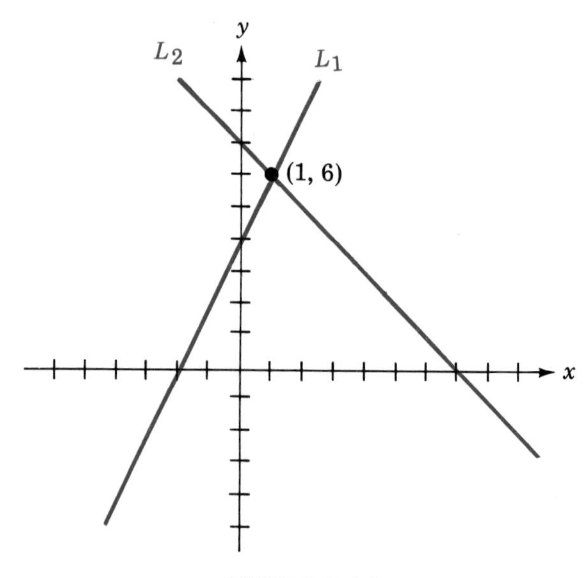

FIGURE 10AA

10.

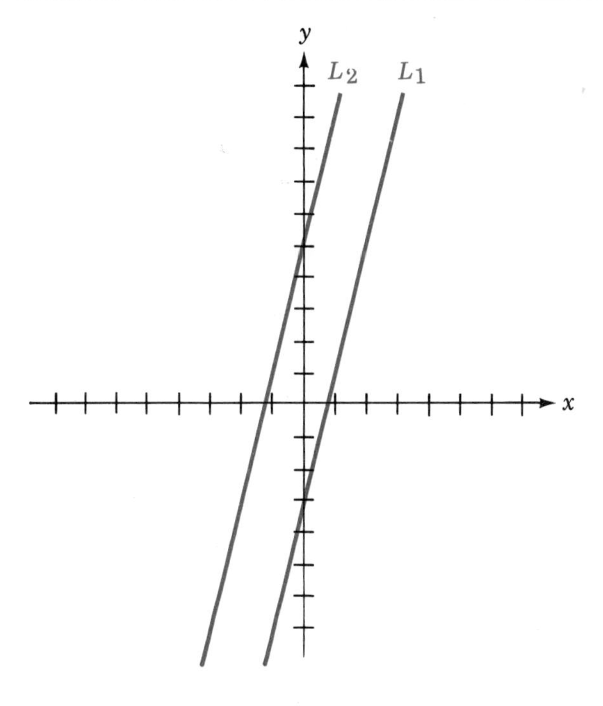

FIGURE 10BB

11.

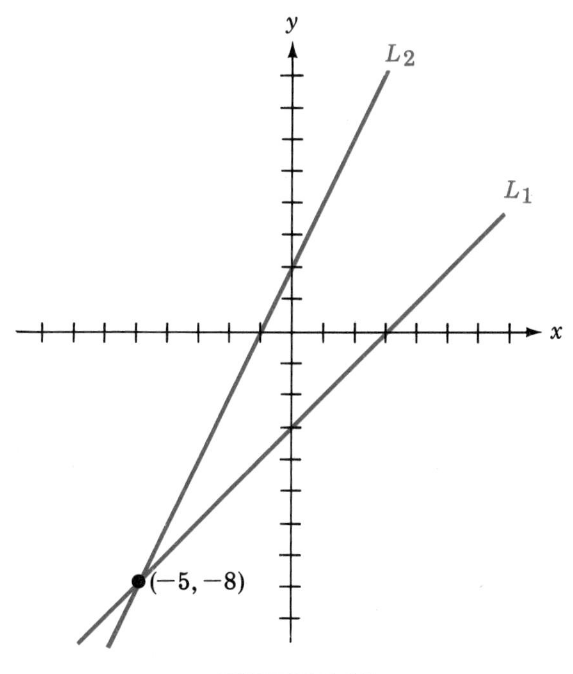

FIGURE 10CC

12. L_3 and L_4

13. $\frac{5}{3}, \frac{13}{3}$

14. (2, 3) Check: $\mathbf{3} + 1 \stackrel{?}{=} 2(\mathbf{2})$ $\quad \frac{\mathbf{3}}{3} \stackrel{?}{=} 3 - \mathbf{2}$

$4 \stackrel{\checkmark}{=} 4 \quad 1 \stackrel{\checkmark}{=} 1$

15. (2, 1)

16. (10, −7)

17. −2

18. 15

19. 32

20. inversely

And these from Chapters 1–9, p. 311

21. 6

22. (a) 1, (b) −5, (c) 7, (d) $\frac{2}{3}$

23. (a) 2, (b) −2, (c) 16, (d) 0

24. $x^2(2x + 3)(2x - 3)$

Try These Exam Questions for Practice, p. 312

1. (a) 3 (b) −3 (c) −1

2. (a) 4 (b) −3

3. $y = -2x + 3$

Check: $2(3) + \mathbf{2} \stackrel{?}{=} 8$

$6 + 2 \stackrel{?}{=} 8$

$8 \stackrel{\checkmark}{=} 8$

4. (3, 2)

$3 - \mathbf{2} \stackrel{?}{=} 1$

$1 \stackrel{\checkmark}{=} 1$

5. (2, −2)

6. $\frac{3}{4}$

CHAPTER 11

11.1, p. 318

1. 4 **3.** 7 **5.** −5 **7.** 9

9. −12 **11.** 40 **13.** 100 **15.** 1000

17. 0 **19.** −3 **21.** 6 **23.** 9

25. x **27.** ab **29.** c^2 **31.** y^5

33. $(xy)^2$ **35.** x^9 **37.** a^{10} **39.** m

41. $-m$ **43.** $-n$ **45.** $-mn$ **47.** n^2

49. 10 inches **51.** 20 inches

53. 8 inches

11.2, p. 325

1. $2 < \sqrt{8} < 3$

3. $4 < \sqrt{23} < 5$

5. $7 < \sqrt{53} < 8$

7. $9 < \sqrt{85} < 10$

9. $11 < \sqrt{125} < 12$

11. $20 < \sqrt{401} < 21$

13. 4.2 **15.** .4 **17.** .3 **19.** 3.5
21. .9 **23.** 1.4 **25.** 3.1 **27.** 6.6
29. 14.4 **31.** 6 inches **33.** 6 inches **35.** 10 inches
37. 4 **39.** 9 **41.** 100 **43.** $\sqrt{5}$

11.3, p. 332

1. $2x$ **3.** $5b^4$ **5.** ab **7.** x^2y^2
9. xyz **11.** ax^5y^2 **13.** $2a^2b^4$ **15.** $6a^2x$
17. $\frac{1}{2}$ **19.** $\frac{6}{7}$ **21.** $\frac{5}{12}$ **23.** $\frac{a}{c}$
25. $\frac{x}{y^4}$ **27.** $\frac{x}{2}$ **29.** $\frac{c^8}{4}$ **31.** $\frac{ab^3}{3}$
33. $\frac{5x^2y^4}{2z^5}$ **35.** (a) $3\sqrt{2}$ (b) 4.2
37. (a) $5\sqrt{2}$ (b) 7
39. (a) $3\sqrt{3}$ (b) 5.1
41. (a) $5\sqrt{3}$ (b) 8.5
43. $a\sqrt{a}$ **45.** $\frac{a^2\sqrt{a}}{b^3}$ **47.** $2a^6\sqrt{3}$ **49.** $ab^2\sqrt{ab}$
51. $\frac{3a^4\sqrt{a}}{b}$ **53.** $7a^2b\sqrt{2b}$ **55.** $\frac{5a\sqrt{2a}}{7b^4}$ **57.** .3
59. .02 **61.** .001 **63.** $9x^4y^2$

11.4, p. 337

1. $2\sqrt{2}$ **3.** $4\sqrt{2}$ **5.** $5\sqrt{6}$
7. $8\sqrt{11}$ **9.** $7\sqrt{5}$ **11.** $3\sqrt{2}$
13. $2\sqrt{2} + 3\sqrt{3}$ **15.** $5\sqrt{7} + 2\sqrt{5}$ **17.** $3\sqrt{2}$
19. 0 **21.** $\sqrt{7}$ **23.** $7\sqrt{3}$
25. $5\sqrt{a}$ **27.** $19y\sqrt{y}$ **29.** $23b\sqrt{a}$
31. $3\sqrt{3a}$ **33.** $12x\sqrt{2x}$ **35.** $6b\sqrt{7ab}$
37. $\frac{2 + \sqrt{5}}{5}$ **39.** $\frac{\sqrt{2} - \sqrt{5} + \sqrt{7}}{3}$ **41.** $\frac{7\sqrt{2} - 5\sqrt{3}}{35}$
43. $\frac{2\sqrt{2} + 3\sqrt{3}}{24}$

11.5, p. 341

1. $\sqrt{10}$ **3.** $\sqrt{15}$ **5.** $2\sqrt{14}$ **7.** 2
9. 7 **11.** 6 **13.** 10 **15.** $2\sqrt{15}$
17. a **19.** b^3 **21.** ab **23.** $10x$
25. $2b^2\sqrt{6}$ **27.** $xy^2z^2\sqrt{x}$ **29.** $6abc^2$ **31.** $2 + \sqrt{6}$
33. $a + \sqrt{ab}$ **35.** 3 **37.** $6(1 + 4\sqrt{3})$ **39.** 47
41. -7 **43.** $38 + 16\sqrt{5}$ **45.** $16 + 12\sqrt{35}$ **47.** $-23 - 5\sqrt{55}$
49. $3 + 2\sqrt{2}$ **51.** $\frac{5}{6}$ **53.** $\frac{2\sqrt{3}}{5}$ **55.** 2

11.6, p. 345

1. 3 **3.** -2 **4.** 2 **7.** $2\sqrt{3}$
9. 7 **11.** $\frac{\sqrt{2}}{3}$ **13.** $\frac{3\sqrt{xz}}{5z}$ **15.** $\frac{4\sqrt{y}}{3z}$
17. 1 **19.** $\frac{3}{10}$

21. (a) $3(1 + \sqrt{2})$, (b) $\frac{1 + \sqrt{2}}{3}$

23. (a) $5(1 + \sqrt{2})$, (b) $\frac{-(1 + \sqrt{2})}{3}$

25. (a) $7(1 + \sqrt{2})$, (b) 7

27. (a) $\sqrt{2}(1 + \sqrt{3})$, (b) $\frac{1 + \sqrt{3}}{2}$

29. (a) $\sqrt{6}(1 - \sqrt{3})$ or $\sqrt{2}(\sqrt{3} - 3)$, (b) $-\sqrt{2}$

31. (a) $\sqrt{ab}(a + \sqrt{c})$, (b) $\frac{a + \sqrt{c}}{ab}$

33. (a) $2(a - b)$, (b) $\frac{1}{5}$

35. (a) $-ab$, (b) 5

37. 1 **39.** 1 **41.** $\frac{5st\sqrt{s}}{12}$ **43.** $\frac{8}{5}$

11.7, p. 350

1. 4 **3.** 1 **5.** -1 **7.** 7
9. -9 **11.** a **13.** a^2 **15.** c^4
17. a^3 **19.** c^{10} **21.** $2a$ **23.** $2b$
25. $10ab^2$ **27.** $8ab^2$ **29.** $\frac{-2}{3}$ **31.** $\frac{5}{4}$
33. $\frac{3}{10}$ **35.** $\frac{b^2}{2}$ **37.** 2 **39.** 1
41. 6 **43.** $-\sqrt[4]{12}$ **45.** No
47. (a) 1, (b) -1, (c) 2, (d) $\frac{1}{2}$

Let's Review Chapter 11, p. *352*

1. 5 **2.** 8 **3.** -30 **4.** x^2
5. $-xy$ **6.** 30 inches **7.** $7 < \sqrt{55} < 8$ **8.** 9.9
9. 8 inches **10.** 11 **11.** $5a$ **12.** a^2b^3
13. $\frac{9}{10}$ **14.** $\frac{2a^2\sqrt{5}}{b^4}$ **15.** $3\sqrt{5}$ **16.** $\sqrt{a}$
17. $17ab^2$ **18.** $\frac{2\sqrt{3}}{5}$ **19.** 10 **20.** x^2y
21. 2 **22.** 2 **23.** 2 **24.** $\frac{2}{5}$

25. (a) $7(1 + \sqrt{2})$, (b) $\frac{1 + \sqrt{2}}{3}$

26. $\frac{4\sqrt{3}}{3}$ **27.** 5 **28.** $2a^3$ **29.** $\frac{-2a}{b^3}$

30. -1

And these from Chapters 1–10, p. *354*

31. $\frac{3}{2abc}$ **32.** .0644

33. (a) 16, (b) 2, (c) 18, (d) 32

34. (a) $>$, (b) $=$, (c) $<$, (d) $<$

Try these Exam Questions for Practice, p. *354*

1. (a) 12 (b) $10x^2$ (c) $\frac{4}{7}$

2. $\frac{4a^2b^4\sqrt{2}}{c^4}$ **3.** $\frac{\sqrt{2}}{4}$ **4.** $18x^3y^2z$

5. $\frac{10}{7}$ **6.** (a) $3(\sqrt{2} - 3)$ (b) $\frac{\sqrt{2} - 3}{2}$

7. 2

CHAPTER 12

12.1, p. 359 (The checks follow.)

1. 0, 4 **3.** 2, 3 **5.** -3, 8

7. $\frac{-1}{2}, \frac{1}{4}$ **9.** 1, 4 **11.** -2, -4

13. 4, -2 **15.** -3, -5 **17.** -3, -6

19. 5 **21.** 3, -7 **23.** 1, -9

25. 1, -2 **27.** $\frac{-1}{2}$, -1 **29.** $\frac{1}{2}$, -4

31. $\frac{-1}{2}, \frac{-3}{2}$ **33.** $\frac{-5}{2}$ **35.** $\frac{3}{2}$, 2

37. 0, 2 **39.** 0, $\frac{1}{2}$

41. $f(x) = 0$ when $x = -5$ and when $x = 3$

43. $f(x) = 0$ when $x = 5$

Checks:

5. *For* -3:

$(-3 + 3)(-3 - 8) \stackrel{?}{=} 0$

$0(-11) \stackrel{?}{=} 0$

$0 \stackrel{\checkmark}{=} 0$

For 8:

$(8 + 3)(8 - 8) \stackrel{?}{=} 0$

$11 \quad (0) \stackrel{?}{=} 0$

$0 \stackrel{\checkmark}{=} 0$

11. *For* -2:
$(-2)^2 + 6(-2) + 8 \stackrel{?}{=} 0$
$4 - 12 + 8 \stackrel{?}{=} 0$
$0 \stackrel{\checkmark}{=} 0$

For -4:
$(-4)^2 + 6(-4) + 8 \stackrel{?}{=} 0$
$16 - 24 + 8 \stackrel{?}{=} 0$
$0 \stackrel{\checkmark}{=} 0$

25. *For* 1:
$2(1^2) + 2(1) - 4 \stackrel{?}{=} 0$
$2 + 2 - 4 \stackrel{?}{=} 0$
$0 \stackrel{\checkmark}{=} 0$

For -2:
$2(-2)^2 + 2(-2) - 4 \stackrel{?}{=} 0$
$2(4) - 4 - 4 \stackrel{?}{=} 0$
$8 - 8 \stackrel{?}{=} 0$
$0 \stackrel{\checkmark}{=} 0$

12.2, p. 363 (The checks are given below.)

1. ± 4 **3.** ± 7 **5.** $\pm\sqrt{6}$
7. $\pm\sqrt{13}$ **9.** $\pm 2\sqrt{2}$ **11.** $\pm 3\sqrt{2}$
13. $\pm 3\sqrt{5}$ **15.** $\frac{\pm 2}{3}$ **17.** $\pm\sqrt{\frac{2}{3}}$ or $\frac{\pm\sqrt{6}}{3}$
19. $3, -4$ **21.** $\frac{6}{5}, \frac{-2}{5}$ **23.** $\frac{4}{3}, -4$
25. $\frac{2 \pm \sqrt{3}}{3}$ **27.** $\frac{-4 \pm \sqrt{7}}{5}$ **29.** $\frac{-5 \pm \sqrt{6}}{3}$
31. $\frac{1 \pm 2\sqrt{2}}{2}$ **33.** $-3 \pm 4\sqrt{2}$ **35.** $1 \pm 14\sqrt{2}$
37. 2 **39.** $\frac{3}{5}$

Checks:

7. *For* $\sqrt{13}$:
$(\sqrt{13})^2 - 13 \stackrel{?}{=} 0$
$13 - 13 \stackrel{?}{=} 0$
$0 \stackrel{\checkmark}{=} 0$

For $-\sqrt{13}$:
$(-\sqrt{13})^2 - 13 \stackrel{?}{=} 0$
$13 - 13 \stackrel{?}{=} 0$
$0 \stackrel{\checkmark}{=} 0$

19. *For* 3:
$[2(3) + 1]^2 \stackrel{?}{=} 49$
$7^2 \stackrel{?}{=} 49$
$49 \stackrel{\checkmark}{=} 49$

For -4:
$[2(-4) + 1]^2 \stackrel{?}{=} 49$
$[-7]^2 \stackrel{?}{=} 49$
$49 \stackrel{\checkmark}{=} 49$

25. *For* $\frac{2 + \sqrt{3}}{3}$:
$\left[3\left(\frac{2 + \sqrt{3}}{3}\right) - 2\right]^2 \stackrel{?}{=} 3$
$[2 + \sqrt{3} - 2]^2 \stackrel{?}{=} 3$
$[\sqrt{3}]^2 \stackrel{?}{=} 3$
$3 \stackrel{\checkmark}{=} 3$

For $\frac{2 - \sqrt{3}}{3}$:
$\left[3\left(\frac{2 - \sqrt{3}}{3}\right) - 2\right]^2 \stackrel{?}{=} 3$
$[2 - \sqrt{3} - 2]^2 \stackrel{?}{=} 3$
$[-\sqrt{3}]^2 \stackrel{?}{=} 3$
$3 \stackrel{\checkmark}{=} 3$

12.3, p. 368

1. (a) 8 (b) 2 (c) $-2 \pm \sqrt{2}$
3. (a) 1 (b) 2 (c) $1, \frac{3}{2}$

5. (a) 13 (b) 2 (c) $\dfrac{3 \pm \sqrt{13}}{2}$

7. (a) 12 (b) 2 (c) $\dfrac{3 \pm \sqrt{3}}{2}$

9. (a) 0 (b) 1 (c) $\dfrac{1}{2}$

11. (a) 320 (b) 2 (c) $\dfrac{5 \pm 2\sqrt{5}}{2}$

13. (a) 5 (b) 2 (c) $\dfrac{-1 \pm \sqrt{5}}{2}$

15. (a) -35 (b) 0

(The checks follow.)

17. $\dfrac{-5 \pm \sqrt{33}}{2}$ **19.** $\dfrac{5 \pm \sqrt{13}}{6}$ **21.** $-5 \pm 2\sqrt{5}$

23. $\dfrac{-3 \pm \sqrt{13}}{2}$ **25.** $-1, \dfrac{-3}{2}$ **27.** $3, \dfrac{1}{3}$

29. 1, 2 **31.** 4, -5 **33.** $-1 \pm \sqrt{5}$

35. $\dfrac{1}{2}, \dfrac{-1}{3}$ **37.** $f(x) = 0$ when $x = -3 \pm \sqrt{3}$

Checks:

17. *For* $\dfrac{-5 + \sqrt{33}}{2}$:

$$\left(\frac{-5 + \sqrt{33}}{2}\right)^2 + 5\left(\frac{-5 + \sqrt{33}}{2}\right) - 2 \stackrel{?}{=} 0$$
$$\frac{58 - 10\sqrt{33}}{4} + \frac{-25 + 5\sqrt{33}}{2} - 2 \stackrel{?}{=} 0$$
$$\frac{58 - 10\sqrt{33} - 50 + 10\sqrt{33} - 8}{4} \stackrel{?}{=} 0$$
$$0 \stackrel{\checkmark}{=} 0$$

For $\dfrac{-5 - \sqrt{33}}{2}$:

$$\left(\frac{-5 - \sqrt{33}}{2}\right)^2 + 5\left(\frac{-5 - \sqrt{33}}{2}\right) - 2 \stackrel{?}{=} 0$$
$$\frac{58 + 10\sqrt{33}}{4} + \frac{-25 - 5\sqrt{33}}{2} - 2 \stackrel{?}{=} 0$$
$$\frac{58 + 10\sqrt{33} - 50 - 10\sqrt{33} - 8}{4} \stackrel{?}{=} 0$$
$$0 \stackrel{\checkmark}{=} 0$$

19. *For* $\frac{5+\sqrt{13}}{6}$:

$$3\left(\frac{5+\sqrt{13}}{6}\right)^2 - 5\left(\frac{5+\sqrt{13}}{6}\right) + 1 \stackrel{?}{=} 0$$
$$3\left(\frac{38+10\sqrt{13}}{36}\right) + \frac{-25-5\sqrt{13}}{6} + 1 \stackrel{?}{=} 0$$
$$\frac{38+10\sqrt{13}-50-10\sqrt{13}+12}{12} \stackrel{?}{=} 0$$
$$0 \stackrel{\checkmark}{=} 0$$

For $\frac{5-\sqrt{13}}{6}$:

$$3\left(\frac{5-\sqrt{13}}{6}\right)^2 - 5\left(\frac{5-\sqrt{13}}{6}\right) + 1 \stackrel{?}{=} 0$$
$$3\left(\frac{38-10\sqrt{13}}{36}\right) + \frac{-25+5\sqrt{13}}{6} + 1 \stackrel{?}{=} 0$$
$$\frac{38-10\sqrt{13}-50+10\sqrt{13}+12}{12} \stackrel{?}{=} 0$$
$$0 \stackrel{\checkmark}{=} 0$$

12.4, p. 372

1. 4 and 5
3. 10 and -7
5. ± 6
7. 5
9. 20 feet by 12 feet
11. 40 feet by 30 feet
13. 6 seconds
15. (a) 1 second, (b) 7 seconds
17. 6 or 10
19. 10 feet

Let's Review Chapter 12, p. 375 (The checks follow.)

1. $-3, -4$
2. 3
3. $\frac{-1}{2}, -3$
4. $\frac{1}{3}, -2$
5. $\pm\sqrt{17}$
6. $\pm 4\sqrt{5}$
7. $1, -2$
8. $\frac{5 \pm \sqrt{2}}{3}$
9. (a) -11, (b) 0
10. $\frac{-5 \pm \sqrt{5}}{2}$
11. $\frac{-1}{2}$
12. $-3 \pm \sqrt{10}$
13. 9 and -5
14. 12 inches by 3 inches

Checks:

1. *For* -3:

$$(-3)^2 + 7(-3) + 12 \stackrel{?}{=} 0$$
$$9 - 21 + 12 \stackrel{?}{=} 0$$
$$0 \stackrel{\checkmark}{=} 0$$

For -4:

$$(-4)^2 + 7(-4) + 12 \stackrel{?}{=} 0$$
$$16 - 28 + 12 \stackrel{?}{=} 0$$
$$0 \stackrel{\checkmark}{=} 0$$

7. *For* 1:

$$[2(1)+1]^2 \stackrel{?}{=} 9$$
$$(3)^2 \stackrel{?}{=} 9$$
$$9 \stackrel{\checkmark}{=} 9$$

For -2:

$$[2(-2)+1]^2 \stackrel{?}{=} 9$$
$$(-3)^2 \stackrel{?}{=} 9$$
$$9 \stackrel{\checkmark}{=} 9$$

10. *For* $\frac{-5+\sqrt{5}}{2}$:

$$\left(\frac{-5+\sqrt{5}}{2}\right)^2 + 5\left(\frac{-5+\sqrt{5}}{2}\right) + 5 \stackrel{?}{=} 0$$
$$\frac{30-10\sqrt{5}}{4} + \frac{-25+5\sqrt{5}}{2} + 5 \stackrel{?}{=} 0$$
$$\frac{30-10\sqrt{5}-50+10\sqrt{5}+20}{4} \stackrel{?}{=} 0$$
$$0 \stackrel{\checkmark}{=} 0$$

For $\frac{-5-\sqrt{5}}{2}$:

$$\left(\frac{-5-\sqrt{5}}{2}\right)^2 + 5\left(\frac{-5-\sqrt{5}}{2}\right) + 5 \stackrel{?}{=} 0$$
$$\frac{30+10\sqrt{5}}{4} + \frac{-25-5\sqrt{5}}{2} + 5 \stackrel{?}{=} 0$$
$$\frac{30+10\sqrt{5}-50-10\sqrt{5}+20}{4} \stackrel{?}{=} 0$$
$$0 \stackrel{\checkmark}{=} 0$$

And these from Chapter 1–11, p. *376*

15. (a) 6 (b) 12 (c) 20
(d) $f(x) = 0$ when $x = -1$ and when $x = -2$

16. (a) -10 (b) -6 (c) $\frac{-45}{4}$
(d) $f(x) = 0$ when $x = -2$ and when $x = 5$

17. (a) 1 (b) $\frac{17}{4}$ (c) $\frac{73}{36}$
(d) $f(x) = 0$ when $x = -3 \pm 2\sqrt{2}$

18. $(5x+1)(x-1)$ **19.** $(5x+2)(2x+1)$

20. 21, 22, 23

Try These Sample Questions for Practice, p. *376*.

1. $7, -2$ **2.** $\frac{1 \pm \sqrt{5}}{2}$ **3.** $\frac{-9 \pm \sqrt{69}}{2}$

4. $\frac{5}{2}, -1$ **5.** 5 and 12

Index

Absolute value, 13, 20–21
function, 251–253, 255
Adding corresponding sides of equations, 297–299
Addition:
of decimals, 168
of fractions, 146–148, 150, 154, 157–158
of like terms, 53–55
and other operations, 19, 40–44
of polynomials, 57–58
of rational expressions, 146–150, 158
of real numbers, 18–28, 31, 58
of roots, 334–337
of unlike terms, 54–55
Addition Property, 179–182
Additive inverse, 12–13, 19–20
Age problems, 206–209, 247
Altitude of triangle, 62, 196
Approximating square roots, 320–323
Area:
of circle, 64
of rectangle, 64, 196
of square, 61–62, 64, 247, 323, 326
of triangle, 62, 64, 196
Arithmetic, 18–48
Associative Law:
of addition, 23–24, 54–55
of multiplication, 28–29, 69–70, 91
for subtraction, 44

Base, 39–40, 69, 71
Base of triangle, 62, 196
Binomial, 56–57, 86–92
Boyle's Law, 308
Branch of hyperbola, 306

Center of earth, 309
Centigrade, 194–95
Change in *x*, 261–262
Change in *y*, 251–262
Check:
for division of numbers, 32
with remainder, 33–34
for division of polynomials, 128–129
with remainder, 132–133
for equations, 176–180, 182, 184, 188–189, 192
for graphical solutions, 286
for quadratic equations, 357, 361–362, 364–365
for systems of equations, 294
Circle, 6–7, 64–65, 196, 241, 247, 303
Circumference (*see* Circle)
Coefficients, numerical, 51, 70, 72, 80, 82, 85, 118
Commission, 190, 221–222, 224–225, 247
Common Factors, 79–80, 82–85
Commutative Law:
of addition, 23–24, 54–55
of multiplication, 28–29, 51, 69–70, 91
Complex fractions, 160–162
Complex numbers, 362
Complex rational expressions, 162–163
Composites, 78
Consecutive integers, 202–204
Constant function, 240
Constant of variation, 301–308
Coordinate axes, 229–231
Coordinates, 231–234
Cost, 222–223, 374
Cross-product, 107
Cube of number, 38
Cube root, 346–348
Curves, 240–241 (*see also* Graphs)

Decimal(s), 6, 165–172
in equations, 189–190
place, 169–170, 321–323

point, 165–170
representation, 322–323
roots of, 332
Degree, 125–126, 131
Denominator:
of fraction, 106, 146–148, 150, 154–158
of rational expression, 111–112, 146–150, 154–159
rationalizing, 343–344
of rational number, 5, 111, 166
Difference, 18
Difference polynomial, 127–133
Difference of squares, 89–92
Digits, 165
Direct variation, 301–304
Discount, 223–226
Discriminant, 366–368
Distance:
from origin, 12–13, 20, 251
problems, 197, 209–215, 237, 242, 247, 301, 309 (*see also* Motion problems)
stretched, 309
unit, 1–2, 229
Distributive Laws:
for like terms, 53–54
for polynomials, 73–76, 84, 86, 183–184
for real numbers, 42–44, 53–54
for roots, 339–340
Dividend, 31–35, 126–133
Division:
of decimals, 170
of fractions, 143–144, 344–345
of integers, 3–6
of polynomials, 74–75, 111–134
with remainder, 131–133
of powers, 71–72, 116–119
process, 126–133
of rational expressions, 143–145
of real numbers, 31–37, 71, 115
with remainder, 32–34
of roots, 342–346
by 0, 34–35, 193
Divisor, 31–35, 126–133
of integer, 77
monomial, 120–123

Equality of fractions, 106–108
Equation of line, 274–283, 291–299
Equations, 176–199
equivalent, 178–179
literal, 191–197
with parentheses, 183–185
quadratic, 355–376
with rational expressions, 185–191
Evaluating:
polynomials, 60–65
rational expressions, 112–115
Even integers, 29, 35, 203–205

Exponents, 39, 40, 69, 71, 78

Factors:
of integer, 77–80, 82
of polynomial, 82–85
of product, 27–30
Factoring, 77–80, 82–85, 89–102
Fahrenheit, 194–195
Formulas, 61–62, 64–65, 194–197, 317
Fractions:
addition and subtraction of, 146–148, 150, 154–155, 157–158
complex, 160–162
and decimals, 165–170
division of, 143–144, 344–345
equality of 106–108
in lowest terms, 108–110
multiplication of, 138–140, 340
and proportions, 185–186
with radicals, 336, 340
rationalizing the denominators of, 344–345
Functions, 237–261, 312, 324–325, 376

Gas, 308
gcd, 79–80, 82–83
Geometry problems:
for circle, 64–65, 196, 247, 303
for rectangle, 64, 196, 369–370, 373, 376
for rectangular box, 197
for square, 61–62, 64, 247, 323, 326
for triangle, 62, 64, 196, 316–319, 323–324, 327
Graphs:
of functions, 248–257
of hyperbolas, 306
of lines, 248–251, 261, 285–289, 291, 299, 302
of parabolas, 253–257
Greater than, 8–10
Greatest common divisor, 79–80, 82–83

Horizontal lines, 229, 270, 272, 280–281
Hundredths, 165, 216
Hyperbolas, 306

Inequalities, 8–12
Infinite nonrepeating decimal, 322–323
Infinite repeating decimal, 166–167, 322
Inputs, 238
Integer problems, 200–205 (*see also* Number problems)

Integers, 1–2, 5, 112, 320
Interest problems, 216–220
Intersection of lines, 284–301
Inverse:
  of number, 12–13, 19–21, 29
  of polynomial, 57–58
Inverse variation, 304–308
Irrational numbers, 6–7, 106–107, 320–323, 334–336
Isolating the common factor, 83–85, 96–97, 100–101, 120–121, 123–124, 343–344
Isosceles triangle, 323–324

Larger than, 8–10
lcd, 154–159, 187–189
lcm, 150–159
Least common denominator, 154–159, 187–189
Least common multiple, 150–159
Left (on number line), 1–2, 4, 8–10, 12, 18–20
Less than, 8–10
Like terms, 52–55, 57
Linear equations, 291–304
Lines, 248–251, 261–304
Line segment, 1, 3–4
Literal equations, 191–197
Lowest terms, 108–110

Machine, function as, 238–239
Marked price, 223–224
Mark-up, 222–223
Mixed number, 5–6, 165, 168
Monomial, 56–57, 120–121
Motion problems, 371, 373
Multiple, 77, 150
Multiplication:
  of binomials, 86–89
  of decimals, 169–170
  of fractions, 138–140, 340
  of monomials, 68–73
  and other operations, 31–35, 40–44, 115
  of polynomials, 68–74, 77, 86–89
  by positive integer, 27–28
  of powers, 68–69
  of rational expressions, 140–141
  of real numbers, 27–31
  of roots, 337–341
  by 0, 51
Multiplication Property, 180–182

Negative numbers, 9–10, 12–13
  and addition, 20, 24
  cube roots of, 347
  and division, 35
  and multiplication, 29–30
  powers of, 39
  square roots of, 314, 362
  on y-axis, 229–230

nth powers, 38–39
nth roots, 346–351
Number line, 1–17, 107–108, 229
Number problems, 369, 372–373 (*see also* Integer problems)
Numerator:
  of fraction, 106
  of rational expression, 111–112
  of rational number, 5, 111, 166

Odd integers, 29, 35, 205
Order of operations, 40–45
Ordered pairs, 231–234, 291–292
Origin, 1, 9–10, 12–13, 20, 229, 232, 252, 302
Outputs, 238

Parabolas, 235–257, 372
Parallel lines, 284, 288–289
Parentheses, 23–25, 40–44, 58, 183–184
Percent, 216
Perimeter of rectangle, 64, 196, 369–370, 373, 376
Perpendicular lines, 229
π, 6–7, 323
Polynomial(s), 56–105
  addition of, 56–57
  definition of, 56
  evaluating, 60–65
  factoring, 80, 82–85, 89–102
  inverse of, 56–57
  least common multiple of, 152–153
  multiplication of, 68–76, 86–89
  subtraction of, 56–57
Positive numbers, 9–10, 12–13
  addition of, 19–20, 24
  and division, 35
  and multiplication, 27–30
  powers of, 39
  on y-axis, 229–230
Power:
  even, 39, 315–316
  of number, 68–69, 71
  odd, 39, 329
  of polynomial, 119
  positive integral, 38
  raising to, 37–42
  of variable, 68–72, 82–83
Pressure, 308
Prime, 77–80
  factors, 78–80, 118, 151
Principal, 217–219
Product, 27–28, 31, 50, 91, 306 (*see also* Multiplication)
Profit problems, 372–374
Proportion, 185–187, 303
Pythagorean Theorem, 316–319, 323–324, 327

Quadratic equations, 355–376
Quadratic Formula, 364–368

Quotient, 31–35, 126–133, 303

Radical sign, 314
Radius:
of earth, 309
of circle (*see* Circle)
Rate of commission, 221–222, 247
Rate, constant (*see* Distance problems)
Rate of mark-up, 222–223
Rational expressions, 111–115, 140–150, 154, 156–159, 162–163 (*see also* Division of polynomials)
Rationalizing the denominator, 343–344
Rational numbers, 3–8, 32, 106, 108, 111, 165–167, 320, 322
Real line, 1–17, 107–108, 229
Real numbers, 1–50, 60, 112, 229–232, 362–363
Rectangles, 64, 196, 309, 369–370, 373, 376
Rectangular coordinates, 229–236
Reduction, 223–224
Remainder, 33–34, 131–133
Right angle, 316
Right (on number line), 1–2, 4, 8–10, 12, 18–20
Right triangle, 316–319, 327
Rise, 261–271, 274
Roots of equation, 176–177
Roots of number, 313–354
Run, 261–271, 274

Sale price, 190, 221–224
Second-degree equation, 355–376
Sign:
of number, 20–21, 24, 58
of variable, 315
Slope, 248, 261–282, 289, 302
Smaller than, 8–10
Solution:
of equation, 176–177, 291–292
of system of equations, 292–293
Spring, 309
Square, 37, 40, 90–92
Square, area of, 61–62, 64, 247, 323, 326
Squaring function, 253–255
Square root(s), 313–345
addition and subtraction of, 334–337
approximating, 320–323
of decimals, 332
division of, 342–346
function, 324–325
in geometry, 323–324
multiplication of, 337–341
of product, 328–329
of quotient, 330–332
Standard form, 126
Subscripts, 261
Substituting in systems of equations, 293–296
Subtraction:
of decimals, 168
of fractions, 146–148, 150, 154 157–158
and other operations, 41–44
of polynomials, 57–58, 158–159
of rational expressions, 146–150, 154
of real numbers, 18–27, 31, 58
of roots, 334–336
Sum, 18
of squares, 91–92
Surface area, 65
Surface of earth, 309
Symmetry, 252
Systems of linear equations, 291–301

Tension, 309
Tenths, 165
Terms, 49–57, 112
Thousandths, 165
Time (*see* Distance problems)
Triangles, 62, 64, 196, 316–319, 323–324, 327
Trinomials, 56–57, 93–102

Variable, 49–52, 60–61, 68–71, 180–181, 191–192, 303, 306
dependent, 237–240
independent, 237–239
Variation, 301–309
Vertex of parabola, 255–256, 372
Vertical lines, 229, 241, 271–272, 281–282
Volume, 65, 197, 308

Weight, 309
Word problems, 200–228, 369–374

x-axis, 229–233, 281
x-coordinate, 231–233, 261, 271–272, 281

y-axis, 229–233, 248, 275, 281
y-coordinate, 231–233, 261, 270, 272, 280, 372
y-intercept, 274–281